Anomalous Magnetic Field of the World Ocean

Anomalous Magnetic Field of the World Ocean

Edited by
Alexander M. Gorodnitsky
Professor of Geology
P.P. Shirshov Institute of Oceanology
Russian Academy of Sciences
Moscow, Russia

CRC Press
Boca Raton   Ann Arbor   London   Tokyo

Library of Congress Cataloging-in-Publication Data

Gorodnitsky, Alexander M.
    Anomalous magnetic field of the World Ocean / edited by Alexander M. Gorodnitsky
      p. cm.
    Includes bibliographical references and index.
    ISBN 0-8493-8937-2
    1. Geomagnetic reversals—Measurement. 2. Magnetic anomalies—Measurement.
    3. Submarine geology—Measurement. 4. Geology, Stratigraphic—Mesozoic.
    5. Geology, Stratigraphic—Cenozoic. I. Gorodnitskii, A. M. (Aleksandr Moiseevich)
QC827.A56 1995
538'.78162—dc20
DNLM/DLC
for Library of Congress   94-6657
                                                                                                                              CIP

    This book contains information obtained from authentic and highly regarded sources. Reprinted material is quoted with permission, and sources are indicated. A wide variety of references are listed. Reasonable efforts have been made to publish reliable data and information, but the author and the publisher cannot assume responsibility for the validity of all materials or for the consequences of their use.

    Neither this book nor any part may be reproduced or transmitted in any form or by any means, electronic or mechanical, including photocopying, microfilming, and recording, or by any information storage or retrieval system, without prior permission in writing from the publisher.

    All rights reserved. Authorization to photocopy items for internal or personal use, or the personal or internal use of specific clients, may be granted by CRC Press, Inc., provided that $.50 per page photocopied is paid directly to Copyright Clearance Center, 27 Congress Street, Salem, MA 01970 USA. The fee code for users of the Transactional Reporting Service is ISBN 0-8493-8937-2/95/$0.00+$.50. The fee is subject to change without notice. For organizations that have been granted a photocopy license by the CCC, a separate system of payment has been arranged.

    CRC Press, Inc.'s consent does not extend to copying for general distribution, for promotion, for creating new works, or for resale. Specific permission must be obtained in writing from CRC Press for such copying.

    Direct all inquiries to CRC Press, Inc., 2000 Corporate Blvd., N.W., Boca Raton, Florida 33431.

© 1995 by CRC Press, Inc.

No claim to original U.S. Government works
International Standard Book Number 0-8493-8937-2
Library of Congress Card Number 94-6657
Printed in the United States of America 1 2 3 4 5 6 7 8 9 0
Printed on acid-free paper

TABLE OF CONTENTS

Chapter 1
Techniques for Marine Magnetic Measurements 1
I. I. Belyaev and A. M. Filin

Chapter 2
Interpretation Procedure of Marine Magnetic Data: Topical Problems 21
G. M. Valyashko, A. N. Ivanenko, G. E. Czerniawski, and S. V. Lukyanov

Chapter 3
Mid-Oceanic Ridges and Deep Oceanic Basins: AMF Structure 67
**V. Yu. Glebovsky, S. P. Maschenkov, A. M. Gorodnitsky, I. I. Belyaev,
A. M. Filin, S. A. Mercuriev, N. A. Sochevanova, S. V. Lukyanov,
G. M. Valyashko, E. A. Popov, and K .V. Popov**

Chapter 4
Geomagnetic Study of Transform Faults 145
S. V. Lukyanov, G. M. Valyashko, and A. M. Gorodnitsky

Chapter 5
Geomagnetic Study of Seamounts 171
V. Yu. Brusilovsky, A. M. Gorodnitsky, and A. N. Ivanenko

Chapter 6
Genesis of Primary Magmatic Titanomagnetite of
Tholeiite Basalt from Mid-Oceanic Ridges 223
A. G. Gorshkov

References 233

Index 249

PREFACE

The anomalous geomagnetic field of the World Ocean is believed to be one of the principal sources of information on the oceanic lithosphere structure and tectonic evolution. The discovery of the striped pattern of geomagnetic anomalies and their bilateral symmetry off the rifts of the mid-oceanic ridges allowed F. Vine and D. Matthews to come up with an enlightening hypothesis (1963). They supposed the existence of an active magnetic layer in the Earth's crust, formed as the result of the sea-floor spreading and the Earth's magnetic field reversals. The first comparison of the elaborated scale of reversals with geomagnetic profiles across the Juan de Fuca Ridge allowed F. Vine and J. Wilson to obtain a well-grounded verification of the hypothesis and to estimate for the first time the rate of spreading (1965). The correlation of the revealed and mapped world systems of magnetic lineations with polarity changes, detected from the magnetization of basalt lava, allowed W. Pitman and J. Heirtzler and others to construct a geomagnetic time scale and to play back the chronology of the geomagnetic field reversals, first for the past 80 million years, and then for 160 million years.

The chronology of the geomagnetic field has been verified by deep oceanic drilling data. The Euler theorem used for estimating poles and angular velocities of oceanic floor opening allowed one to apply the system of linear paleomagnetic anomalies for numerical analysis of plate kinematics. The latter was used as a basis for various positional paleoreconstructions of the oceans and continents in the Mesozoic and Cenozoic eras.

Thus, the first stage in the study of the magnetic field of the oceans was completed in principle. It realized the techniques and know-how options of the time. They are as follows:

1. Geomagnetic survey (in particular, measurement of the full vector modulus of the geomagnetic field) was characterized by a moderate accuracy: standard crossover error of the survey was about 15 nT. Such a moderate accuracy was due to insufficient account of the distorting effect of geomagnetic variations in the open ocean. This hindered the study of the fine spatial structure of the anomalous magnetic fields (AMF) and their geological nature.

2. Geomagnetic measurements were carried out only on the ocean surface, i.e., at a distance of 3 to 5 km above the magnetic sources, which made it possible to obtain only smooth integral patterns of the AMF.

3. There was practically no necessary information available on the composition of magnetic anomalies, thickness of the active magnetic layers in the oceanic crust, and magnetic parameters of rocks comprising it. This hindered the resolution of the inverse magnetometric problem in the ocean due to the scarcity of specific data on the geometry of magnetic sources and their magnetization.

This volume, a monograph devoted to the study of the AMF of the ocean and their relations to the geological structure and tectonic evolution of the oceanic lithosphere, reflects the beginning of the second stage in the geomagnetic study of the World Ocean by the Russian researchers. The principal trends in the research of the stage in question are (1) elaboration of new techniques and know-how in the geomagnetic study of the ocean; (2) geomagnetic mapping and study of the fine spatial structure of geomagnetic field anomalies and simulation of the spreading kinematics for the individual regions of the World Ocean; and (3) study of the structure of magnetic anomalies and their correlation with the petromagnetic structure of the oceanic crust.

This work is based on the results of geomagnetic surveys and studies of the oceanic crust carried out recently by Russian specialists in various regions of the World Ocean. The latest know-how and technique development, including the results obtained by the towed and man-operated submersibles used for bottom geomagnetic measurements and petromagnetic testing, are also presented in the monograph.

The authors contributing to the monograph are leading specialists in geomagnetic studies from the P. P. Shirshov Institute of Oceanology, Russian Academy of Sciences; St. Petersburg Division of the Institute of the Earth's Magnetism, Russian Academy of Sciences; and the Research Institute of Ocean Geology, Geological Committee of Russia.

The first section of the monograph illuminates the progress in the Russian techniques and know-how concerning geomagnetic measurements in the ocean. Considerable attention is paid to the latest development of differential sea magnetometers and the experience in the gradient magnetic survey with measurement of the course gradient that allows one to avoid the distorting error from magnetic variations. The first results of the Russian-borne bottom measurements with the towed probe are presented here. The principal original technology of computer processing and interpretation of geomagnetic survey data on the ocean are also given in the work. The original technique for magnetic simulation of the seamount topography is proposed and justified.

Considerable attention is paid to the state of progress in the chronology of geomagnetic anomalies of the oceanic crust.

The pivot of the monograph is the result of extensive geomagnetic surveys carried out recently by Russian specialists in the zones of mid-oceanic ridges and deep basins of the Atlantic, Pacific, and Indian Oceans. The presented kinematics simulations and geomagnetic profiles provide new information on the plate kinematics and on the deep structure of the active magnetic layer of the oceanic crust.

A comprehensive analysis of the results of magnetic simulation of transform faults and petromagnetic study of the rocks from the dredging data gave evidence of a complex AMF structure in the said zones and made it possible to reveal considerable effects of magnetic sources sited in the lower layers of the oceanic crust. The data from magnetic simulation and petromagnetic study of the lowermost oceanic crust testifies that the former suppositions about the active magnetic layer structure should be revised.

Paleooceanology, no doubt, has a great interest in the paleomagnetic data on the seamounts of the Atlantic, Pacific, and Indian Oceans, as well as of the Tyrrhenian Sea. The data illustrate the great advantages of magnetic simulations of seamounts when estimating the kinematics of the underlying oceanic lithosphere and investigating the nature of basalt volcanism.

Physical processes of the natural remnant magnetization of the oceanic crust and also the genesis of the primary titanomagnetite are considered in the monograph.

Interpretation of a large scope of experimental data obtained recently from extensive geomagnetic surveys over oceanic territories and from the petromagnetic study of the oceanic crust rocks is based on the modern geodynamic models and suppositions about the evolution of the oceanic crust and the Earth as a whole.

The manuscript was prepared with the financial support of the Russian Fund of Fundamental Research (Project Code 93-05-9218).

Alexander M. Gorodnitsky

THE EDITOR

Alexander M. Gorodnitsky, Ph.D., D.Sc.
Professor of Geology, P. P. Shirshov Institute of Oceanology, Russian Academy of Sciences, Moscow, Russia

Dr. Gorodnitsky was born on March 20, 1933, in Leningrad, Russia. In 1957 he graduated from the Geophysical Department of the Leningrad Mining Institute. He earned his D.Sc. degree (Doctor of Geology and Mineralogical Sciences) in 1982 at the P. P. Shirshov Institute of Oceanology. His subject of study was geology of the oceans and seas and his thesis was entitled "Structure of the Oceanic Lithosphere and Creation of the Seamounts." In 1988 Dr. Gorodnitsky earned his Ph.D. degree at the Geological Department of the Moscow State University. There his specialty was geophysics and his thesis was entitled "Geophysical Methods for Exploration of the Seamounts."

During the years 1957 to 1969, Dr. Gorodnitsky served at the Institute of Geology of the Arctic, Russian Ministry of Geology, Leningrad, in the capacity of Geophysicist-Researcher (1957 to 1969) and as Chief of the Sea Geophysical Laboratory (1969 to 1972). He then moved to the P. P. Shirshov Institute of Oceanology, Russian Academy of Sciences, Moscow, where he served as Senior Researcher (1972 to 1985), Chief of the Geomagnetic Laboratory (1985 to present), and Professor of Geology (1991 to present) specializing in geology of the oceans and seas.

Dr. Gorodnitsky is currently engaged in theoretical and experimental investigations of the structure of the oceanic lithosphere: its thickness and tectonic evolution; creation of the seamounts and islands; structure of anomalous magnetic fields (AMF) of the World Ocean; and the nature of the magnetic anomalies. He has participated in 22 scientific expeditions to various regions of the World Ocean, including diving with submersibles to the seamounts in the North Atlantic, Pacific, and Mediterranean. Dr. Gorodnitsky has authored more than 220 scientific works, including 7 monographs and books. Since 1992 he has been a member of the National Academy of Nature Sciences.

CONTRIBUTORS

I. I. Belyaev
Oceanology Institute
Moscow, Russia

V. Yu. Brusilovsky
Oceanology Institute
Moscow, Russia

G. E. Czerniawski
Oceanology Institute
Moscow, Russia

A. M. Filin
Oceanology Institute
Moscow, Russia

V. Yu. Glebovsky
VNIIOkeangeologiya
St. Petersburg, Russia

A. M. Gorodnitsky
Oceanology Institute
Moscow, Russia

A. G. Gorshkov
Oceanology Institute
Moscow, Russia

A. N. Ivanenko
Institute of Volcanic
  Geology and Geochemistry
Petro-Pavlovsk-Kamchatsky, Russia

S. V. Lukyanov
Oceanology Institute
Moscow, Russia

S. P. Maschenkov
VNIIOkeangeologiya
St. Petersburg, Russia

S. A. Mercuriev
Earth Magnetism and
  Radiowaves Institute
St. Petersburg, Russia

E. A. Popov
Oceanology Institute
Moscow, Russia

K. V. Popov
Oceanology Institute
Moscow, Russia

N. A. Sochevanova
Earth Magnetism and
  Radiowaves Institute
St. Petersburg, Russia

G. M. Valyashko
Oceanology Institute
Moscow, Russia

Chapter 1

Techniques for Marine Magnetic Measurements

I. I. Belyaev and A. M. Filin

CONTENTS

I. Instruments and Techniques for Scalar and
   Component Geomagnetic Measurements 1

II. Instruments and Technique for Gradient
   Geomagnetic Measurements 3

III. Gradiometer Survey for Removal of Geomagnetic Variations
   and in the Study of the Fine Spatial Structure of the Ocean's AMF 10
   A. The Barents Sea 11
   B. The Northeastern Pacific 13
   C. Seamounts of the Tyrrhenian Sea 14

I. INSTRUMENTS AND TECHNIQUES FOR SCALAR AND COMPONENT GEOMAGNETIC MEASUREMENTS

Measurements of the Earth's total magnetic field intensity, or scalar measurements, have become the priority in marine magnetic research. This research uses mainly proton and optically pumped magnetometers. Proton magnetometers are being widely used due to operational simplicity and reliability of their sensors and higher measurement accuracy. The latest models of proton magnetometers are equipped with special processors that allow statistical processing of the signal. The processors increase considerably the sensitivity and accuracy of proton magnetometers.

Marine optically pumped magnetometers, namely, KM-2, KM-2M, and KM-2U were used widely 10 to 15 years ago (Krasil'nikov et al., 1975; Lyubimov and Perfilov, 1980). Later, they were almost entirely replaced by proton magnetometers. At present, only the magnetometer KM-2U is used. Its technical parameters are given in Table 1. Recently, absolute accuracy of optically pumped magnetometers has been increased greatly. They are not only highly competitive with proton magnetometers, but in some cases are superior to the latter. Therefore, in the near future, optically pumped magnetometers may be widely used in marine research, in particular, for gradient measurements.

Most of the Russian modern proton magnetometers have sensitivities of 0.1 nT and better and measurement accuracies of 0.1 to 0.5 nT. Table 1 shows the technical characteristics of the Russian modern scalar magnetometers. However, the increased accuracy of marine magnetometers and the satellite radio navigation systems used did not stimulate any sharp increase in the precision of magnetic surveys. This can be explained by the shortage of industrially produced marine magnetic variation base stations (MVS) and also by certain difficulties arising in the course of their exploitation during surveys. Therefore, the usual accuracy of surveys in the ocean falls within the range of 10 to 30 nT and more.

Table 1 **Marine magnetometers**

Magnetometer and type	Model	Sensitivity (nT)	Range K=1.000 (nT)	Sampling time (s)	Accuracy (nT)	Readout A=analog D=digital	Ref.
Proton precession	MBM-1	0.1 1	20-100k	10,30,60 2	±0.5 ±1	A & D	Livotov et al., 1979
	MPM-5M	0.1	20-70k	1,3,5,10,20,60	±1	A & D	Afonyashin et al., 1984
	MMP-2M	0.01	20-100k	10	±0.15	A & D	Uglov et al., 1986
	MPM-7	0.1	20-85k	5,10,20,60 2	±0.5 ±1	A & D	Zagurcky et al., 1988
	MAMP-0.1	0.1	20-100k	>4 2,3	±0.5 ±1	A & D	Balyberdin et al., 1987
Cesium-vapor	KM-2U	0.01-10	15-30k 30-70k	0.02-60	±5 ±3	A & D	Lyubimov and Perfilov, 1980

To measure the geomagnetic field components optically pumped and fluxgate magnetometers are used. The main difficulty in the elaboration of component magnetometers is the creation of high-accuracy stabilizing systems, which help to set and maintain the position of magnetosensitive or supplementary elements (of a ring system, for instance). The most popular are pendulum stabilization by gimbals and gyro-stabilization.

The magnetic declination D is determined usually by the measurements of the angle between the geographic azimuth (determined with a gyrocompass) and the magnetic azimuth (determined with the orientation sensor of a fluxgate device).

The main portion of marine component magnetic measurements was made by the St.Petersburg Division of the Institute of the Earth's Magnetism and Radio Waves Propagation (Russian Academy of Sciences) aboard the vessel *Zarya*. The measurements were taken with the special shipboard acquisition system including the active gyro-vertical SU-2, digital fluxgate magnetometer, optically pumped H-magnetometer, developed airborne magnetometer AMM-13, towed proton magnetometer, and microcomputer system "Micron" (Tsutskarev et al., 1986). Other Russian geophysical institutions carry out component measurements only episodically.

When geomagnetic measurements are made from the ocean surface, a significant part of information useful for geological interpretation is lost because of the large distance between magnetic sources and the surface of measurements. This reduces the application of magnetic methods to problems such as investigations of the fine lateral magnetic field structure near magnetic bodies and the search for local magnetic objects near the sea bottom (connected with ore deposits).

The deep-sea towed system "Zvuk-6" was created in the P. P. Shirshov Institute of Oceanology, Russian Academy of Sciences (Onischenko et al., 1986). This system includes the proton magnetometer MPM-6 developed at the "Rudgeophysica" Design Bureau in cooperation with the Institute of Oceanology (IO RAS) for near-bottom total magnetic field measurements (Belyaev et al., 1987a). The magnetometer consists of a measuring device, operator's terminal, analog chart recorder, fish with sensor, alarm disconnection system, and a 25-m towing cable of KNG-2 type. The magnetometer is shown in Figure 1. The technical specifications of the instrument are measuring range, 25,000 to 80,000 nT; sensitivity, 0.1 nT; accuracy, ±0.5 nT; sampling rate, 4, 10, 30, 60 s; registration, analog and digital. The operator's terminal is connected to the measuring device by a telemetric system. The weight of the fish with sensor is 19 kg and it is: 180 mm in diameter and 800 mm in length. The weight of magnetometer with cable is about 70 kg.

Figure 1 Deep-tow proton magnetometer model MPM-6: 1, fish; 2, measurement device; 3, operator's terminal; 4, alarm disconnection system; 5, sensor.

Tests in the northeastern Atlantic were carried out in the vicinity of the Cruiser Seamount to develop a technique for bottom geomagnetic measurements (Yastrebov et al., 1991). The data acquisition and processing results are given in Chapter 2.

The creation of the pilot-operated submersible vehicles (POVs) allowed direct geomagnetic measurements at the sea bottom to be carried out under both mobile and stable conditions, accompanied by sampling of oriented specimens and drill cores directly from the analyzed objects. This is a new step in the survey accuracy, which allows the solution of a number of fundamental and applied problems.

The two "MIR" POVs with operating depths of 6 km, used by IO RAS are equipped with component magnetometers JH-16/RR. These magnetometers are designed for measuring the total magnetic field intensity in the orthogonal reference system (X, Y, Z) together with the magnetic declination. They have sensors of the fluxgate type. The range of ±100,000 nT is divided into 200 equal subranges of ±2,000 nT, each with sensitivity of 1 nT, and into subranges of ±20,000 nT, each with the sensitivity of 10 nT.

II. INSTRUMENTS AND TECHNIQUE FOR GRADIENT GEOMAGNETIC MEASUREMENTS

Proper allowance must be made for the geomagnetic field time variations, which introduce large errors into the measurements, in particular, in the high-latitude auroral and near-equatorial zones, since time variation is one of the principal factors affecting the precision of marine magnetic measurements.

The spectrum of variations is wide and overlaps completely the spectrum of geologically meaningful anomalies. Variation amplitudes lie within the range of several parts of a nanotesla to several hundred nanoteslas, and in the case of a magnetic storm, they may reach 2000 to 3000 nT.

It is obvious that the proper allowance for temporal magnetic field variations must be made in high-precision marine magnetic surveys. At present, both direct and indirect techniques are used to measure temporal variations.

Direct techniques are used to measure variations in marine MVS, and so are the data of coastal MVS following extraction of temporal magnetic field variations generated in the magnetosphere and ionosphere (MIMF).

However, due to the dynamic proporties and electrical conductivity of the media, marine MVS observations are not easily performed. Thus, if a MVS is installed in the near-surface water layer (buoy stations), the drift of the station produces an error in measurements. When the station is at the bottom (bottom stations) screening of the short-period variations occurs due to the thick layer of water, which displays considerable electrical conductivity. As a result, there is a discrepancy in the records of the buoy and bottom stations.

The current direct techniques of measuring geomagnetic variations are based on the linear interpolation of data MVS from located in the region of the magnetic survey; the techniques are primarily used at the survey site because the error in the linear interpolation of MIMF increases as the distance between the reference MVS and the survey point increases.

As we know, an anomalous gradient zone of field variation occurs at the shore/ocean boundary and in the seamount regions, which causes the shift of the variation phase and a sharp change in the amplitude. This hinders considerably the usage of coastal MVS to account for the temporal variations in marine surveys.

Along with direct techniques for measuring temporal variations, indirect techniques are also elaborated and used by the results of the magnetic survey proper. These are based on the analysis of the discrepancy at the point of the intersection of profiles (Gordin, 1980; Pylaeva and Roze, 1981). The indirect techniques permit one to find with a certain reliability only the long-period disturbance of the magnetic field of the Earth. However, the reliability also depends considerably upon the navigation quality. A frequency range of 10 to 60 min is hardly open for the indirect techniques. This forced one to search for measurement modes free of the perturbation effect of geomagnetic variations, one of which is marine gradiometry — recently developed to advantage in some countries.

The gradiometry technique is based on the synchronous measurements of the Earth's total magnetic field intensity (TMFI) or its components at two or more points at some distance from each other. The ratio of the total field difference to the distance between points allows one to estimate the derivative of the field with respect to the vector joining the points of measurements. The field gradient $T(r)$ characterizes the rate of field variations with respect to the vector r and may be expressed as:

$$\frac{dT}{dr} = \lim_{\Delta r \to 0} \frac{T_r - T_{(r+\Delta r)}}{\Delta r} \approx \frac{\Delta T}{\Delta r} \quad (1)$$

where $\Delta T = T_r - T_{(r + \Delta r)}$ is the total field differential between two sensor positions spaced Δr apart; Δr is small compared to the distance to the nearest source of magnetic anomalies. In practice, the following condition should be observed:

$$\Delta r \leq (0.1 \text{ to } 0.2) r_m \quad (2)$$

where r_m is the distance to the source of magnetic anomalies.

Registration of the total field differential ΔT is performed by gradiometers. Three principal techniques of gradient measurements are elaborated in accordance with the position of sensors with respect to the direction of the vessel, namely, longitudinal, transverse, and vertical. Measuring the vertical gradient seems to be most interesting, but is difficult to perform in marine conditions. In the present work we consider the longitudinal configuration for measurements of the magnetic field gradient in the direction of vessel movement.

The gradiometer reading in the field difference channel, with no account for the instrument and procedure errors, could be represented as

$$\frac{\Delta T}{\Delta X} = \frac{T_1 - T_2}{\Delta X} = \frac{[T_1(x) + \delta T_1(t)] - [T_2(x + \Delta X) + \delta T_2(t)]}{\Delta X} \tag{3}$$

where T_1 is the combined field value measured by the first gradiometer sensor, which includes the constant $T_1(x)$ and variable $\delta T_1(t)$ magnetic field components; T_2 is the same value for the second sensor separated from the first by a distance ΔX, called the base of measurements. Since the sources of magnetic field time variations are in the Earth's ionosphere and magnetosphere and are separated from the Earth's surface by several hundred kilometers, which is considerably greater than the base of measurements ΔX,

$$\Delta X \ll r_v \tag{4}$$

where r_v is the distance to the source of variations, the effects of time variations on each gradiometer sensor are equal. Therefore,

$$\delta T_1(t) = \delta T_2(t) \tag{5}$$

and expression (3) could be rewritten as

$$\frac{\Delta T}{\Delta X} = \frac{T_1(x) - T_2(x + \Delta X)}{\Delta X} = G(x) \tag{6}$$

This expression shows the principle of the gradient measuring technique. When both conditions (2) and (4) are fulfilled, the gradiometer measures the derivative of the geomagnetic field with respect to the given direction and these measurements are free of the geomagnetic variation effect.

If information on the reference value of the total field $T(0)$ at some point P_0 and continuous recording of the gradiometer differential channel $\Delta T(l)$ along the direction of the vessel travel are available, it is possible to determine the constant component of the magnetic field at any point of profile $P(t)$ as (Roze and Markov, 1984):

$$T(p) = T(0) + \frac{1}{l_0} \int_{P_0}^{P(t)} \Delta T(l) \, dl \tag{7}$$

where l_0 is the gradiometer base.

The reference value of $T(0)$ could be determined either from marine MVS measurements or by indirect techniques at the survey site.

Upon integrating the value of the gradiometer differential channel, we obtain magnetic field values free of magnetic variations. By this means differential measurements characterize only the constant magnetic field. Information on the distribution of time variations along the profile may be obtained by subtraction of the calculated values $T(p)$ from the data of one of the gradiometer channels:

$$\delta T(t) = T_1(p) - \left[T(0) + \frac{1}{l_0} \int_{P_0}^{P(t)} \Delta T(l) dl \right] \quad (8)$$

The principal advantage of the gradiometer technique is the opportunity to obtain magnetic field data free of magnetic time variations. The separation of the total field into constant and variable components occurs with the precision of a linear constituent. To exclude the latter, it is necessary, from time to time, to return to the reference point, i.e., the MVS (Gordin et al., 1986).

The first marine magnetic gradient measurements were attempted in the former USSR with a set of two fluxgate magnetometers (Novysh et al., 1974).

Theoretical and modelling investigations performed by E. N. Roze and R. B. Semevsky showed the advantages of the gradient technique (Roze, 1973; Semevsky, 1976; Semevsky et al., 1977; Roze, 1978; Roze and Semevsky, 1986). Beginning with the latter half of the 1980s, gradient measurements were carried out with a set of two modular proton magnetometers; also, gradiometers were elaborated to measure longitudinal gradient, namely, the proton gradiometer DPM-1 (Filin et al., 1985) and the optical helium gradiometer KMMG-1 (Uglov, 1984).

At present, several gradiometers have been developed, namely, DPM-2 (Figure 2), MMP-2D, and MAMP-D (Figure 3).

Table 2 presents the technical parameters of the Russian-origin marine gradiometers used at present.

The gradiometry technique has some specific features and depends to some extent upon the problems stated. We can list some of them:

1. study of horizontal gradients of the stationary magnetic field of the Earth (EMF);
2. estimation of the anomalous magnetic field (AMF) of the Earth reconstructed by the integration of the gradient;
3. recognition of the EMF time variations.

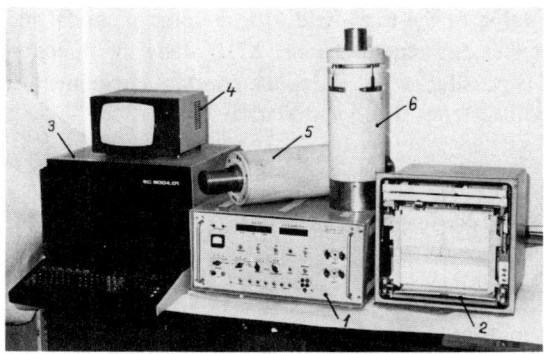

Figure 2 Proton gradiometer model DPM-2: 1, measurement device; 2, dual-channel strip chart recorder; 3, tape recorder model UPDML EC-9004; 4, monitor of tape recorder; 5, forward "Slave" fish; 6, rear "Master" fish.

Table 2 **Marine magnetic gradiometers**

Gradio-meter and type	Model	Sensi-tivity (nT)	Range k=1.000 (nT)	Sampling time (s)	Accu-racy (nT)	Readout A=analog D=digital	Ref.
Proton preces-sion	DPM-2	0.1	20-70k	3,5,10,20, 60	±0.5	A & D	Belyaev and Filin, 1990
	MMP-2D	0.01	20-100k	10	±0.15	A & D	Uglov et al., 1986
	MAMP-D	0.1	20-75k	4-99	±0.5	A & D	Balyberdin et al., 1987

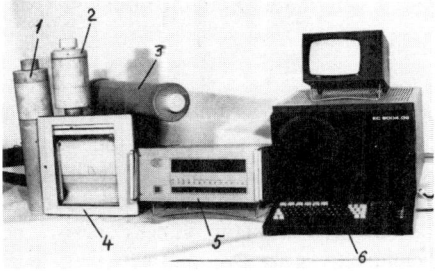

Figure 3 Proton gradiometer mode MAMP-D: 1, rear Master fish; 2, forward Slave fish; 3, previous amplifier; 4, analog chart recorder; 5, measurement device; 6, tape recorder.

A twosensor system is usually used in gradiometry works. The system is realized in two ways, namely, towing by one and by two cables.

The base of measurement depends upon the problems to be solved and upon the amplitude-frequency characteristics of the AMF of the given region, as well as upon the accuracy of gradiometer data. In each case, there is an optimal base length (Markov, 1986):

$$l_0 = \rho \sqrt[6]{2\tau\sigma_\epsilon^2 / 3F\sigma_a^2} \qquad (9)$$

where ρ is the correlation radius; τ is the sampling rate; F is the minimum period of the signal allowed by the low-frequency filter; σ_ϵ is the variance of a random irregular noise; σ_a is the variance of the AMF.

According to estimations, the base length may vary from 50 to 400 m.

Numerous measurements of the horizontal gradient in the deep parts of the Atlantic, Pacific, and Indian Oceans, as well as of the Barents Sea, showed that the base value should lie within the range of 100 to 200 m. When measurements are performed in shelf regions, at mid-oceanic ridges, or at seamounts reaching the ocean surface, the base varies within the range of 50 to 100 m.

When the gradiometer is towed, the fishes with sensors are at various depths and, thus, the factual length of the base should vary: the lower the speed of the vessel, the greater the variations.

The sampling rate depends upon the problem to be solved and upon the character of the AMF analyzed. When studying the stationary magnetic field, the measurement cycle is determined from:

$$v\tau \le x_d / 7 \text{ to } 10 \qquad (10)$$

where v is velocity of the vessel; x_d is the minimum base length of the magnetic anomaly for the survey area, which should be calculated from 7 to 10 data readings. According to Roze and Markov (1984) the sampling rate is determined from:

$$\tau_0 \le \tau < 0.1 \rho / v \qquad (11)$$

where τ_0 is the correlation radius of the lowest-frequency harmonics of the noise in the differential measuring channel which exceeds the instrumental noise in amplitude; ρ is a spatial correlation radius for the measured field gradient.

When analyzing the pattern of EMF time variations in the area studied, the sampling rate is determined from the ratio:

$$\tau \le \tau_{min} / 7 \text{ to } 10 \qquad (12)$$

where τ_{min} is the minimum period of variations.

From the given ratio and the need to have several measurements of the gradient when it exceeds the length of the base, the sampling rate in the ocean should be up to 10 to 20 s and should not exceed 5 s in the shelf areas.

The gradiometer survey error includes both systematic and random parts. Instantaneous error of the EMF, $\varepsilon(t)$, may be calculated as (Roze and Semevsky, 1986):

$$\varepsilon(t) = at + b + \eta(t) + \xi(t) \qquad (13)$$

where $a = \delta v / l_0$, δ is the net absolute error of the gradiometer differential channel measurements; v is velocity of the vessel; l_0 is the gradiometer base; b is the combined absolute error of the reference value determination at each profile; $\eta(t)$ are random errors due to the measurement of longitudinal gradients at the base with a finite length, which are stationary in the case of statistical homogeneity of the anomalous EMF; $\xi(t)$ are random, nonstationary errors originating from the integration of random errors in the gradiometer differential channel.

The principal errors in gradient measurements result from: (1) instrumental error, (2) instability of the instrument base, and (3) deviational error.

Since the gradiometer is used for measuring the finite field differences, its systematic error may be determined from the difference between systematic errors of each modular channel, which is 0.5 to 1 nT for modern proton magnetometers. The random error of the gradiometer is 0.1 to 1.0 nT.

The peculiar error in gradiometer measurements is due to the instability of the instrument base and its deflection from the general direction of the towing vessel. Experimental data (Uglov and Lygin, 1988) showed that the horizontal base dimensions do not vary considerably (not more than 2%) when the towing velocity changes from 5 to 9 knots. The main part of the error due to the changing parameters of the gradiometer fishes towing system configuration results from the effect of the EMF, vertical gradient. The relative vertical positions of the fishes depend considerably upon the vessels velocity, the towing length of the fish, and upon the fish and cable types.

When performing a gradiometer survey, constant velocity of the vessel is one of the principal parameters. The most stable velocity that can be achieved varies within ±1 knot. At greater velocities, false anomalies may originate from the changing relative vertical positions of the fishes and from growth of vertical gradient influences, particularly in high-anomaly fields. One should also bear in mind that when the vessel changes its heading, the towing depth of the fish changes due to a decrease in the velocity. It takes some time to stabilize the fish in a vertical position after the vessel turns. This time depends upon the vessel type, its velocity, and the weather conditions and is approximately 5 to 7 min.

One of the most important effects in gradiometer surveys is the effect of the magnetic field of the vessel, called deviation effect. It depends upon the mass of the vessel, distance between the fish and the vessel, the self-magnetic field of the fish and the sensor transducer, and also upon the area studied. When integrating, the error from the deviation causes great errors in reconstructing the EMF. As a rule, it is almost impossible to reduce the deviation to 0.3 to 0.5 nT. Thus, when performing a gradiometer survey, the deviation should be determined for each area studied and be considered in the data processing. The experimental data, obtained in the Atlantic and Indian Oceans by R/V *Dmitry Mendeleev* (length, 110 m; displacement, 6900 metric tons), are presented below. Figure 4a shows the function of the magnetic effect of the vessel upon the towing length of the fish. Measurements were taken in two headings (10° and 190°), when deviation was maximum. As is clear from the figure, the effect of the vessel is small when the towing length exceeds 400 m. Thus, to avoid the magnetic effect of the vessel, the towing length should be 4 to 5 times greater than the length of the vessel itself.

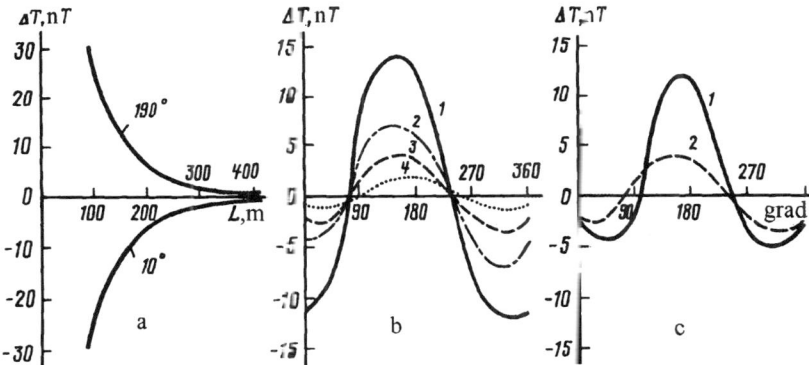

Figure 4 Magnetic effect of towing vessel.

Figure 4b shows the deviation curves at various towing lengths (from 150 to 300 m). Figure 4c illustrates the influence of the area studied upon deviation. These deviation curves were obtained in the areas with various inclinations of EMF (25° and 65°). It is clear that maximum deviation differs from region to region by a factor of three.

Therefore, when the profiles are long, in particular in the longitudinal direction, it is difficult to account for the deviation effect. That is also true for the deviation changes during the site survey, where the heading of the vessel is maintained with a definite error. If the total deviation effect is great (over 2 to 3 nT), the yaw of the vessel may cause considerable deviations along the profile line. Some ways to solve this problem have been proposed. Roze and Markov (1984) proposed a procedure for determining the real gradiometer base and for direct estimation of the azimuth deviation errors. This procedure provides for a higher accuracy compared to the traditional procedure called a "star" (measurements along eight profiles at different headings, which cross at one point near the center of the area studied), but it depends strongly upon geomagnetic variations. Another procedure, worked out at Yuzhmorgeo Company and Moscow State University is more sophisticated (Leibov et al., 1986), and, according to the authors, is highly efficient only when no high-frequency variations occur.

It is also possible to account for a strong magnetic effect of the vessel by using a multi-sensor longitudinal towing system, the theoretical basis of which was considered in a number of papers (Tomoda and Fuimotu, 1982; Roze and Semevsky, 1986; Leibov et al., 1988). In 1990, during the 20th cruise of R/V *Akademik Mstislav Keldysh*, a three-sensor longitudinal gradiometer system was used in the experimental works of the IO RAS. The system is characterized by equal bases of measurements, of 150 m each, between the nearest and intermediate, and also between the intermediate and far fishes. Results of the three-sensor gradiometer data processing are considered in Chapter 2.

Length of the profile and survey scale. There are no restrictions on the profile length while measuring the distribution of the longitudinal gradient along the regional profile. However, when estimating the stationary AMF or variable field δT, the length of the profile is restricted since summation of errors occurs in the process of integration. Errors of the integration should not exceed the given limit, defined by the required accuracy of the survey. Error variance in the recognition of the stationary field varies for a nearly linear function at lengthy profiles, and the filtering of the initial data cannot reduce significantly the error for the restoration of the constant magnetic field (Markov, 1986).

The survey scale, or the grid size, is specified by the problems to be solved by the survey. The minimum distance between profiles is calculated from (Fastovsky, 1989):

$$d \geq (2 \text{ to } 3)\sigma_n \qquad (14)$$

where σ_n is a mean square error of the navigation system.

The distance between profiles is specified with respect to the stated geological problem, bearing in mind the most-complete three-dimensional description possible of the expected anomalies.

Estimating the gradiometer survey accuracy is a complicated problem since the values to be measured depend on the survey direction. Some authors (Lashkov et al., 1988) proposed estimating the gradiometer survey accuracy by comparing the repeated profiles. However, this does not ensure a comprehensive accuracy estimate for the entire survey. Therefore, it seems expedient to estimate the survey by comparing integrated values at the intersection points of profiles (Markov et al., 1981).

III. GRADIOMETER SURVEY FOR REMOVAL OF GEOMAGNETIC VARIATIONS AND IN THE STUDY OF THE FINE SPATIAL STRUCTURE OF THE OCEAN'S AMF

By 1990, the laboratory of geomagnetic research of the P. P. Shirshov Institute of Oceanology had performed numerous gradient hydromagnetic surveys during 11 cruises in various regions of the World Ocean with the instruments MPM-5M, DPM-1, and DPM-2. The total amount of gradient magnetic data is over 60,000 lines per kilometer.

The first measurements of the longitudinal gradient were performed with the gradiometer DPM-1 in 1977 during the combined geological-geophysical survey at

Baikal Lake (Belyaev et al., 1975). Several illustrations of gradient magnetic surveys are given below.

A. THE BARENTS SEA

In 1987, the integrated geophysical studies were carried out in the Barents and Kara Seas during the 19th cruise of R/V *Professor Shtockman*, including echo-sounding, continuous seismic profiling (CSP), and magnetic survey. Geomagnetic survey was performed with the gradiometer DPM-2. Since the bottom of the Barents Sea is covered by thick sediments and has a rather quiet magnetic field, and the velocity of the vessel was not high, the base for longitudinal gradient measurements was 150 m. The first fish of the gradiometer DPM-2 was towed 300 m astern, whereas the second was 450 m astern (the vessel length being 69 m). The azimuth deviation during the survey did not exceed the noise level of 0.5 to 0.7 nT as determined in the course of processing. The sampling rate was 5 s, which allowed the mapping of the magnetic field every 20 to 30 m.

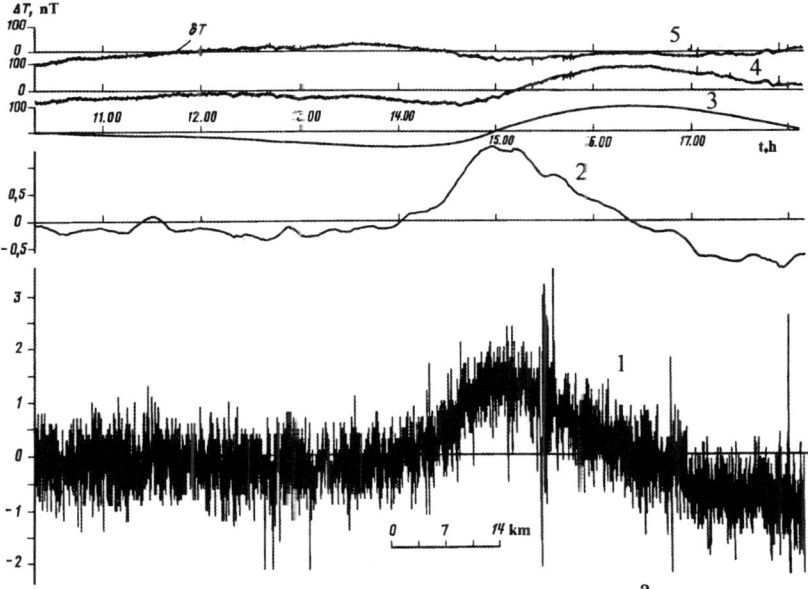

Figure 5 Gradient measurement processing of results obtained during (a) a quiescent variation day and (b) a magnetic storm: 1, measured longitudinal gradient on base 150 m; 2, optimal filtering result of longitudinal gradient; 3, computed total field (integrated gradient); 4, measured total field; 5, computed magnetic variation.

Figure 5 shows the results of the processing of the total field intensity and its longitudinal gradient. As follows from the picture, the gradient data are complicated by a high-frequency noise of 0.57 nT. Results of the filtering obtained by applying the adaptive families showed that a high-frequency mode is excluded from the primary data. In the filtered gradient field the anomalies of up to 1.5 nT with periods of about 20 km are recognized. They are complicated by local anomalies with smaller amplitudes and periods of no more than 2 km. From CSP and geological sampling data, these local anomalies are of geological origin and are identified with the wash swells of the subsided banks or with glacial transportation of detrital rock material.

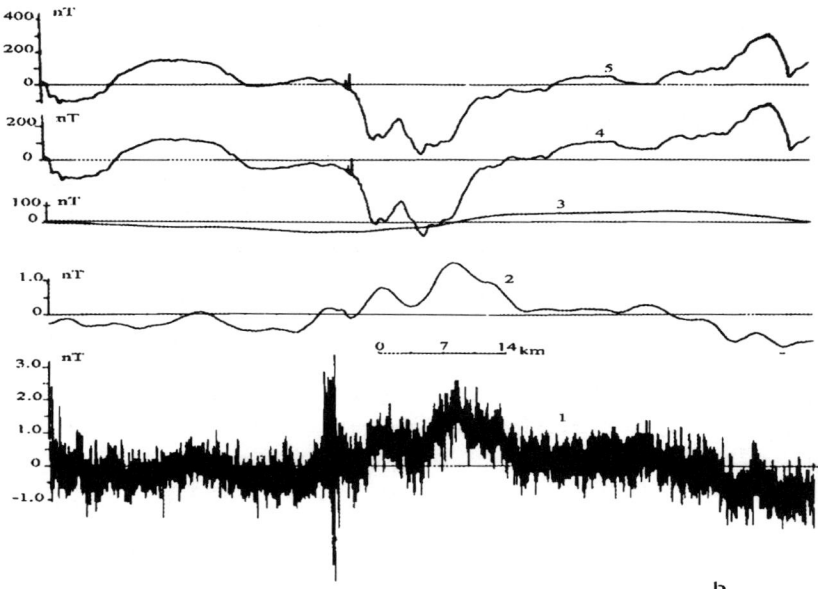

Figure 5 (continued).

It follows from the processing results that this solution of the filtering problem, resolving anomalies of two classes, allows the numerical high-precision integration of the longitudinal gradient data. When no deviation effect occurs in the gradient data, the applied technique of processing allows successful processing of great data massives. In fact, a curve of the integrated field in Figure 5a conforms to that of the measured field. Since there are no marine MVS in the study area, the primary value of the field, $T(0)$, was assumed to be zero. From the records of the coastal magnetic observatory *Loparskaya*, located in the vicinity of the city of Murmansk, that day was marked by quiet variations.

Quite a different situation is observed in Figure 5b. Integrated and measured curves of the total field differ sharply from each other. High-amplitude aperiodic variations of the magnetic storm type with amplitudes of up to 500 nT were observed on that day. Records of the coastal magnetic observatory *Loparskaya* confirm the occurrence of a magnetic storm at that time.

Analysis of the survey data allows the supposition on the existence of two types of anomalies within the Barents Sea area, namely, long-period with periods of 30 to 60 km and amplitudes of 50 to 70 nT, caused by deep magnetic objects, and short-period local anomalies with periods of 0.3 to 0.7 km and amplitudes of several nanoteslas, confined to the shallow waters (Valyashko et al., 1990a).

From the experience of American scientists, gradient surveys can be used for the exploration of hydrocarbon deposits on the ocean shelf in high-latitude areas since this technique allows the registration of small-amplitude magnetic anomalies in the presence of magnetic field time variations (Cunningham et al., 1987).

Since the Barents and Kara Seas have a high oil and gas exploration potential, short-period, small-amplitude magnetic anomalies recognized there should be analyzed thoroughly.

Aperiodic time variations of the magnetic storm type have been determined with the gradiometer DPM-2 in the auroral zone of the Barents and Kara shelves. Together with the data of on-land magnetic observatories and buoy and bottom MVS, this material may be used to study the origin and distribution of the sources of the ionospheric variations of the EMF and also makes it possible to embark on a study of the fine spatial AMF structure, which was hardly possible before. The instruments and techniques used in the data processing allowed the recognition of magnetic anomalies with amplitudes of 0.5 nT with the presence of time variations up to 500 nT.

B. THE NORTHEASTERN PACIFIC

Experimental surveys were carried out with the gradiometer DPM-1 in 1982 during the 29th cruise of R/V *Dmitry Mendeleev* (Filin et al., 1985). The Northeastern Pacific, in particular the region of magnetic lineations along 42°N profile, was chosen as the study area. Fishes were towed at distances of 280 and 450 m astern.

The experiment was aimed at testing the efficiency of the proton gradiometer DPM-1 in studying the fine spatial AMF structure in the presence of sources of small-amplitude anomalies as far as 4 to 5 km below the survey plane. The gradiometer survey was performed in the region where lineations from 9 to 32 were registered earlier. Gradient measurements helped to distinguish clearly the fine structure of anomaly lineations. Figure 6 (upper graphs) shows a fragment of a summary plot of an AMF and longitudinal gradient at a latitudinal profile in the vicinity of linear anomalies 13 and 15. Despite the fact that these short-period anomalies are seen in the graph of the AMF, they have magnitudes of no more than 50 nT. Their confident recognition is difficult in the event of time variations. Meanwhile, the fine structure of the AMF within one period of the same magnetic polarity is recognized with sufficient reliability in the graph of the gradient.

Some years ago, Leibov and Mirlin considered the problem of identification of small, short-period magnetic anomalies in the vicinity of the East Pacific Rise, where the spreading rate reaches 5 cm per year and more and where magnetic anomalies 6 to 15 have regular structure (Leibov and Mirlin, 1979). The authors resolved 19 short-period anomalies in the gap between the earlier identified anomalies 6 and 15. Almost all these anomalies were traced in other regions of the Pacific and Atlantic. Using the inverse filtering procedure to investigate the origin of the magnetic field fine structure between anomalies 6 and 15, the age of which is 23 to 37 million years (Oligocene), the authors showed that the recognized short-period anomalies may correspond to short inversions of the EMF.

The gradiometer survey carried out on the 29th cruise of R/V *Dmitry Mendeleev* showed with more confidence the geological origin of small-amplitude (25 to 30 nT) anomalies, the existence of which was doubted earlier because it was impossible to account for time variations. The graph of the longitudinal gradient is shown in Figure 6. Short-period anomalies are clearly seen within the negative part of linear anomaly 13, their periods ranging from 3 to 9 km and amplitudes reaching 20 nT/km. These anomalies do not seem to be related to the magnetic field variations but are of geological origin.

Analysis of the gradient data showed that in the event that the horizontal gradient is measured at a distance of 4 to 5 km over the field source with a base of 170 m within the areas of standard oceanic crust and typical linear anomalies, its value reaches 10 nT/km, which is sufficient for its reliable registration. Therefore, the observed values of the horizontal gradient may be used in direct and inverse problems.

The fine spatial structure of the magnetic field is most pronounced in the areas corresponding to the epochs of constant magnetic field polarity, marked by the numbered linear magnetic anomalies.

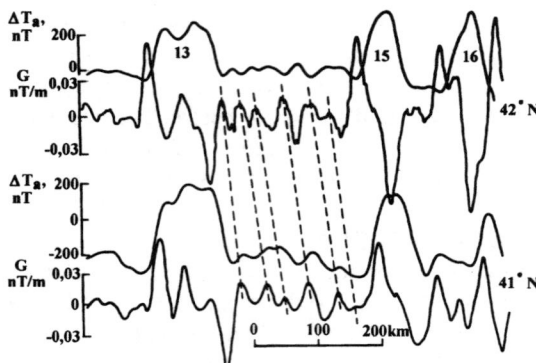

Figure 6 Fine spatial structure of the anomalous magnetic field in the Northeastern Pacific: ΔT_a, anomalous magnetic field; G, longitudinal gradient.

It follows from the analysis of the data on the anomalous structure of the longitudinal gradient that the width of the transient zone between the blocks of direct and inverse magnetization is constant irrespective of the duration of magnetic lineation polarity. The width of this zone is about 5 to 8 km when spaced 5 km, on average, from the magnetic sources. The magnitude of the longitudinal gradient anomalies reaches its maximum in these zones.

The gradiometer system MPM-5M (two scalar magnetometers working simultaneously) was used for measurements on the 42nd cruise of R/V *Dmitry Mendeleev* in 1988. The survey covered the latitudinal profile 60 miles south of the profile made during the 29th cruise of the same vessel. Figure 6 (lower graphs) shows the curves of the AMF and its longitudinal gradient. Despite considerable distance between the profiles, short-period anomalies of small amplitudes correlate well in the gradient graphs.

Thus, the gradiometer survey in the Northeastern Pacific made it possible to study in detail the fine spatial structure of the AMF in the area between linear anomalies 9 and 15. The reliability of short-period, small-amplitude magnetic anomalies of global origin could be proved by gradiometer measurements and they may be used in compilation and refinement of magnetic geochronological scales.

C. SEAMOUNTS OF THE TYRRHENIAN SEA

Seamounts, which are usually of volcanic origin, are some of the most promising objects for geomagnetic study by the gradiometer technique. The magnetization inhomogeneity of igneous rocks composing these mountains is reflected directly in the fine spatial structure of the AMF (FSS AMF).

A detailed gradiometer survey was performed in 1984 during the 7th cruise of R/V *Vityaz* at the Vavilov Seamount within the test area of 39°47′ to 39°55′N and 12°32′ to 12°44′ E (see Chapter 5). The survey followed the net of mutually perpendicular latitudinal and meridional profiles spaced 0.5 mile (latitudinal) and 1 mile (meridional) apart. The measurement base for the longitudinal gradient was 200 m.

The mean square error of the geomagnetic survey at the test area was ±10.5 nT. The total length of a detailed gradient survey was 181 miles (Filin et al., 1988).

Two maps of the longitudinal gradient field have been compiled from the results of the gradient measurements: the map of G_x based on the meridional profiles data (Figure 7a), and the map of G_y based on the latitudinal profiles data (Figure 7b). As can be

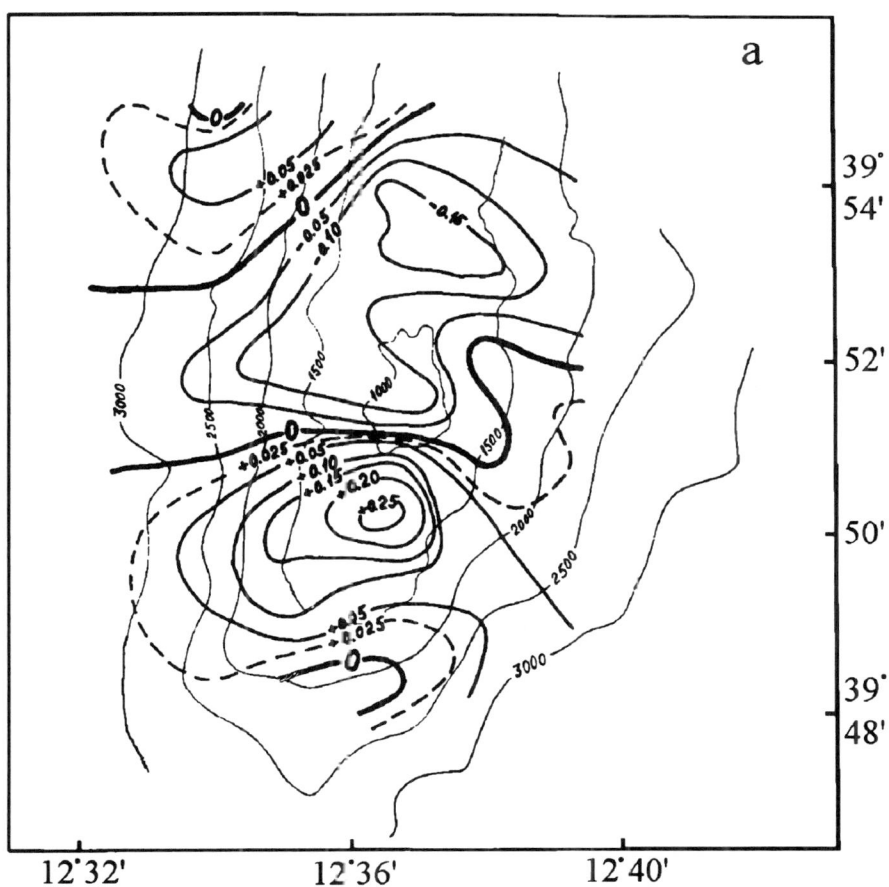

Figure 7 Intensity maps of longitudinal gradient over (a) meridional and (b) latitudinal profiles of the Vavilov Seamount: 1, longitudinal gradient (nT/m); 2, bathymetric contours (m).

seen from Figures 7a,b the largest longitudinal gradient anomalies — up to 0.2 to 0.25 nT/m — were obtained on the meridional profiles. The zone of the largest gradients G_x (Figure 7a) is confined to the southern slope of the mountain dissected by deeply cut canyons with depths of 1000 to 1200 m. Over the center of the mountain summit and the top of its western slope, a negative G_x zone reaching 0.1 to 0.15 nT/m is registered. These zones are isometric in plan and usually have a sublatitudinal strike.

In their turn, the largest values of the latitudinal gradient G_y (Figure 7b) reach +0.15 and –0.15 nT/m and are confined to the mountain summit with a minimum of 725 m, where G_y equals –0.15 nT/m, and also to the steep western slope where the value of +0.15 nT/m is registered at the bottom, at a depth of 1600 m.

The map of latitudinal gradients shows a submeridional strike of anomalies, which seems to be related to linear magnetic features of the same strike. This agrees well with geological data (Selli et al., 1977), according to which the largest volcanoes of the Central Tyrrhenian Sea are confined to the submeridional faults with the 15°-azimuth strike.

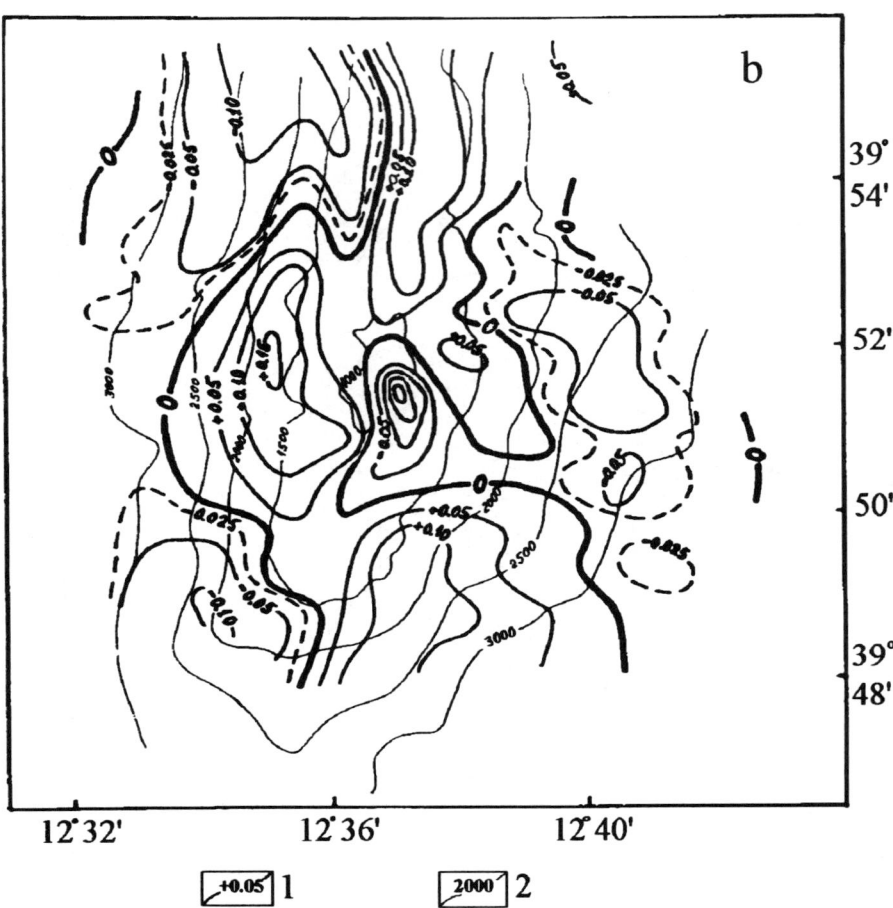

Figure 7 (continued). Intensity maps of longitudinal gradient over (a) meridional and (b) latitudinal profiles of the Vavilov Seamount: 1, longitudinal gradient (nT/m); 2, bathymetric contours (m).

The maps of the longitudinal magnetic field gradients data taken in meridional and latitudinal directions, together with total field intensity and bathymetry maps, were used in constructing magnetic models of the Vavilov Seamount (Filin et al., 1988).

The maps of the orthogonal constituents of a horizontal gradient better reflect graphically the FSS AMF. At the same time they depend strongly upon the chosen direction of profiles. The most easy-to-grasp picture of magnetic objects on a plane configuration is presented in the map of the total horizontal gradient intensity compiled for the Vavilov Seamount (Figure 8). The intensity is determined from the longitudinal gradient taken from the maps of meridional and latitudinal gradients according to the formula:

$$G_{xy} = \left|(G_x^2 + G_y^2)^{1/2}\right| \qquad (15)$$

The gradients are estimated in the interprofile space after linear interpolation of the longitudinal gradients data. The intensity map of a total horizontal gradient gives information about the boundaries and structure of anomalous zones and the geometry of the sources. Four anomalous zones are distinguished in Figure 8. They are confined, respectively, to the mountain summit, and to its southern, northern, and western slopes. As follows from the figure, such a map correlates well with the contrast of the bottom topography to the total-field topography.

Since there are measured data of the longitudinal gradient intensity in two orthogonal directions, we have attempted to present the results of the gradiometer survey as a map of the total horizontal gradient vector (Figure 9). Such a method of magnetic mapping is widely used in detailed on-land geomagnetic survey.

The vector map for a horizontal gradient is compiled on the basis of a joint analysis of gradient surveys according to the systems of meridional and latitudinal (or submeridional and sublatitudinal) profiles, forming an orthogonal square grid. Despite the great number of profile intersections (60), additional values of longitudinal gradients were determined on the isodynamic maps of the longitudinal gradient by interpolation in the interprofile space to better reflect the structural characteristics of the Vavilov Seamount AMF; thus, the number of profile intersections was increased to 266. The vector direction was estimated by two orthogonal constituents and was determined as the angle φ,

$$\varphi = \arctan G_x / G_y \tag{16}$$

The sizes of the arrows in Figure 9 correspond to the horizontal gradient values in nanoteslas per meter.

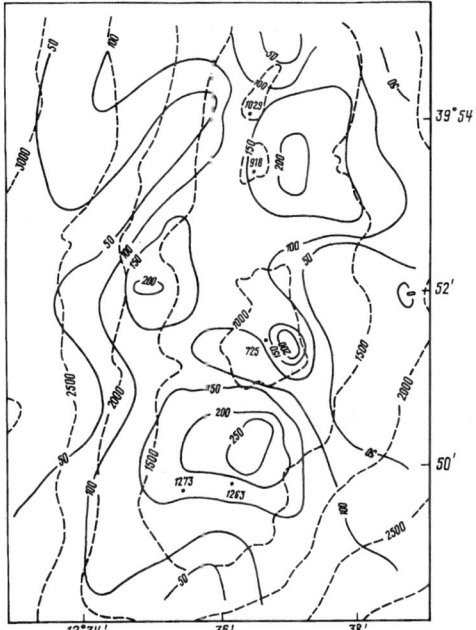

Figure 8 Intensity map of total horizontal gradient G_{xy} of the Vavilov Seamount: continuous line, total horizontal gradient (nT/km); broken line, bathymetric contours (m).

Analysis of the vector map shows two interesting areas: first, where the vectors run to a single center, and second, where the vectors run off the fixed line. In the first case, the vector directions indicate strictly the position of the mountain summit. This allowed the supposition that the vectors are oriented towards a main volcanic crater. Such a supposition agrees well with geomorphological data and TV observations of the summit of the mountain by the towed apparatus "Zvuk". The second area, delineated in Figure 9 by a dashed line, is interpreted as a submeridional fault. This was verified by observations from the towed apparatus "Zvuk-Geo" during the 16th cruise of R/V Akademik Mstislav Keldysh.

The described technique of gradient area survey and mapping seems to be efficient and has greater resolution compared to traditional magnetic mapping for exploration purposes and for recognition of local magnetic bodies.

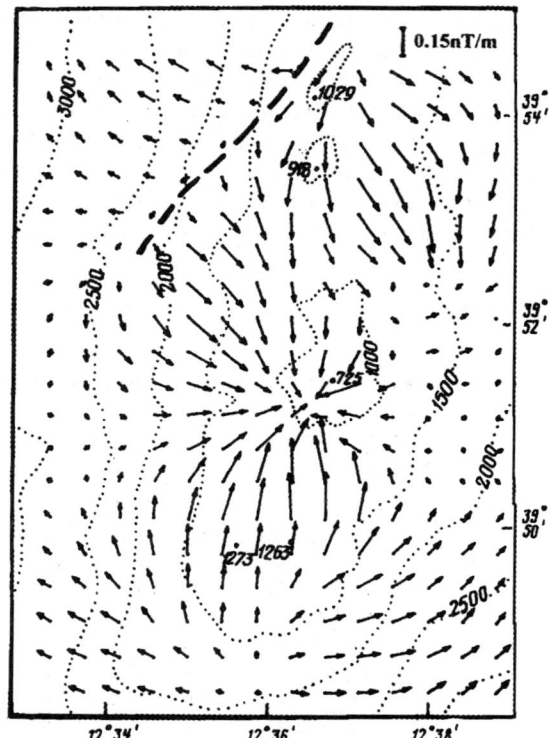

Figure 9 Vector map of total horizontal gradient of the Vavilov Seamount: arrows, vector of total horizontal gradient; dotted lines, bathymetric contours (m); broken line, assumed fault.

A similar survey with the gradiometer DPM-2 was carried out over the Magnaghi Seamount, located west of the Vavilov Seamount, during the 16th cruise of R/V Akademik Mstislav Keldysh in 1988 (see Chapter 5). The geophysical survey was performed over the net of orthogonal submeridional and sublatitudinal profiles extending 1.5 to 2 miles from each other. The gradient measurement base was 100 m. The mean square error of the survey was ±9 nT.

A vector map of the total horizontal gradient was also compiled for the Magnaghi Seamount (Figure 10). As follows from analysis of the map, the vectors indicate the existence of a local magnetic object of high magnetization off the mountain summit, near the rise of the western slope, at a depth below 3000 m. The above, as well as the absence of any direct correlation between the AMF and the Magnaghi Seamount topography, evidences a complex multi-stage process of volcano formation and the processes of secondary superimposed volcanism, to which intrusion of younger magmatic high-magnetized bodies near the foot of the formed volcanic feature is related.

The remaining area is characterized by orthogonally oriented vectors of the horizontal gradient along two predominant directions, one of which is subparallel to the mountain crest strike (15° to 20° NE), whereas the second is perpendicular to it. Comparison of the results of the gradient survey with the materials of the submersible study of the mountain slopes by "MIR" POVs and a geomorphological survey allows the supposition that the orthogonal vectors indicate the position of crossing faults, in the respective directions, probably controlling the successive stages of volcanic activity.

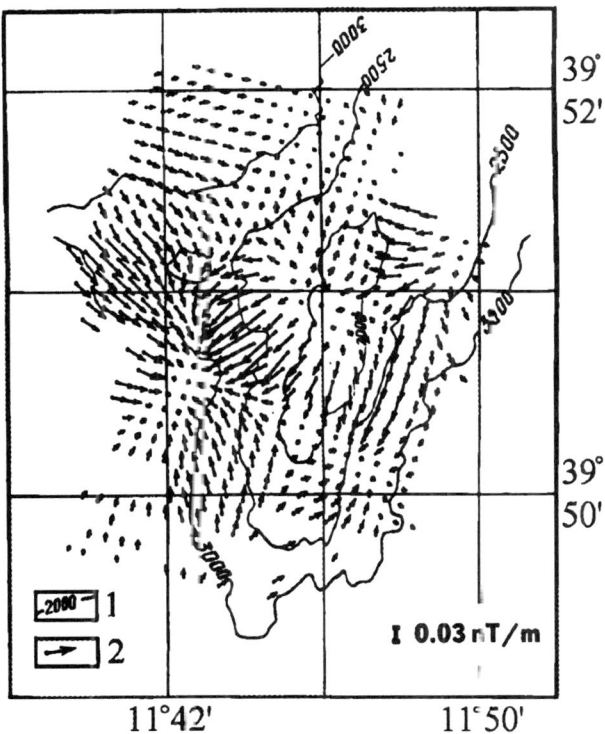

Figure 10 Vector map of total horizontal gradient of the Magnaghi Seamount: 1, bathymetric contours (m); 2, total horizontal gradient vector.

Summarizing the results of the use of magnetic gradiometers to study the fine spatial structure of the seamount AMF, one should note the following. Maps of isodynamic lines of the longitudinal gradient measured in two orthogonal directions are useful when performing magnetic simulations of seamounts since it allows one to localize within a test area the regions of maximum magnetization and to specify the boundaries of areas with direct and inverse magnetic polarity. The intensity map of a total horizontal gradient and its vector diagrams may be used for plane graphical reflection of a number of principal geological-geomorphological structures in the ocean; in particular, the isometric ones.

Chapter 2

Interpretation Procedure of Marine Magnetic Data: Topical Problems

G. M. Valyashko, A. N. Ivanenko, G. E. Czerniawski, and S. V. Lukyanov

CONTENTS

I. Gradiometric Survey 22
   A. Three-Sensor Scheme 22
   B. Experimental Data 22
   C. Processing Procedure 24
   D. Processing Results 27
   E. Conclusion 29
II. Bottom Measurements. Processing Procedure and Interpretation
   of Results 29
   A. Formulation of the Problem 31
   B. Generalization of the Results 35
III. Inverse Problems of Marine Magnetometry: Procedure of a Generalized
    Linear Inversion 38
   A. Reparameterization and Amalgamation of the Model 41
   B. A Uniqueness Solution of an Inverse Problem 44
   C. Adaptive Reparameterization 45
   D. Practical Results 46
IV. Magnetic Field Simulation for Seamounts 48
   A. Formulation of the Problem 48
   B. Solution of a Direct Magnetometry Problem for Seamounts 55
   C. Determining the Direction of Magnetization from
      the Magnetic Field Spectrum and Transformations 58
   D. Use of a Transfer Function (Admittance) between
      the Topography and Magnetic Field 64

A unique approach to the processing and interpretation procedures for the marine magnetic data was developed at the laboratory of geomagnetic studies of the P. P. Shirshov Institute of Oceanology during the last three years.

The approach embraced a wide range of problems of marine magnetic surveys, in particular, three-sensor gradiometer and bottom magnetic surveys. The possibility of applying the generalized linear inversion technique to the solution of inverse problems of magnetometry was considered. Resolution of parameters is of special concern: this problem is vital when addressing the geological models, when, in particular, it is necessary to reparameterize the model and adaptively reformulate the initial problem. Different ways of the interpreting the data from marine magnetic site surveys over seamounts are considered. A simulation technique for the magnetic field of sea mountains is given.

This approach requires the application of an adaptive regularization strategy to linearly incorrect problems that ensures the preservation of medium- and high-frequency spectrum segments of valid signals and reduces, in some cases, the relative error from 15 to 30% down to 7 to 10%. The strategy essential also for solving other problems is considered in the present work.

The procedures described are illustrated by abundant practical examples. The simulation technique for seamount magnetic fields is presented in Chapter 6, whereas the technique of an adaptive reparameterization in magnetization reconstructions, in a two-dimensional formulation, was used to interpret magnetic anomalies within transform faults (Chapter 4).

I. GRADIOMETRIC SURVEY

A. THREE-SENSOR SCHEME

At present, marine magnetic surveys use more and more widely the gradiometeric technique to ensure the accuracy of the survey and to exclude the effect of temporal variations of the magnetic field. However, even now, one cannot exclude a number of errors that are present in the gradiometeric survey data and that hinders the application of quantitative interpretation and thus reduces the quality of the materials obtained.

The procedural basis for gradiometeric surveys in the ocean, as well as the approach to the processing of these data, were formulated in the works of E. I. Roze, R. V. Semevsky, and I. M. Markov (Roze, 1978; Roze and Markov, 1984; Semevsky, 1981). Further development of the processing procedure and its practical introduction into gradiometeric surveys in the ocean is reflected in several works (Roze and Markov, 1984; Roze and Semevsky, 1986; Leibov et al., 1988; Valyashko et al., 1990a). Theoretical aspects of multisensor measurements were considered earlier, mainly as a generalization of the experience of two-sensor observations (Tomoda and Fuimoto, 1980; Roze and Semevsky, 1986; Leibov et al., 1988). We failed to discover any published material on either measurements performed according to a three-sensor scheme or descriptions of their processing. We believe that one of the possible ways to improve the efficiency of the gradiometric technique is the use of a multisensor, in particular, three-sensor systems.

B. EXPERIMENTAL DATA

Experimental three-sensor observations with 150-m bases were carried out in April 1990 in the Arabian Sea during the 20th cruise of R/V *Akademik Mstislav Keldysh*. The measurements were taken with synchronized instruments, namely, MAMP-D and MPM-5M (Balyberdin et al., 1987; Afonyashin et al., 1984). The sensor closest to the vessel was connected to the instrument MPM-5M and was towed at a distance of 280 m from the vessel. The second and third sensors were connected to the differential magnetometer MAMP-D and were towed at a distance of 430 and 580 m, respectively. The towing scheme is given in Figure 1b. The vessel was 110 m long and was equipped with an integrated navigation system, which was controlled by a special computer. The mean speed of the vessel was about 10 knots. Measurements were taken every 10 s and were registered by a computer acquisition system together with the current coordinates and the heading of the vessel.

Two profiles in the western Arabian Sea were chosen for processing. The region is located within the oceanic crust, which is about 60 million years old and has a pronounced system of magnetic lineations. However, profiles were oriented almost

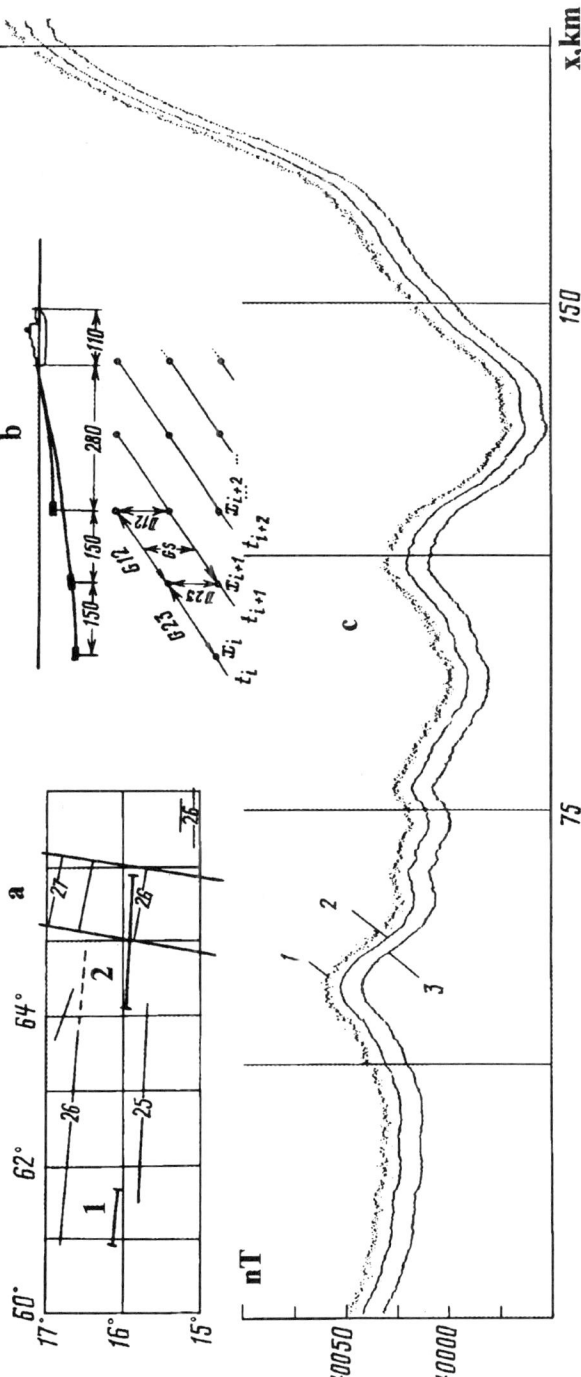

Figure 1 Results of the experiment. (a) Scheme of profile pattern: 1, 2, numbers of profiles; 25-27, numbers of linear magnetic anomalies. (b) Scheme of three-sensor measurements. (c) Magnetic field measured by three channels at profile 2; numbers correspond to channel numbers.

parallel to their strike. The first profile is placed in a 25r, the crust being 57.5 million years old, while the second one is placed in a 26r, the crust being about 59 million years old (Figure 1a). Figure 1c shows the results of the measurements of the magnetic field by three channels at profile 2. Local anomalies about 30 km long with amplitudes of up to 30 nT and a long-period one 60 km long with amplitude up to 200 nT were resolved. The normal magnetic field trend is 140 nT. The noise of the sensor closest to the vessel was considerably higher compared to that of the other two.

The principal distinction between a three-sensor system and a two-sensor one is the registration of the difference of magnetic field gradients measured at one and the same point by two pairs of sensors with equal base. This parameter is called the superposed gradient. Field data processing and analysis show that the use of superposed gradients allows one to estimate the influence of the vessel (deviation) and the parameters of a stationary magnetic field.

C. PROCESSING PROCEDURE

The magnetic field values measured by an n-channel system can be represented in the form

$$F_n(x,t,k,h,r...) = T(x) + V(t) + D(k,r,z,h...) + \varepsilon + \delta \tag{1}$$

Here x is a horizontal distance, t is time, k is the vessel's heading, h is the towing depth, r is the distance from the vessel, $T(x)$ is the stationary magnetic field in the point of x, $V(t)$ is the temporal magnetic field variation, $D(k, r, x, h...)$ is the composite deviation effect, dependent, among other agents, on the vessel's magnetic mass and the self-deviation of the sensors, ε is sum of other errors due, in particular, to the base instability and orientation (Semevsky, 1981), and δ is a noncorrelated noise.

In order to estimate the unknown value of a noncorrelated noise, the discriminant-functional technique was used (Strakhov and Valyashko, 1976). The algorithm of adaptive regularization (Valyashko and Strakhov, 1984; Strakhov and Valyashko, 1984) was used to eliminate the noise from Equation 1. The adaptive algorithm was chosen because of its higher accuracy and, more importantly, because it ensures maximum safety for high and medium-frequency modes of magnetic fields, which are the combined effect of geological objects and the noise of different origin.

The vertical magnetic gradient at sea could be several times higher than the horizontal one. Thus, the data were corrected for the difference in sensor towing depths, which reached 20 m.

Considering the equal bases of measurements and, for now, neglecting the various errors, the three-channel data in the time t can be represented in the form

$$\begin{aligned} F1(x_{i+2}) &= T1(x_{i+2}) + V1(x_{i+2}) + D1(x_{i+2}) \\ F2(x_{i+1}) &= T2(x_{i+1}) + V2(x_{i+1}) + D2(x_{i+1}) \\ F3(x_i) &= T3(x_i) + V3(x_i) + D3(x_i) \end{aligned} \tag{2}$$

Here, for the first channel, $F1(x_{i+2})$ is the measured field, $T1(x_{i+2})$ is the stationary magnetic field, $V1(x_{i+2})$ is the temporal field variation, and $D1(x_{i+2})$ is the deviation of sensor 1. The notations for sensors 2 and 3 are similar.

Let us consider the next three-sensor differences.

The gradients $G12$ and $G23$ measured by the first and second sensor pairs, respectively, for the same time t are given by

$$G12(x_{i+2}) = g12(x_{i+2}) - d12(x_{i+2})$$
$$G23(x_{i-1}) = g23(x_{i+1}) - d23(x_{i+1})$$
(3)

For the first pair of sensors, $g12(x_{i+2})$ is the true longitudinal gradient related to the point x_{i+2} and $d12(x_{i+2})$ is the difference in deviation of channels 1 and 2; for the second pair of sensors, $g23(x_{i+1})$ and $d23(x_{i+1})$ are defined in a similar manner.

The superposed fields, the differences $D12$ and $D23$ of field values measured at the same point for the first and second pairs of sensors, respectively, are given by

$$D12(x_{i-2}) = v12(x_{i+2}) - d12(x_{i+2})$$
$$D23(x_{-1}) = v23(x_{i+1}) - d23(x_{i+1})$$
(4)

For the first pair of sensors, $v12(x_{i+2})$ is the variation difference, and $d12(x_{i+2})$ is the deviation difference; notations $v23(x_{i+1})$ and $d23(x_{i+1})$ are written in a similar manner.

Let us introduce the superposed gradient GS, using Equation 4 as an analogy:

$$GS(x_{i+2}) = G12(x_{i+2}) - G23(x_{i+2}) = d12(x_{i+2}) - d23(x_{i+2})$$
(5)

From Equations 4 and 5 the following equation can be derived

$$D12(x_{i+2}) = GS(x_{i+2}) + D23(x_{i+2})$$
(6)

The superposed gradient does not depend on either the stationary magnetic field or the temporal variations, because it is the pure characteristic of the deviation. Further, we shall omit indexes if possible

Figure 2 shows the graphs of the superposed field and superposed gradient. The amplitude and general configuration of graphs GS and $D12$ are similar, but differ considerably from $D23$. Accounting for the character of GS, it is possible to identify the determining contribution of the deviation in the three-sensor differences. It is known that the magnetic field of the vessel can be approximately represented as the field of a magnetic dipole. Thus, the deviation difference attenuation at a fixed base can be estimated by a dipole derivative,

$$A_m = \frac{3}{r^4}\sqrt{1 + 2\cos(90° - \varphi)}$$
(7)

where φ is the geographic latitude and r is the distance from the vessel. The vessel field attenuation is proportional to r^4, and it can be assumed that $GS >> D23$. From this and Equation 6, it is possible to use GS as the initial estimate for $D12$.

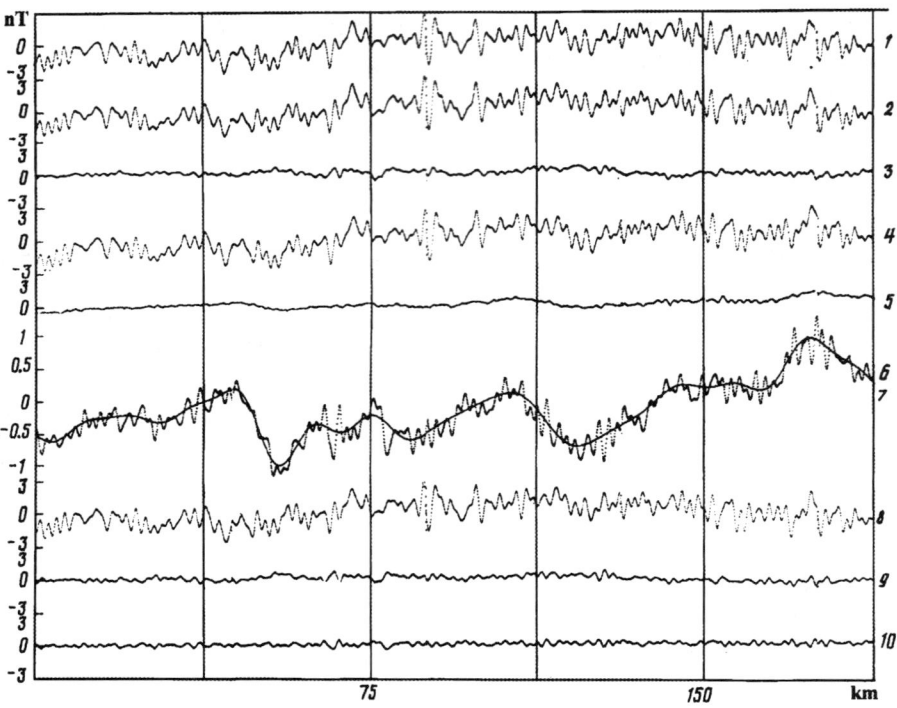

Figure 2 Differential characteristics. Downwards: (1-5) after filtering of noncorrelated noise; (6, 8, 9) after iteration; (7, 10) after band filtering; (1) GS, (2) D12, (3) D23, (4) G12, (5) G 23, (6) g12* and g23*, (7) g12* filtered, (8) d12*, (9) d23*, (10) error of orientation and self-sensor deviation.

To refine this estimation, we use the variation properties data. It follows from analysis of these data that high-frequency geomagnetic field variations have small magnitudes. According to the data of Vanyan and Butkovskaya (1980) and Campbell (1976), the variations with 0.5 periods are less than 1 nT, and during high solar activity, they do not exceed 5 to 10 nT. Furthermore, according to estimations by N. Pal'shin (personal communications), high-frequency variations attenuate while passing through the water medium. They lose from 40 (1-min period) to 10% (15-min period) of their magnitude. Therefore, we suppose that in the majority of cases, the deviational component predominates in the differences $D12$ and $D23$ at least within the 0.5-h interval, whereas the variational component has a negligibly small contribution in the phase spectrum of the superposed field. This time interval will be considerably wider only in the case of long-period variations. Magnetic storms and some bay-shape variations are the exception.

Such account for this *a priori* information allows the construction of the following convergent iterational process $(GS \rightarrow 0)$ to obtain estimates of $g12$, $g23$, $v12$, $v23$, $d12$, and $d23$, which can be solved by a simple iteration method. Let $GS(\omega)$ be the amplitude spectrum of the superposed gradient and $\theta_{D23}(\omega)$ the phase spectrum of the superposed field for the second pair of sensors. Thus, for the k-th iteration

$$d23_k = A_{opt} \int_{-\pi/2}^{\pi/2} [GS_{k-1}(\omega)\exp(\theta_{D23}(\omega))] \exp(j\omega x) d\omega$$

$$\begin{aligned} d12_k &= d23_k + GS_{k-1} \\ g23_k &= g23_{k-1} - d23_k \\ g12_k &= g12_{k-1} - d12_k \\ GS_k &= g12_k - g23_k \end{aligned} \quad (8)$$

It should be noted that none of the components of the expression depends on the variation magnitude.

To obtain the estimate of the deviation difference due to the influence of the vessel, $d12^*$, the current values of $d12_k$ should be summed. The estimate of the longitudinal gradient $g12^*$, which is free of any deviations, can be obtained as the output of the iterative process. Furthermore, from Equation 4 we can obtain the variation increment for the time the vessel passes the base length

$$v12^* = D12 - d12^* \quad (9)$$

The factor A could be best adjusted to the data by using the "gold section" method. Assuming that the variation increment $v12$ is smaller than the deviation difference, we studied the variance of $v12^*$ as a function of factor A. The constructed function has a distinct minimum (Figure 3a). The optimal value of factor A was found as $A_{opt} = \arg\min F_p[A]$, where p is the interval of allowable values of A. Computer experiments showed that even if the variations are 20 times higher than the valid signal, the function $F(A)$ will have a singularity at the point A_{opt}.

D. PROCESSING RESULTS

Figure 2 shows the graphs of $g12^*$ and $g23^*$. They almost coincide, but vary at two points. The difference corresponds to the base of 150 m. The curves also include a high-frequency harmonic 0.5 nT in magnitude, which obviously is not connected with the valid signal. When constructing the iterational process, we did not consider "other" errors in Equation 1, only variations and deviations. To find the source of high-frequency errors, we compared the autocorrelation functions of the gradient and the derivative of the vessel's heading (Figure 3b). A high-frequency component with a period of 7 to 10 min coincides with the periodical heading changes. As shown in one work (Semevsky, 1981), these changes cause both the orientation error of the longitudinal gradient and the error due to the self-deviation of the sensors. The sum of these two errors was removed from longitudinal gradient data by frequency filtering. The autocorrelation function of $g12^*$, after removal of all errors, is shown in Figure 3b; filtering results along with the errors are shown in Figure 2.

The graphs of $d12^*$ and $d23^*$ are shown in Figure 2. They include both a high-frequency component with a period of 5 to 10 min and a low-frequency component with a period of 0.5 h and more. Note that the magnitude of $d12^*$ is significantly higher than that of the longitudinal gradient, and the magnitude of $d23^*$ is comparable to it.

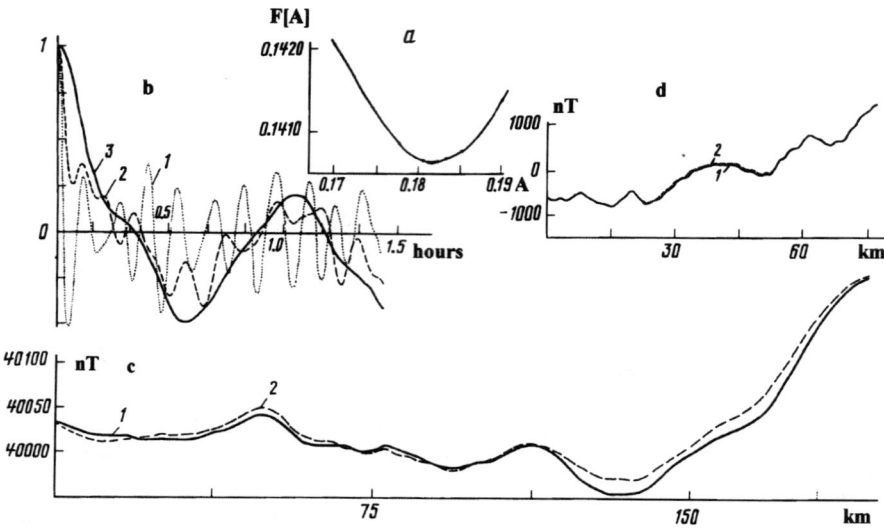

Figure 3 (a) Choice of optimal coefficient A by "golden section". (b) Comparison of autocorrelated function (ACF): 1, ACF of the heading derivative; 2, ACF of $g12^*$; 3, ACF of $g12^*$ after filtering. (c) Results if integration: 1, measured field; 2, integrated field. (d) Experiment with introduced variation: 1, introduced magnetic storm; 2, derived variation.

To reconstruct the anomalous stationary magnetic field and temporal variations obtained as the result of iterations, the estimates of the longitudinal gradient and variation increment should be integrated. The specific features of the integration process were discussed by Roze (1978) and Leibov et al. (1988). In the absence of magnetovariational measurements the values of variations and anomalies are unknown at the starting point of the profile. Thus, the integration of the gradient can provide only relative values of anomalies and not the linear field trend and variations.

The linear constituent of the stationary field is usually connected with the normal field and can be calculated with the IGRF coefficients. We have introduced the normal field during integration in the form of a gradient, taking as a zero level the value of the first measurement of the farthest channel, which is the least influenced by the vessel. The variations were calculated as the difference between the measured and integrated fields. This made it possible to account for a linear component.

Figure 3c gives the integrated field in comparison with the measured one. The main anomaly is of 200 nT. The field is complicated by local anomalies 30 km long and up to 30 nT. The variation recognized has an alternating sign, though the positive values prevail; the main periods are from 0.5 to 3 h and more, and the magnitude reaches 20 nT. Note that the general character and magnitude of the integrated field are consistent with the fine structure of the oceanic magnetic field, which can be recorded from the ocean surface. The magnitude and behavior of temporal variations are usual for this region on quite days. The accuracy of field reconstruction at profiles 1 and 2 is about 1% and 5%, respectively.

To check the efficiency of the algorithm under conditions of high variations, the real magnetic storm recorded at the Borok observatory was artificially introduced into the real observed field of profile 1. Figure 3d shows the result of the comparison of the introduced and calculated variations in the form of the difference between original and integrated fields. The introduced variation has a trend of 2000 nT and local peaks of up to 500 nT with periods from 15 to 30 min. The separated variation is similar to the introduced one; the difference in the central part is due to the presence of a natural variation in the original data.

E. CONCLUSION

The proposed processing technique for three-sensor systems was tested and showed good results both in fields with low and those with medium anomalies. It became possible to extract the deviational part of a signal and to separate temporal variations from the stationary field. However, due to the lack of MVS and profile intersections, the accuracy of variation extraction was determined indirectly. Registration of variations, along with three-sensor measurements, is necessary to obtain accurate estimates. The procedure of calculations is simple enough. It may be divided into three stages: (1) filtering of a noncorrelated noise, (2) recognition and elimination of the deviational influence of the vessel, using *a priori* information on the variation behavior, and (3) elimination of other errors and integration of gradient estimates and variation increments. Analysis of the results shows that the deviational error appears to be the main one in gradiometric measurements. The change from the traditional two-sensor system to a three-sensor one, however, allows the very accurate elimination of this error along with the others from the measurements using the pure deviational observable — the superposed gradient — with the wider spectrum of temporal magnetic field variations.

II. BOTTOM MEASUREMENTS. PROCESSING PROCEDURE AND INTERPRETATION OF RESULTS

The efficiency of bottom geomagnetic surveys in deep-sea regions of the ocean is illustrated convincingly by several American works (Spiess and Mudie, 1970; Atwater and Mudie, 1973; Macdonald, 1977; Mudie et al., 1972). No such surveys were carried out in the former Soviet Union until recently. This fact constrained considerably the choice of problems which could be solved by marine magnetometry and did not allow either experimental study of the fine spatial structure of an anomalous magnetic field near its sources or the search for local magnetic objects on the ocean bottom, which are believed to be confined to ore deposits.

To solve this problem, an instrumentation system has been developed and tested at the P. P. Shirshov Institute of Oceanology, Russian Academy of Sciences, in the department of bottom research with the assistance of the laboratory of geomagnetic research. This system is appropriate for bottom geomagnetic surveys and simultaneous measuring of depth-to-the-bottom by the deep-sea towed vehicle "Zvuk-6" (Onischenko et al., 1986; Yastrebov et al., 1991). The instrumentation system combines an echo sounder, two side-scan sonars, an acoustic profiler, the proton magnetometer MPM-6, and stereophoto and TV systems. The sea bottom images obtained with sonars and acoustic profilers are registered on the facsimile systems "Inei-P". The total magnetic intensity data are registered in both digital and analog forms. The sampling rate of magnetometer is 4 s. The magnetometer fish is towed at a distance of 18 m astern the vehicle by a special cable. The altitude of "Zvuk-6" over the sea bottom is controlled by a deck winch in accordance with the echo sounder data.

Test surveys were carried out in the Northeastern Atlantic to probe the instrumental system "Zvuk-6" and to work out the procedure and technique for bottom magnetic measurements. The area was 2×2 km and was located on the northeastern slope of the Cruiser Seamount, on the plateau with bottom depths of 1500 to 1800 m. The navigational reference of the instrument during the complex bottom survey was performed by the hydroacoustic navigation system (GANS SNP-20) operated and controlled by the computer system IVK developed at IO RAS (Onischenko et al., 1986). IVK IO RAS together with SNP-20 coordinated the position of the towed vehicle, with an error of 9 m, in the reference system connected with the bottom transponders (Nafikov et al., 1988; Russak, 1986). The position of the latter was referred to geographical coordinates by a satellite navigational system. Simultaneously with the "Zvuk-6" operation, the diving depth of the apparatus was determined with an accuracy of 0.5% of the measured depth and the distance from the bottom was determined with an accuracy of 0.4 m, which allowed estimation of the depth along the towing line. Joint analysis of the hydrolocation mosaic compiled from the results of a sonar survey of the bottom topography and also from a phototelevision research shows basic rock outcrops which form an uplift up to 300 m high in the northern part of the area studied. Slopes of the uplift are covered by thin sandy-clayey sediments (several tens of centimeters).

A unique procedure was developed in the laboratory of geomagnetic research to process the results of the bottom geomagnetic survey, which considers the specific character of the latter. In fact, geomagnetic surveys performed by bottom magnetometers have some peculiar features. The principal difference from normal hydromagnetic surveys carried out over the sea surface is that the altitude of the meter over the bottom varies even along the survey profile. The range of that altitude diversity depends mainly upon the bottom topography raggedness and towing conditions. On average, it varies from several hundred to tens of meters. However, such small variations compared to the general depth of the ocean bottom still affect considerably the results of magnetic field measurements. The point is that the measurements are taken near the source of high-intensity magnetic anomalies. Therefore, anomalies appear which are not of geological origins. They result exclusively from the variation of the sensor altitude over the bottom of the ocean, which may be identified with the top of the active magnetic layer. The amplitudes of such "false" anomalies may reach several hundred nanoteslas.

Bottom hydromagnetic surveys, similar to surface ones, are performed along the grid of profiles. However, the area covered by bottom surveys is usually considerably less. Meanwhile, the grid of survey profiles should be denser since medium- and high-frequency components of magnetic fields increase sharply when measured near their sources, and the spectra of such fields are more complicated if compared to that of the anomalies registered on the ocean surface. However, the profile geometry of the survey is retained. The mean discreteness of such measurements is several meters.

The procedure considered for bottom geomagnetic data processing incorporates two stages. The first one is the reduction of the survey data to a surface with a constant altitude over the bottom (for instance, to the surface conformal to the bottom topography) or to the reference plane with a constant depth (to the plane parallel to the sea surface). The second stage involves the calculation of the transformants of the field reduced to a given surface and solution of some problems of quantitative interpretation. Here, we shall present only the final result of the problem of calculation of effective magnetization of the magnetoactive layer with known geometry. Various algorithms of its solution are given in several works (Bott, 1973; Shouten and McCamy, 1975; Strakhov and Lapina, 1976; Parker and Huestis, 1974 a,b). We assumed as a basis the approaches described in Strakhov and Lapina (1976) and Strakhov (1981) for a two-dimensional case and in Bott (1973) and Strakhov (1981) for a three-dimensional one. Thus, all problems can be solved both in a two- and three-dimensional formulation according to the data provision and representativity of the bottom geomagnetic survey.

When the processing procedure was being developed, the following requirements were taken into account: specific features of the bottom survey, use of a unique mathematical approach, criteria of survey accuracy definition, and value of random error in the measurements.

At the first stage, a modified two-dimensional version of the marine magnetic data processing was used due to a profile-mode of the survey — measurements are concentrated along profiles at close intervals (Strakhov and Valyashko, 1977; Valyashko, 1981). At the second stage, the problems were solved in a two- and three-dimensional formulations.

A. FORMULATION OF THE PROBLEM

Let us choose the reference system. For the case of two-dimensional transformations, axis x is to the right-hand side and axis z is going upward. In the case of three-dimensional transformations, axis x runs eastward, axis y goes northward, and axis z goes upward. Further, we assume that the Earth's geomagnetic field intensity $f_\delta(x, y, z)$ measured by a bottom magnetometer is an additive sum of the valid signal $f(x, y, z)$ and of a random non-correlated noise δf:

$$f_\delta = f + \delta f \qquad (10)$$

This is also true for $f(x, z)$ at $y = 0$. Let us assume that geomagnetic field variations have been freviously removed or are negligibly small. Further, we believe that we know the interval N in the observation profiles such that the standard deviation of noise within it is a constant, and we assume it is known when formulating the problem. Let us assume that the value $z = -H$ is known, where $H > 0$ is the depth of the field singularity which is nearest to the surface of observation and belongs to the U-interface (water/Earth's crust). The latter may be assumed to be the sea bottom surface in the first approximation. Taking into account that measurements are performed near the U-interface, we assume that $H << N$. Functions $f(x, z)$ and $f(x, y, z)$ are harmonic in the outer sphere of the sources. The problem is formulated as follows: find out values of the field and its transformants in the vertical semiplane, or in the semispace, for z lying within $(-H < z < +\infty)$, with respect to the given functions $f(x, z)$ or $f(x, y, z)$.

On the whole, the problems of transformation were solved by an approximation method using a discrete cosine Fourier transform. We shall present discrete convolution operators of field transformations. For two-dimensional transformations:

$$f_\delta(x,z) = \sum_{n=0}^{N} a_n^\delta \exp\left(-z_n \frac{n\pi x}{N}\right) \qquad (11)$$

Here, a_n^δ are the cosine coefficients of the Fourier series for the field $f(x, z)$, reduced to a uniform grid along coordinate x. For three-dimensional transformations, we used the expression:

$$f_\delta(x,y,z) = \sum_{n=0}^{N}\sum_{m=0}^{M} a_{n,m}^\delta \exp\left[-z_{n,m}\sqrt{\left(\frac{n\pi}{N}\right)^2 + \left(\frac{m\pi}{M}\right)^2}\right], \qquad (12)$$

where $a_{n,m}^\delta$ are the cosine coefficients of the double Fourier series for the field $f(x, y, z)$

reduced by Equation 11 to the given plane and then interpolated into a regular grid by coordinates x and y.

The random noise δf occurring in measurements is determined by a correlation discriminator prior to applying the corrections for altitude diversity (Strakhov and Valyashko, 1976; Strakhov, 1977; Valyashko, 1981).

To solve the incorrect problems of potential field transformations and the abovementioned problems of quantitative interpretation, we used the algorithms of an adaptive regularization of linearly incorrect geophysical problems. In particular, we used the regularizing family based on the orthogonal principle introduced in the paper by Strakhov and Valyashko (1981).

The calculation procedure was as follows. Since the measurements were of a profile type, the values f occurring in measurements were determined by a correlation discriminator. The data measured at various altitudes over the sea bottom were reduced to a given depth using Equation 11 and algorithms of an adaptive regularization. Values $f(x, z)$, obtained at a given plane, were interpolated into a regular (x, y) grid, and then further transformations were performed for field $f(x, y, z)$ in a two- $(y = 0)$ or three-dimensional formulation, with respect to the representativity of the survey performed. The regularization parameter was determined for each transformation.

However, the application of usual interpolation algorithms developed for the plane to the data located, for instance, on the surface conforming to the bottom topography, is restricted by the degree to which the relief is rugged (Aronov, 1976). The estimations we performed showed that the bottom topography is smooth over the area studied, and errors for the interpolation are one order less than survey errors after data reduction to the surface of constant altitude over the sea bottom. Therefore, the corresponding corrections were not applied. Completing the description of the calculation procedure, we shall mention the approximation technique proposed by Aronov (1970a,b; 1976). The essence of the transformations applied to the processing of gravity data implies that the field values measured on the irregular surface are to be approximated by a set of artificial sources, and then all transformations of the field are reduced to the solution of a direct magnetometry problem. In fact, such an approach is efficient in gravity data processing. The vector character of the magnetic field sources requires specification of the angular parameters of the magnetization vector as early as the stage of approximation. These parameters themselves are often subjects of investigations. Therefore, our approach, when no hypotheses on the possible origin of the magnetic field are introduced at the stage of the processing, seems to be more general. We shall note that, in principle, the technique of reducing data to a surface of constant altitude was first formulated by Andreev in 1947 (Andreev, 1947; 1949).

Let us proceed with the analysis of the data obtained. Figure 4 shows a scheme of bottom survey profiles. Twelve profiles were made within the area studied. The most elongated are over 2 km. The mean discreteness of measurements of the total magnetic field intensity, sea bottom depth, and altitude of the towing apparatus over the bottom is 5 to 10 m. Profiles cover the study area rather evenly and have 16 intersection points.

Maps of the bottom topography over the area are shown in Figure 4. In the northern part of the area, there are three small uplifts with depths of about 1500 m and relative elevations up to 300 m. In the central part of the area, depths may reach 1700 m. In the southern part, depths increase to 1800 m. There, a local depression of 1854 m is recorded as well. Thus, the depths change within the area by about 350 m. We believe that isometric projection of the bottom topography with a 15-m-deep contour interval,

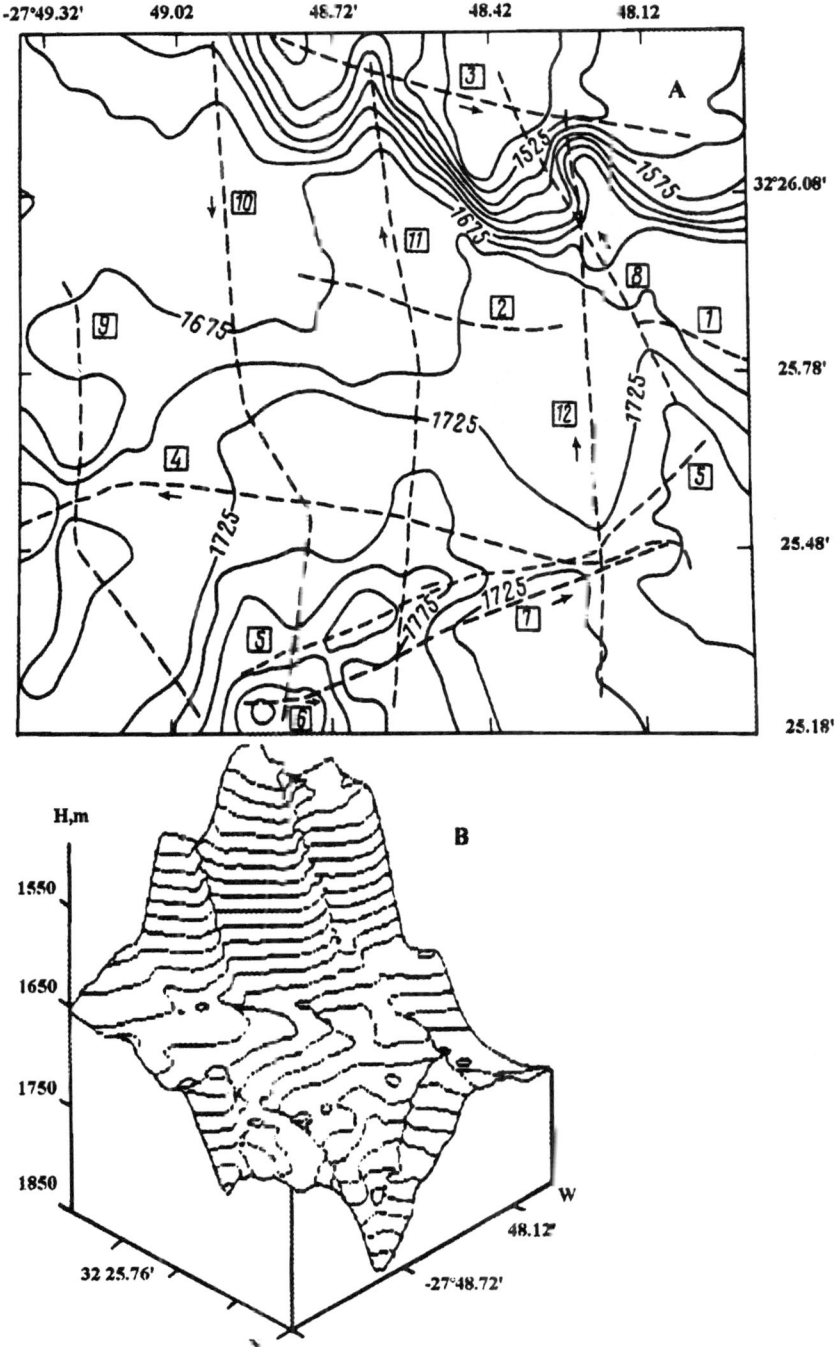

Figure 4 (A) Bathymetric map of the area studied. Isobaths are spaced 25 m apart. Dashed lines show survey profiles. Arrows show survey direction. Numbers in squares are profile numbers. (B) The same bathymetric map in isometric projection. Isobaths are spaced 15 m apart.

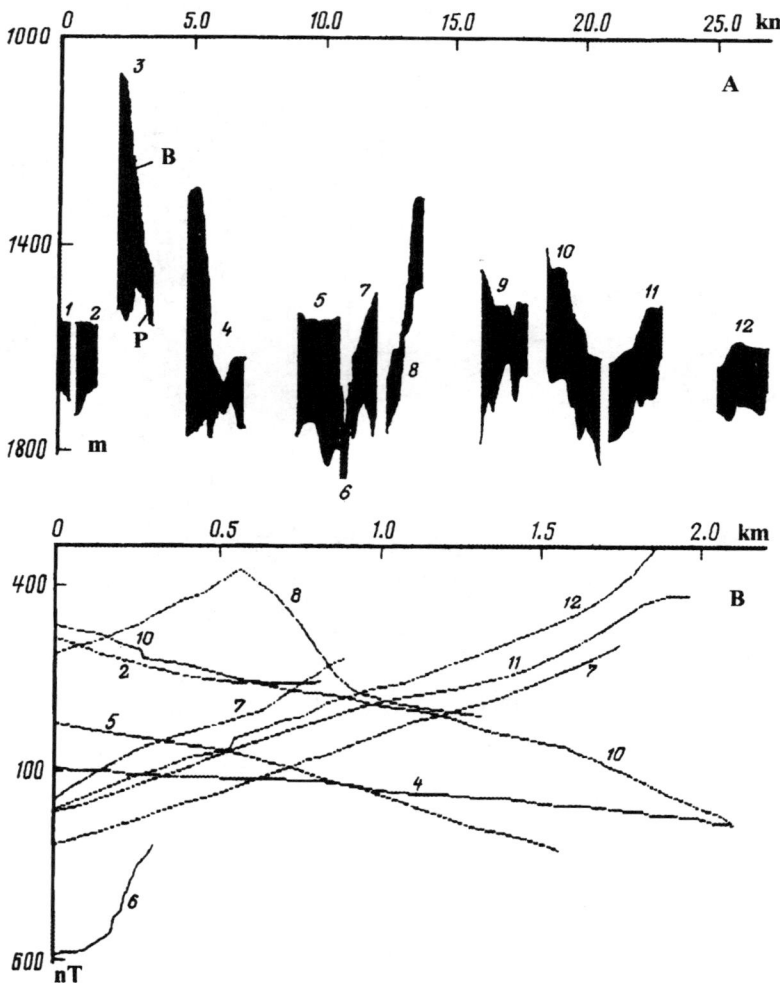

Figure 5 (A) Matched graphs of towing depth of "Zvuk-6" apparatus and bottom topography along survey profiles and (B) measured anomalous magnetic field along profiles. B, altitude of apparatus over the bottom; P, bottom topography. Numbers by graphs correspond to profile numbers. Data are spaced along survey direction.

shown in Figure 4b, is the most pronounced one. As follows from the maps given there, the topography of the area is a slope fragment typical of the northeastern part of the plateau of the Cruiser Seamount.

Figure 5A shows the combined graphs of bottom topography and towing depth of "Zvuk-6" along the survey profiles. Analysis of these data shows that the average altitude of the fish over the bottom is 150 to 200 m. In profiles 3 and 4 these values vary from 450 to 100 m, and from 30 to 200 m along profile 8. The diversity of altitude during the survey over the given area resulting from changes in the towing depth does not exceed 450 m. The minimum depth surveyed was 1800 m. Therefore, a depth of 1000 m was chosen as the common plane to which the field was reduced. The constant altitude of the surface conformable to the bottom topography, to which measurement results were also reduced, was assumed as 300 m, taking into account the values of the altitude diversity and mean towing altitude over the bottom. At this depth, transformations to the lower half-space were performed only for some parts of profiles 3 and 4.

Analyzing the results of geomagnetic measurements, we note that the age of the oceanic crust in the vicinity of the Cruiser Seamount, at the eastern slope of which the area studied is located, seems to be about 80 million years. The normal magnetic field was calculated from the IGRF-85 coefficients with the survey depth taken into account, and a constant of 250 nT was added to the obtained value. Figure 5B shows the set of the most elongated profiles. Analysis of this data shows that the periods of the main anomalies registered by the bottom survey exceed 2 km. The minimum magnetic field intensity is registered in the southern part of the area along profile 6: it reaches –600 nT at a distance of 30 to 60 m from the bottom to the apparatus. The maximum value of 45 nT is registered along profile 8 at a survey height of about 80 m. The figure shows that the greatest gradient was registered at meridional profiles 10 and 12. However, a number of local anomalies with periods of about 1 km and smaller are registered, for instance, at profiles 3 and 8.

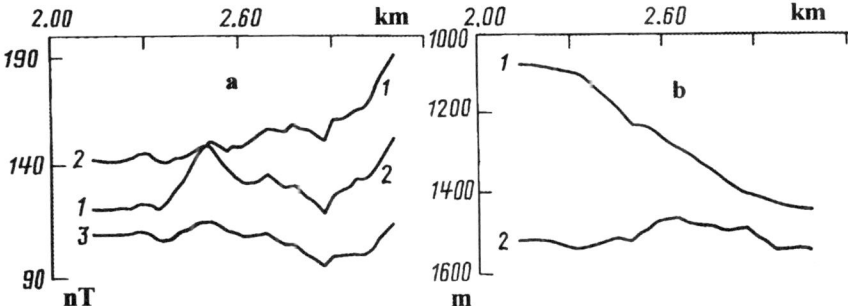

Figure 6 (a) Introduced correction for different depth of survey: 1, measured values; 2, adjusted to altitude of 300 m, conformed to bottom topography; 3, adjusted to plane at depth of 1 km. (b) Matched graphs of apparatus position and bottom topography at profile 3: 1, altitude of apparatus over bottom; 2, bottom topography. Data are spaced along the survey direction.

Figure 6a, b shows results of the data reduction to a bottom-conformable surface and to a common reference plane. Thus, at profile 3, the towing altitude over the bottom varies from 450 to 88 m. In the observed field, the magnetic field intensity varies from 120 to 190 nT. When reduced to a bottom-conformable surface, the field gradient becomes inverted and thus stresses the local uplift in the bottom topography with a relative elevation of 150 m. In the field reduced to the reference plane, the noted tendencies are preserved: the gradient is also inverted, but the field intensity decreases considerably and is only 90 to 100 nT. Thus, corrections change considerably the configuration of anomalies.

B. GENERALIZATION OF THE RESULTS

Figure 7 shows the maps of the anomalous magnetic field reduced to the bottom-conformable surface and to a common reference plane. To estimate the accuracy of the obtained results we used traditional procedures to estimate the root mean square error using the profile intersection points. The formal use of this procedure for the initial data gives an error of 153 nT, or 27%. Reduction to the bottom-conformable surface reduces the error to 26 nT, or 14%, whereas the error after reduction to a reference plane at a depth of 1 km is 18 nT, or 11%. These values attest that the applied correction procedure for a diverse-altitude survey provides better accuracy, and the obtained data may be used in further quantitative interpretation.

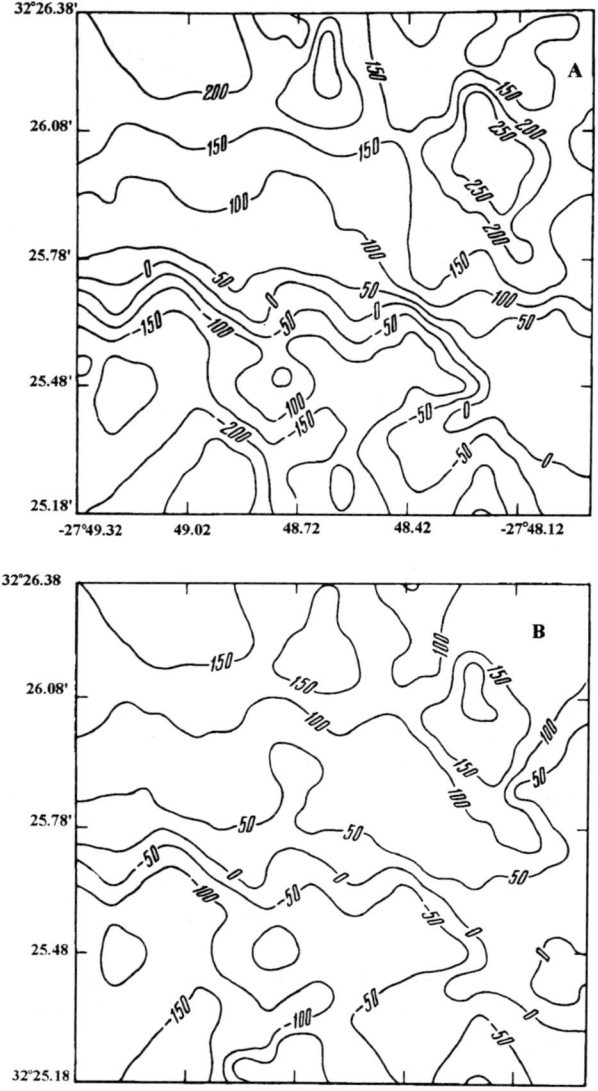

Figure 7 AMF map adjusted to the surface conformed (A) to bottom topography 300 m over the bottom, and (B) to plane at depth of 1 km. Isodynams are spaced 50 nT apart.

In the map of the data, reduced to a bottom-conformable surface, anomalies with amplitudes below 800 nT are resolved. The gradient varies in the meridional direction. In the northeastern part of the area, the isometric anomaly with amplitude up to 300 nT is an exception. It is confined to the slope of the local uplift in the bottom topography. In the southern part of the area, the negative local anomaly with a northeastern elongation should be noted. Its contours correlate with the 1775 m isobath. On the whole, we can conclude that there is good correlation between the bottom topography and magnetic field anomalies. At a depth of 1 km (Figure 7B), anomaly amplitudes do not exceed 200 nT and the pattern of anomalies does not change principally, though local components decay. Taking into account the introduced constart in the level of a normal field and good positive correlation between the principal forms of the bottom topography and anomalous magnetic field, we may suppose that the active magnetic layer of that part of the Cruiser Seamount completed its formation in an epoch of normal polarity of the Earth's magnetic field.

Figure 8 shows the magnetization intensity in a layer 5 km thick. The field reduced to a bottom-conformable surface was taken as the original one. The obtained magnetization intensity was about 1 A/m and agrees well with the mean magnetization intensity for seamounts of the Cretaceous Canary Basin. A contour line configuration attests to sufficiently good resolution of the applied algorithm and the expediency of using data of geomagnetic bottom surveys when solving such problems.

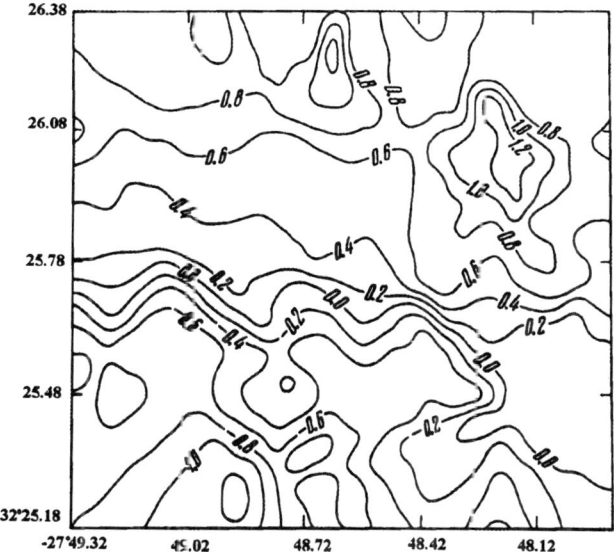

Figure 8 Scheme of effective magnetization intensity within the study area. Isodynams are spaced 0.2 A/m apart.

In conclusion, we state that the above is mainly methodical in character since the dimensions of the area studied are so small and, besides, the goal of the work was to test the bottom survey procedure. Meanwhile, the first step made in the development of the Russian-origin technique of bottom magnetic surveys using the towed vehicle, as well as the procedure developed and estimations performed, show that this approach has high potential in the study of the magnetoactive layer and of the geological structure of the bottom along with other methods of researching the bottom.

III. INVERSE PROBLEMS OF MARINE MAGNETOMETRY: PROCEDURE OF A GENERALIZED LINEAR INVERSION

Many marine magnetometry problems, for instance, the reconstruction of magnetization from the measured field in models of various orders, allow finite-dimensional approximation and are reduced to a system of M linear equations with N unknowns:

$$Ax = y \qquad (13)$$

Reliability of the results, when such systems are to be solved numerically, depends primarily upon the dimensions and properties of matrix A. The number of equations M in System 13 is determined by the number of chosen measurement points of the magnetic field and also by the quality of the geomagnetic survey, field features, computer capacity, and choice of an interpreter. Up to several thousand points of the primary field may be necessary for an adequate representation of magnetic anomalies. The number of unknowns N depends upon the type of the problem, the method of approximating and anatomizing the model object. The number N may be rather large since tens or even hundreds of elementary volumes are used to perform the dissection necessary to reconstruct the magnetic field of complex-origin objects.

We note that usually selection of points of measurements and dissection of the object are completely nonformalized procedures and depend upon the experience and intuition of a researcher. For instance, when dissecting the object, Bott (1973) and Bhattacharyya (1980) do not recommend choosing horizontal dimensions of the body less than one half of its depth to the upper edge, because stability of the system decreases sharply with such choices. Such decreases occur because anomalous effects of bodies similar in shape and position hardly differ at the measurement points, and matrix columns become almost linearly dependent.

Poor determination of great systems of linear equations, obtained as the result of digitization of gravitational- and magnetometric inverse problems, is common (Verlan' and Sisikov, 1986; Gordin et al., 1981). Therefore, to obtain plausible results it is necessary to regularize such systems.

To solve System 13 we use the procedure of generalized linear inversion at the basis of a singular-value decomposition of the matrix system (Jackson, 1972; Wiggins, 1972). We chose this method because singular-value decomposition is a powerful calculation instrument in the case of a poorly determined and degenerated system of linear equations (Louson and Henson, 1986; Forsight et al., 1980). It permits us not only to construct regularized solutions in accordance with the optimum criterion chosen, but also to carry out a deeper analysis of the inverse problem properties; i.e., it allows us to detect the cause of its poor determination, ensure quantitative characteristics of limit resolution and informative data density, choose a well-grounded level of regularization, perform reparameterization of the model, and investigate the uniqueness of the solution.

Let us write a singular-value decomposition of matrix A from System 13

$$A = U \Sigma V^T \qquad (14)$$

where $U_{M \times N}$, $V_{N \times N}$ are orthogonal and $\Sigma_{N \times N} = \text{diag}\,(\sigma_1,..., \sigma_r, 0, ... 0)$ diagonal matrices, $\sigma_1 \geq \sigma_2 \geq ...\sigma_r > 0$ are singular values of matrix A, and $r = \text{rank } A$ is the order of A. Columns of matrix U form an orthonormal basis in the observational space $Y = E^M$, whereas columns of matrix V form an orthonormal basis in the solution space $X = E^N$.

The normal pseudo-solution (solution with a minimum norm ensuring minimum error of closure) is found by applying the pseudo-inverse matrix

$$A^+ = V\Sigma^+ U^T \qquad (15)$$

to the right part of System 13:

$$x^+ = A^+ y = V\Sigma^+ U^T y \qquad (16)$$

where

$$\Sigma^+ = \text{diag}(\sigma_1^{-1} ..., \sigma_r^{-1}, 0, ... 0) \qquad (17)$$

The purpose of a pseudo-inverse operator A^+ in Equation 16 is reduced to a simple interpretation following the sequence of matrix products in the formula. Product $\mathbf{g} = U^T \mathbf{y}$ determines vector $\mathbf{g}$ of a "generalized observation", components of which may be considered as coefficients of decomposition of $\mathbf{y}$ along the orthogonal system of matrix U columns, i.e., the "spectrum" of observations. The first r "generalized" observations are multiplied by values inverse to singular values of matrix A — "gain factor coefficients", "spectrum" of a pseudo-inverse operator — whereas the remainder $(M-r)$ are reduced to zero. As a result, we obtain the vector $\mathbf{p} = \Sigma^+ \mathbf{g}$ of "generalized" parameters ("spectrum" of the solution), which are correlated with the desired unknown by orthogonal transformation as $\mathbf{x}^+ = V\mathbf{p}$.

The procedure for constructing A^+ using Equation 15 was first elaborated by Lanczos (1961) and is called "natural inversion". It is equivalent to the pseudo-inverse by the Moore-Penrouse technique (Albert, 1972). Later Wiggins (1972) and Jackson (1972) proposed modifying it and considering nonzero singular values $q \leq r$ in Equation 17, because the introduction of small singular values into A^+ causes sharp increases of the norm and variance of the solution. This follows, apparently, from the expressions for the norm of the solution, norm of the residuals, and variance of the solution:

$$\left\| \mathbf{x}^{(q)} \right\|^2 = \sum p_i^2 = \sum (g_i / \sigma_i)^2 \qquad (18)$$

$$\left\| \mathbf{r}^{(q)} \right\|^2 = \left\| A\mathbf{x} - \mathbf{y} \right\|^2 = \sum g_i^2 \qquad (19)$$

$$\text{var}(x_k^{(q)}) = s^2 \sum (V_{ki} / \sigma_i)^2 \qquad (20)$$

The norm and variance of the solution show monotone increase with growing q, while the norm of the residuals decays monotonously. General variance of parameters determined as a trace of their covariance matrix also increases monotonously with growing q:

$$\text{var}(\mathbf{x}^{(q)}) = \text{Tr}[\text{cov}(\mathbf{x}^{(q)})] = s^2 \sum 1/\sigma_i^2 \qquad (21)$$

Here, s^2 is a variance of observations (it is unknown as a rule and is estimated within the framework of the same model by the residuals).

The matrix $R = A^+A = VV^T$ characterizes the resolution of unknowns: the closer it is to the unique matrix, the better is the resolution of the parameters. This can be explained by the fact that pseudo-solution $\mathbf{x}^+$ and unknown real solution $\mathbf{x}$ are related by:

$$\mathbf{x}^+ = R\mathbf{x} \qquad (22)$$

That is, R characterizes the linear relation of the obtained solution with the real one. The smaller the diagonal and the greater the nondiagonal elements R, the less pronounced are real parameters which merge amongst themselves.

The matrix $S = AA^+ = UU^T$ is called a matrix of the information density. It indicates what part of the given measurement participates in the construction of a pseudo-solution. If the former is close to the unique matrix, it indicates that all measurements are used properly for parameter estimation. Matrices R and S in the Lanczos technique are closer to the unique matrices than any others (Jackson, 1972). However, the choice of the value $q = r$ may not work due to the great variance of the solution, and we would have to sacrifice the resolution to improve the stabilities of the results.

The choice of an appropriate regularization parameter is a pivotal question in the theory of inverse problems. Specific recommendations on the choice of regularization level are given in a classical Tikhonov technique and its modification, namely, ridge regression (Hoerl and Kennard, 1970), oriented toward the statistical estimation of parameters; for instance, the error of closure principle is widely used (Tikhonov and Arsenin, 1986).

The number q, which is a pseudo-rank of matrix A and which determines the dimension of a subset of the greatest singular values, included in the pseudo-inverse operator $A^+ = V_q \Sigma_q^+ U_q^T$, is used as a regularizator in the procedure of a generalized linear inversion. It follows from Equations 18 to 21 that decay of q causes, on one side, decrease of $\|\mathbf{x}^+\|$, var x_k^+ and var $(\mathbf{x}^+)$, and on the other, increase of the residuals, shift of the solution (Marquardt, 1970), and loss of resolution. The process of constructing the optimal pseudo-solution can be best simulated as a procedure of successive growth in the dimensions of a pseudo-inverse operator due to the introduction of new singular values and simultaneous control of the chosen characteristics of the model. Several approaches can be proposed for a motivated choice of the regularization level with due consideration of the properties of observations and solutions:

1. Constraints on the norm of the residuals (error of closure principle). The number q is to be such that a monotonously decreasing residuals norm in Equation 19 becomes less than the given value Ms^2, where s^2 is a mean square error (mse) of the field mesurements.
2. Constraints on the norm of the solution. The norm of the solution, monotonously increasing in Equation 18, should not exceed some limit, for instance, $N|x_{\max}|$, where $|x_{\max}|$ is an *a priori* known (from physical consideration) maximum of any parameter. The value $|x_{\max}|$ may be also used for restrictions on the maximum for any model parameter.
3. Constraints on the maximum variance of each parameter or the total of all parameters. The number q is to be such that values in Equation 20 or 21 do not exceed some level determined from the assumption that the desired variance allows some practical interpretation of the results.

4. Forsight et al. (1980) proposed considering only those singular values whose ratio to the maximal of them is not less then the relative error in the initial data.
5. The analog of the Strakhov-Valyashko adaptive regularization (1984) may be obtained if we extract successively the maximum of the modulo numbers from the generalized observations $g = U^T y$ and introduce the corresponding singular values in the inverse operator. The solution obtained by adaptive regularization in accordance with the error of closure principle has a minimum number of degrees of freedom, because the anomaly can be explained by the minimum set of generalized parameters.
6. By analogy with the Winer theory of the optimal filtration of signals, it is possible to construct a pseudo-inverse operator with the spectrum of singular values as:

$$\sigma_k^+ = \frac{s_k}{s_k^2 + G_N^2 / g_k^2} \qquad (23)$$

where $k = 1,...r$, G_N, is a white noise variance which has a regular decomposition over the columns of matrix U.

It is clear from this, by no means complete, list of approaches for choosing the regularization level, that the technique of generalized linear inversion makes it possible to obtain regularized solutions possessing certain desired characterstics. In the end, the interpreter should himself determine what properties of the model best suit the essense of the problem formulated and the initial data, and on this basis decide which method of regularization is optimal.

The Tikhonov regularization technique implies that the matrix $(A^T A + \lambda E)^{-1}$ is to be calculated for various values of λ. Expressions for this matrix, compact and convenient in the calculation, may be also obtained from a singular decomposition. It is important that the singular decomposition of matrix A is calculated only once. Louson and Henson (1986) give the corresponding formulas for a pseudo-inverse matrix, solution and residuals norms, variance of the solution, etc. necessary to analyze the behavior of the system for various values of the regularization parameter λ.

A. REPARAMETERIZATION AND AMALGAMATION OF THE MODEL

To achieve the objective of a detailed model of the medium's structure at the stage when the model is to be chosen, we have to assume an overly detailed dissection of the volume into elementary bodies. This decreases the stability of the problem and implies the simulation of extremely smoothened, poorly resolved regularized solutions. Since this simulation is aimed at obtaining the information for geological-geophysical data interpretation, a geological researcher would obviously prefer a simple and clear model with a small number of anomalous bodies to a more detailed, almost continuous model, in which it would be necessary to specify constantly the resolution degree for each real point in the space. Therefore, the question arises: Is there any simple inner structure of the analyzed phenomenon? In particular, is it possible to explain properly the observed anomaly by a minimal set of objects? Reparameterization seems to be the substitution of a model with a small number of anomalous bodies whose inner physical parameters are constant for a complex medium whose description requires a great number of variables.

Reparameterization is aimed at changing the initial poorly determined model with regular detail and low resolution individual domains for an amalgamated steady model with good resolution.

In linear regression analysis such an approach corresponds to the problem of the best regression (Seber, 1980). In this approach, in any subset of regressors one regressor should be found that would gain an error of closure statistically similar to that of a complete model. The procedure of step-by-step regression and exhaustion of all regressions possible is labor-consuming and, sometimes, leads to diverse results (Dreiper and Smith, 1986).

Our principal model uses vectors of N variables to explain the behavior of M observations. If matrix A is of incomplete rank or is poorly determined, the effective number of descriptive parameters will be less; it will be equal to q, for instance. Thus, a natural question arises: How can we select (from N unknowns) the variables $x_{i_1}, x_{i_2}, ..., x_{i_q}$ which can explain properly the behavior of y and, at the same time, be calculated from the residuals of the system? In other words, the residuals of the corrected model should be statistically compatible with the residuals of a complete model. If the model matrix is of complete rank, this problem can be solved partially in a simple way: the information on correlations between the variables is contained in the estimate of their covariance matrix $\sigma^2 (A^T A)^{-1}$. We can get the corresponding coefficients of the parameter correlations by normalizing this matrix. If for some pair of these parameters the correlation coefficient modulo is close to 1, this evidences a close linear relation between these parameters. Therefore, within the given model, they cannot be determined separately, since a variation in one of them will cause a compensation change in the other. When correlated variables have the same physical meaning, for instance, it may be magnetization of the neighboring blocks, high correlation between them may indicate that they belong to one larger object. Unfortunately, this works only for a pair of correlated variables, and it is not so productive if we consider a set of numerous variables close to the linear relation.

In the technique of generalized linear inversion, when we choose the number q of singular values participating in a pseudo-inverse by Equation 16, we divide the space of model X and observations Y into the pairs of orthogonal subspaces:

$$X = V_q \cup V_0, \quad V_0 = V_q^\perp \qquad (24)$$

$$Y = U_q \cup U_0, \quad U_0 = U_q^\perp \qquad (25)$$

where V_q and U_q are linear spheres of the first q eigenvectors $v_1, ... v_q$ and $u_1, ... u_q$. Any set of parameters describing V_q has an "active" character: it may be obtained when a pseudo-inverse operator A^+ acts on some observation $y_0 \in Y$. On the contrary, if $x_0 \in V_0$, it is called a "passive" set of parameters and it cannot be obtained (at the given A^+) at any observation y.

Analogously, observations belonging to U_q are active, because they participate in the solution simulation, whereas $y \in U_0$ do not affect in any way $x \in V_q$.

From these properties of the models and observations as well, they can be organized with respect to their "active" subspaces V_q and U_q (Ivanenko, 1988). One of the properties of the resolution matrix R is that it projects X onto the subspace V_q. From its symmetry and idempotent character it follows that $R^2 = V_q V_q^T V_q V_q^T = V_q V_q^T = R$. The matrix of information density S has the same characteristics in observations space. It projects the data onto the subspace of active observations. Using the projection properties of these matrices, it is possible to calculate the angles between any model **x** or observation **y** and the corresponding subspace. The smaller is the angle, the better the model or data "matches" the structure of the given inverse problem.

The angle between the vector ξ and subspace L is determined as

$$\arccos(\|\text{Pr}_l\,\xi\| \,/\, \|\xi\|) \tag{26}$$

Let us write the decomposition of model X on the orthonormalized basis

$$\mathbf{x} = \sum x_i \xi_i \tag{27}$$

where $\xi_i = (0, 0..., 0, 1, 0, ..., 0)$ is the i-th unique vector of the basis. ξ_i determines in E^N the direction along which the i-th parameter of model **x** varies. The projection of ξ_i onto V_q will be

$$\text{Pr}_{V_q} \xi_i = R\,\xi_i = \mathbf{r}_i \tag{28}$$

where $\mathbf{r}_i$ is the i-th column of matrix R. From this we obtain:

$$\cos \theta_i = \|\mathbf{r}_i\| / \|\xi_i\| = (\sum r_{ik}^2)^{1/2} = r_{ii} \tag{29}$$

where θ_i is an angle between the direction of the i-th parameter of the model and subspace V_q. Completely analogously:

$$\cos \varphi_j = \|\mathbf{s}_j\| = s_{jj} \tag{30}$$

where φ_j is an angle between the direction of the j-th measurement and active subspace U_q.

Note that formally $\cos \theta_i$ and $\cos \varphi_j$ coincide with the multicoefficient of the correlation, if we consider the problems of "matching" using least-square technique:

$$\xi_i = V_q\, \alpha \tag{31a}$$

$$\eta_j = U_q\, \beta \tag{31b}$$

that is, if we try to find the best linear estimate for the i-th parameter direction (j-th observation) in the active subspace $V_q(U_q)$. Therefore, it is possible to check formally by F-statistics the hypotheses on the equality of $\cos \theta_i$ and $\cos \varphi_j$ to zero.

$$F_1 = \frac{\cos^2 \theta_i / q}{(1 - \cos^2 \theta_i)/(N - q - 1)} \tag{32a}$$

$$F_2 = \frac{\cos^2 \varphi_i / q}{(1 - \cos^2 \varphi_i)/(M - q - 1)} \tag{32b}$$

with q, $N-q-1$, and q, $M-q-1$ degrees of freedom, respectively. The variables and observations for which F-statistics exceed critical values have small angles with active subspaces and affect considerably the process of pseudo-reversal. For the remainder, we must either modify the model or exclude them from the problem because these variables cannot be determined reliably in the given model, and the observations bear no considerable information. Decrease in the problem dimensions and increase of its stability result from such a modification.

To distinguish clusters of nonresolved like parameters and substitution of them for amalgamated characteristics of the model, a special procedure has been proposed (Ivanenko and Semenets, 1989). It turned out that it is possible to find such a linear combination of poorly resolved parameters, which has better resolution. A group of l nonresolved parameters forms some subspace $L = L(\mathbf{x}_1, \mathbf{x}_2, ..., \mathbf{x}_l)$ in X. We shall look for a vector of unique length $\mathbf{x}^1 = \alpha_1 \mathbf{x}_1, \alpha_2 \mathbf{x}_2, ..., \alpha_l \mathbf{x}_l$ from L, $\|\alpha\| = 1$, which has the best resolution in that subspace. For this purpose we choose coefficients a_k, $k = 1, ... l$ from the conditions that the length of projection $\mathbf{x}^1$ onto the active subspace V_q is maximum. Accounting for the projecting properties of matrix R, we obtain:

$$\left\| \mathrm{Pr}_{V_q} \mathbf{x}^1 \right\| = \left\| R\mathbf{x}^1 \right\| = \left\| R\sum \alpha_k \mathbf{x}_k \right\| = \left\| R_l \alpha \right\| \tag{33}$$

where R_l is formed by corresponding l columns from R. From the definition

$$\max \| R_l \alpha \| = \max \frac{\| R_l \alpha \|}{\| \alpha \|} = \| R_l \|$$

and is equal to the maximum singular number R_l. Thus, the desired coefficients α_k are determined by the eigenvector corresponding to the maximum singular value of the singular decomposition R_l.

The procedure of automatic reparameterization may be performed in a more rapid and simple way: among the blocks with poor resolution we combine neighboring ones whose same parameters overlap at the level-of-confidence interval. The merging of blocks does not require the recalculation of a direct problem because it may be performed by a simple summing up of the corresponding columns of matrix A.

B. A UNIQUENESS SOLUTION OF AN INVERSE PROBLEM

The problem of a uniqueness solution of an inverse problem in magnetometry with a discrete formulation while using the technique of generalized linear inversion may be solved almost automatically. Any vector of the null-space of matrix A (linear sphere of eigenvectors of matrix V with negligibly small singular numbers) will be an annihilator or, in other words, a distribution of magnetization which does not form any field at the points of measurements (Parker and Huestis, 1974). An annihilator is orthogonal to a pseudo-solution. Numerical experiments have been performed for two- and three-dimensional variants of an inverse problem with a scalar magnetization. It followed that:

1. In the models with poor resolution, the annihilator is distributed in poorly resolved blocks and may distort strongly the interpretation in such areas. However, such

annihilators form nonzero fields beyond the measurement points, i.e., their occurrence is determined by the properties of the data and the choice of the model.

2. For the well-determined models with detailed description of the field and high resolution, a nontrivial annihilator should be found as the projection of a regular unique magnetization onto a null-space. The projection is found via the projecting properties of the matrix of a "resolution defect" $R_0 = I - R_q$. The choice of a unique magnetization is motivated by the fact that a regular magnetization is an annihilator for an ideal problem, i.e., for a flat parallel layer.
3. The annihilator is determined by the accuracy of a constant factor whose value is restricted by the physical considerations: (1) the field formed by an annihilator should not exceed some value, for instance, 0.5%, of the anomaly span; (2) the maximum (average) value of the annihilator should not exceed the maximum (average) value of a pseudo-solution.
4. By summing up the annihilator and then extracting it from the pseudo-solution, we obtain the restrictions on the family of possible general solutions of the problem.

When the stable solution is found, the model may be further amalgamated in accordance with the character of the solution obtained. The blocks which remain unsolved may be either combined with the neighboring ones, or some restrictions may be set on them arising from the general magnetization behavior. Neighboring blocks with close values are combined. Restrictions on poorly resolved parameters are given as linear relations between them which are introduced as supplementary equations in the principal system of equations describing the model. Such a procedure is equivalent to the usage of the additional *a priori* information and leads to the simulation of a regularized solution (Jackson, 1979).

C. ADAPTIVE REPARAMETERIZATION

The above methods of reparameterization have some element of formal approach to the problem, because the obtained solvable combinations of parameters do not bear any pronounced physical sense. It is possible to overcome this difficulty when changng for the procedure of compilation of regulated solutions in the simulation class with given characteristics. The latter are introduced to the model when restrictions are given which aggravate iterationally the requested characteristics of simulations. Such an approach to reparameterization is called adaptive.

When reconstructing the magnetization in the active magnetic layer from the observed field, there originates a problem of the simulation of the simplest model of the layer. In the case of an excessive number of blocks comprising the model, the solution with a quasi-piecewise-constant magnetization will correspond to that model. Our objective is to obtain solutions of piecewise-constant magnetization distribution — due to smoothening as a result of regularization — with a minimum number of magnetization changes from one model block to another. Though convergence of the supposed iteration has not been proved by the mathematics, the examples considered later and conclusions described in the previous chapters show the practical value of the proposed approach.

Not so long ago, Constable et al. (1987) proposed the construction of "smooth" solutions of inverse problems by considering restrictions for derivative solutions called the "Okkama inversion".

Restrictions for the norm of the first or second derivative of the solution are introduced in the linear inverse problem described by the system of equations $Ax = y$. In the result of this procedure the solutions become "smooth". (The procedure is an obvious generalization of the on-line-approximation ideas.) In the final-differential formulation, the restrictions have the form of discrete analogs of differential functionals of the first or second order. We note that minimization of the considered problem by the Constable regulating functional is equivalent to transition to the extended model due to the inclusion of equations describing final-differential restrictions in matrix A (Louson and Henson, 1986).

As a result the system of equations of the model has the form:

$$\begin{pmatrix} A & x \\ \lambda D & x \end{pmatrix} = \begin{pmatrix} y \\ 0 \end{pmatrix} \tag{34}$$

where D the band matrix describing the final differences in the magnetization increment in the neighboring blocks; λ is a parameter of regularization considering the relative weight of restrictions and accuracy in the solution choice. We shall consider the case of restrictions only for the first derivative.

Let us consider the modified Problem 34.

$$\begin{pmatrix} A & x^{(n)} \\ \lambda W^{(n)} D & x^{(n)} \end{pmatrix} = \begin{pmatrix} y \\ 0 \end{pmatrix} \tag{35}$$

where $W^{(n)}$ is a diagonal matrix considering the relative weight of restrictions about the equality of magnetization in neighboring blocks; n is the iteration number.

Since our goal is to obtain the quasi-piecewise-constant solution, we shall redistribute those weights at each step of the iteration so as to intensify restrictions where the solution is quasi-constant and to weaken them where the magnetization has a high gradient at the transition to the neighboring block of the model.

For that purpose we shall use the equipment developed especially for a robust estimation of parameters in the procedure of linear regression (M adjustment) used for the adaptive inclusion of previously unknown differences in the observation errors (Steiner, 1980). This can be achieved by redistribution of weights when analyzing the model residuals; in our case, when analyzing the sharp change in magnetization from one block to another. The weights are redistributed by the formula

$$W_i^{(n)} = \frac{1}{\sum_{i=2}^{N}(x_{i+1}^{(n)} - x_i^{(n)})^2 + (\varepsilon^0)^2}$$

where ε^0 can be found from the minimization functional proposed by Steiner (1980). In our problem we normalize these weights so that their sum is equal to the number of restrictions.

From our experience, three to four iterations are sufficient to obtain the solution with the required parameters.

D. PRACTICAL RESULTS

Examples of the magnetization reconstructions in the layer with real geometry using adaptive reparameterization and without the latter are given in Figure 9a for the Northeastern Pacific basin within magnetic lineations 13 and 15. In the field observed, the amplitudes of those anomalies do not exceed 200 nT. The bottom topography is even typical of a basin with an average depth of 5 km and deeper. Sedimentary thickness exceeds 150 m. The sequence of the anomalies is characterized by the fine structure of the field which can be clearly mapped by a gradient survey. The existence of these local anomalies seems to be related mainly to the changes in the stress of the paleofield and is verified by a number of surveys.

According to a magnetic chronological scale for the magnetic field inversions anomalies 13 and 15 consist of three chrones, namely, 13.1n, 13.r, and 13.n, which continued for 0.26, 0.07, and 0.38 million years, respectively, along with 15.ln, 15.r, and 15.n, which continued for 0.26, 0.03, and 0.15 million years respectively. For comparison, the Figure shows the solution in the case when an experienced interpreter dissects the model into relatively large blocks without the use of a computer. Neither the fine structure of the magnetic field, nor inverse chrones 13.r and 15.r are reflected in the magnetization distribution. It follows from the figure that the solution has a general character and can reflect the chronology of magnetic lineations only in general. A qualitatively different picture is obtained when using adaptive reparameterization. All normal and inverse chrones present in the scale are distinguished clearly in the magnetization distribution within magnetic lineations 13 and 15. Five short-period magnetization fluctuations are mapped between anomalies 13 and 15 which correspond to the fine structure of the observed field, their amplitudes reaching 1 A/m. It should also be noted that general tendencies in the distribution of the magnetization are preserved if compared to the manual dissection of the model.

Figure 9b shows the results of estimates along the meridional profile crossing the Heezen transform fault located in the Southern Pacific. In the bottom topography, the fault is expressed as a giant scarp with trough depth down to 6.1 km, the summit of the near fault ridge being located at the depth of 600 m. As seen from the figure, the anomaly over the near-fault ridge reaches 1000 nT. Data of bathymetric measurements were taken for the top of the active magnetic layer (sedimentary layer is practically absent there). The thickness of the active magnetic layer is 5 km. The magnetization distribution pattern over the near-fault ridge summit is specific. There the annihilator values reach 0.2 A/m, whereas in the trough area they increase to 1.8 A/m. This evidences the dominating role of the layer geometry in the considered part of the fault. The confidence interval exceeds 1 A/m. From the estimates, the input of the topography in the formation of the anomaly over the near-fault ridge is over 600 nT. Effective magnetization does not exceed 2 A/m at the fault flanks; they mark the position of lineations 2A from the side of the southern fault slope and number 4, from the northern one. The profile follows the strike of lineations at an acute angle to them. For comparison, similar to the previous example, manual dissection of the model into elementary bodies was performed. The solution has a generalized character as well. No peculiar features even within the near-fault ridge are observed. The most distinct differences from the previous estimates are recorded within the near-fault ridge and trough of the fault. Comparison with the results of petromagnetic analysis of the dredged rocks showed that the most intensively magnetized rocks comprise the upper part of the near-fault ridge but not the bottom of the trough. This better agrees with the solution obtained by adaptive reparameterization of the model.

We believe the examples given illustrate well the advantages of the adaptive approach compared to the traditional one. However, it should be noted that due to the specific character of hydromagnetic surveys, the most informative are profile measurements. Therefore, when results of the interpretation of profile measurements are involved in the solution, the latter is improved considerably.

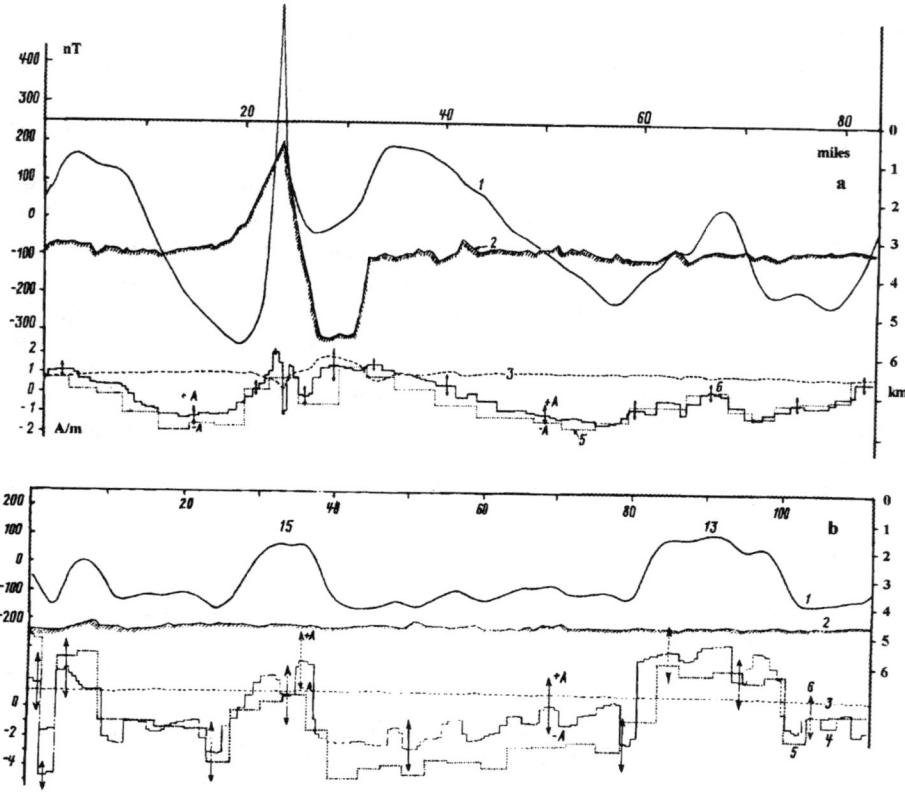

Figure 9 Effective magnetization estimated by adaptive reparameterization algorithms: (a) Northeastern Pacific basin; (b) Heezen fracture zone; 1, observed magnetic field; 2, bottom topography; 3, annihilator; 4, 5, effective magnetization distribution with and without adaptive reparameterization; 6, confidential interval.

IV. MAGNETIC FIELD SIMULATION FOR SEAMOUNTS

A. FORMULATION OF THE PROBLEM

Geomagnetic research of seamounts is carried out in two aspects, namely, paleomagnetic and structural-geological aspects.

The paleomagnetic aspect (seamount method) is aimed at finding out the intensity and vector of a seamount magnetization from the magnetic field and topography of the latter with the subsequent estimation of the coordinates of the virtual paleomagnetic pole. These results are used in paleomagnetic reconstructions for seamounts of known age to determine the kinematics of the lithosphere-bearing plates. The paleolatitude of the formed seamounts bears the information about the latitudinal migration of lithosphere plates, while declination of the magnetization's horizontal component from the meridional one reflects the rotational component of their motion. Comparison of the position of virtual paleomagnetic poles of the undated seamounts with the curves of the apparent drift of paleomagnetic poles of the lithosphere plates allows the dating of those mounts.

The structural-geological approach accents the simulation of a geometrical model for the sources of magnetic anomalies that should reflect the specific features of the interior geological structure of an underwater volcano, stages of its formation and evolution, as well as its rock composition.

The first papers devoted to experimental geomagnetic research and numerical simulation of magnetic anomalies of seamounts are dated back to the middle of the 1960s, namely, the period of intensive geological-geophysical and, in particular, magnetic studies of the World Ocean floor and the period when basis for the plate tectonic concept was formulated. Vacquier (1962) and then Talwani (1965) were the first to formulate the problem of determining the magnetization intensity of a homogeneously magnetized seamount from its topography and magnetic anomaly using the least squares technique. They proposed the method of approximating the bottom topography to solve direct problems. On the basis of this technique, and within the framework of a paleomagnetic approach, magnetization of a series of seamounts was determined (Richards et al., 1967; Vacquier and Uyeda, 1967), and, in particular, a remarkable (30 to 40 latitudinal) northward drift of the post-Cretaceous Pacific plate has been verified (Francheteau et al., 1970).

More complex simulations of block models have been attempted since the 1970s for there is often no way to choose the appropriate field within the framework of the simplest model of homogeneous magnetization, and remnant anomalies are of nonrandom character (Kaminsky and Simovsky, 1976; Harrison, 1971; Sager et al., 1982). Systematic description of this technique may be found, for instance, in the monographs *Plate Tectonics* (Le Pichon et al., 1973) and *Seamounts* (1978). Block models have illustrated in theory and in practice that coexistence of blocks in the body of a seamount where the magnetization differs in intensity and direction may affect considerably the accuracy of determination of paleopoles (Lumb et al., 1973). This effect is particularly critical in the case of lateral inhomogeneities in the distribution of magnetization (Blakely and Christiansen, 1978). At the same time, Harrison (1971), simulating a seamount with a nonmagnetized summit (the Atlantic), concluded that the influence of a vertical inhomogeneity in magnetization upon paleomagnetic parameters of the model was rather weak.

During the last decade techniques were devoted mainly to simulations of models with more complex geometry (McNutt, 1986; Sager, 1984; Ueda, 1985) due to the availability of more efficient and visual procedures for approximating the topography and solving a direct problem (Plouff, 1976). Numerous models were simulated for seamounts, comprised of several large blocks of various magnetizations and, sometimes, various directions of the latter (Kodama and Uyeda, 1979; Sager and Keating, 1984). They seem to better interpret the measured anomaly. However, such interpretation does not seem to be always adequate since it implies some arbitrary approach, i.e., the method of tests and errors. The data revised in such a manner were used as a basis for the reconstruction of the apparent migration of the Pacific plate paleomagnetic pole starting from the middle of the Cretaceous (Sager and Pringl, 1988).

Some authors (Sager, 1984; Gardner et al., 1984) used the procedure of transforming the field upwards for subsequent inversion by the least-squares method to reduce the effect of local anomalies (caused by near-surface sources) upon the estimates of paleocharacteristics. Unfortunately, for a correct interpolation of the field upwards — to amplify long-period anomalies which are the main source of paleoinformation — the detailed knowledge of the field over a large area is necessary, which usually exceeds the real dimensions of a standard geophysical survey area for seamounts.

Once again it became clear that there is no way to obtain reliable results for seamounts with complex inner structures by traditional techniques of simulation when results were published of direct paleomagnetic measurements taken at seamounts such as Kobb (Merrill and Burns, 1972), Suiko (Kodama and Uyeda, 1978; Kono, 1980), and Oshima volcano (Kodama and Uyeda, 1979).

Recently, Parker and others (Hildebrand and Parker, 1987; Parker, 1988; Parker et al., 1987) attempted developing a new interpretation procedure for seamount magnetic fields. This procedure is a developed Backus-Gilbert procedure for solving inverse problems (Backus and Gilbert, 1968) and is reduced to a joint determination of a mean homogeneous and minimum heterogeneous magnetization inside a given seamount body and allows the reproduction of the observed anomaly with the necessary accuracy with due respect to the noise level. The procedure, however, implies great expenses since it is necessary to use supercomputers. One of the principal aspects of the Backus-Gilbert theory — solvability of parameters — has not yet been investigated and this prevents the transition to any geological model.

The review illustrates that the range of the geomagnetic studies of seamounts is enlarged and models become more and more complex as the procedures of data processing and interpretation improve. It has revealed the obvious demerits of a traditional approach: low detail, insufficient solvability, and subjective choice of the models. An incomplete account of the existing data has stimulated the elaboration of a more flexible and powerful technique of interpretation using the recent achievements of numerical techniques for solving geophysical problems.

Magnetic simulations of seamounts are related to inverse problems of geophysics. It is typical of such problems that parameters of the object which is of interest to us are beyond our reach or cannot be measured directly; therefore, we can draw conclusions about them only from some indirect parameters, i.e., anomalies which are related functionally to the parameters in question. Mathematical and algorithmal formulation of the rules describing that functional relation provides for the solution of a direct problem and allows simulation of the measured fields at the given parameters of the model. Inversion of the functional relation implies the search for unknown parameters of the medium from the real observations and is believed to be the subject of an inverse problem.

We present the principal functions determining the relation between the measured magnetic anomaly over a seamount and unknown magnetization of that object. Suppose, that we know from some geological-geophysical studies the volume V accomodating the sources of the magnetic anomaly in question. This volume is to be determined so that its upper boundary coincides with the relief of a seamount, whereas the lower boundary coincides with the top of the basalt layer of the underlying lithosphere. Accounting for the possible inner sources of anomaly in the lithosphere itself, due to volcanic features of the sea volcano (magmatic columns, chambers, etc), it seems possible to determine the lower boundary with some safety margin to include the entire area of the possible occurrence of anomalous magnetization. A simple supplementary determination of magnetization by zero values beyond the area of its real occurrence will retain the expressions written below.

We assume that the magnetic anomaly at the point of observation P with coordinates $\mathbf{r} = (x, y, z)$ is caused by the distribution of bulk magnetization $\mathbf{I}(\mathbf{r}_0) = (I^x(\mathbf{r}_0), I^y(\mathbf{r}_0), I^z(\mathbf{r}_0))$; $\mathbf{r}_0 = (\xi, \eta, \zeta)$ in the body taking up volume V in the lower half-space. Expression for a magnetic potential at point P is as follows:

$$U(\mathbf{r}) = \int_V (\mathbf{I}(\mathbf{r}_0)\nabla_0|\mathbf{r}-\mathbf{r}_0|^{-1})dV = -\int_V (\mathbf{I}(\mathbf{r}_0)\nabla_p|\mathbf{r}-\mathbf{r}_0|^{-1})dV \qquad (36)$$

The vector $\mathbf{T}_a(\mathbf{r})$ of the anomalous magnetic field corresponds to a gradient of magnetic potential: $\mathbf{T}_a(\mathbf{r}) = -\nabla_p U(\mathbf{r})$.

Proton magnetometry is usually used in geomagnetic surveys to measure geomagnetic intensity, which is a sum of the total intensity of the normal magnetic field of the Earth (EMF), $\mathbf{T}_E$, and of the anomaly field $\mathbf{T}_a$. Let the unit vector $\mathbf{t} = (\lambda, \mu, \nu)$ determine the direction of the EMF.

Then, measuring $|\mathbf{T}_E + \mathbf{T}_a|$ and computing $|\mathbf{T}_E|$ (for instance, from an IGRF model), it is possible to estimate the field $\Delta \mathbf{T}_a(\mathbf{r}) = |\mathbf{T}_E(\mathbf{r}) + \mathbf{T}_a(\mathbf{r})| - |\mathbf{T}_E(\mathbf{r})|$, which at $|\mathbf{T}_a| \ll |\mathbf{T}_E|$ coincides approximately with the gradient of magnetic potential along the vector direction $\mathbf{t}$:

$$\Delta T_a(\mathbf{r}) \approx -\frac{d}{dt}U(\mathbf{r}) = -\left(\lambda\frac{\partial}{\partial x} + \mu\frac{\partial}{\partial y} + \nu\frac{\partial}{\partial z}\right)U(\mathbf{r}) \qquad (37)$$

Further, we assume vector $\mathbf{t}$ to be free of the coordinates of observation points within the area of magnetic survey over an individual seamount. This is possible since dimensions of anomalous field areas over seamounts do not exceed several tens of kilometers, and changes of EMF direction within such areas are several parts of the degree.

Differentiating Equation 37 we obtain:

$$\Delta T_a(\mathbf{r}) = \int_V \mathbf{I}(\mathbf{r}_0)^T [W(\mathbf{r},\mathbf{r}_0)] \mathbf{t}\, dV = \int_V \mathbf{t}^T [W(\mathbf{r},\mathbf{r}_0)] \mathbf{I}(\mathbf{r}_0)\, dV \qquad (38)$$

Matrix $[W(\mathbf{r},\mathbf{r}_0)] = (\nabla_p \nabla_p^T) \rho^{-1}$, where $\rho = |\mathbf{r} - \mathbf{r}_0|$ describes the tensor with elements

$$w_{11} = w_{xx} = \frac{\partial^2}{\partial x^2}\rho^{-1} = [3(\xi - x)^2 - \rho^2]/\rho^5$$

$$w_{12} = w_{21} = w_{xy} = w_{yx} = \frac{\partial^2}{\partial x \partial y}\rho^{-1} = [3(\xi - x)(\eta - y)]/\rho^5$$

$$w_{13} = w_{31} = w_{xz} = w_{zx} = \frac{\partial^2}{\partial x \partial z}\rho^{-1} = [3(\xi - x)(\zeta - z)]/\rho^5$$

$$w_{22} = w_{yy} = \frac{\partial^2}{\partial y^2}\rho^{-1} = [3(\eta - y)^2 - \rho^2]/\rho^5$$

$$w_{23} = w_{32} = w_{yz} = w_{zy} = \frac{\partial^2}{\partial y \partial z} \rho^{-1} = [3(\eta - y)(\zeta - z)]/\rho^5$$

$$w_{33} = w_{zz} = \frac{\partial^2}{\partial z^2} \rho^{-1} = [3(\zeta - z)^2 - \rho^2]/\rho^5 \tag{39}$$

Equation 38 reflects the theoretical relation between the magnetization distribution and the magnetic anomaly. It may be considered as an integral equation of the first order with respect to vector $\mathbf{I}(\mathbf{r}_0)$, $\mathbf{r}_0 \in V$ at given values of $\Delta \mathbf{T}_a(\mathbf{r})$, $\mathbf{r} \in \mathcal{D} \subseteq R^2$. Such a formulation of an inverse problem is very general since its solution, which implies just determining $\mathbf{I}(\mathbf{r}_0)$ at $\mathbf{r}_0 \in V$, allows us to answer all questions arising in the course of seamount magnetism, of both paleomagnetic and structural-geological characters.

A general solution of paleomagnetic problem is derived directly from Equation 38 by calculation of the direction of the obtained magnetization vector estimate $\mathbf{I}(\mathbf{r}_0) = [I^x(\mathbf{r}_0), I^y(\mathbf{r}_0), I^z(\mathbf{r}_0)]$. This direction is determined unambiguously by a unit vector $\mathbf{j}(\mathbf{r}_0) = \mathbf{I}(\mathbf{r}_0)/|\mathbf{I}(\mathbf{r}_0)|$ or by declination and inclination angles

$$D(\mathbf{r}_0) = \arctan[I^y(\mathbf{r}_0)/I^x(\mathbf{r}_0)]$$

$$I(\mathbf{r}_0) = \arctan\left[I^z(\mathbf{r}_0)/[I^x(\mathbf{r}_0)^2 + I^y(\mathbf{r}_0)^2]\right]^2 \tag{40}$$

The knowledge of these angles allows us to study the distribution of magnetization directions, to determine positions of virtual paleomagnetic poles, and perform other procedures of paleomagnetic analysis.

The general solution of a geological-structural problem is related primary to the vector of the magnetization intensity behavior:

$$J(\mathbf{r}_0) = |\mathbf{I}(\mathbf{r}_0)| = [I^x(\mathbf{r}_0)^2 I^y(\mathbf{r}_0)^2 + I^z(\mathbf{r}_0)^2]^{1/2} \tag{41}$$

This behavior reflects the inner structure of the object analyzed, as well as tectonic, geological, physical-chemical, and other conditions of rock magnetization and evolution.

It is known from the theory that inverse problems of the potential in such a general formulation as stated in Equation 38 do not have a single solution. V. N. Strakhov (1978) showed that, for this problem, it is possible to find equivalents from outer field magnetization distributions dependent upon the random vector function at which weak restrictions are applied. Such multivaluedness of the solution is stronger compared to that of the general inverse problem of gravimetry.

Stability of the general inverse problem to the errors of the input data does not exist because it is an integral equation of the first type and belongs to the class of incorrect problems. This implies the use of the regularization procedure when dealing with numerical solutions.

For numerical solution of the problem we choose the dissection of volume $V = U\{V_i\}$, $i=1, \ldots N$, and rewrite Equation 38 with respect to it:

$$\Delta T_a(\mathbf{r}) = \sum_{i=1}^{N} \int_{V_i} \mathbf{I}(\mathbf{r}_0)^T [W(\mathbf{r},\mathbf{r}_0)] \, t \, dV \qquad (42)$$

We believe that, within the accuracy of practical goals, $\mathbf{I}(\mathbf{r})$ inside V_i can be substituted by a mean value: $\mathbf{I}(\mathbf{r}) \approx \mathbf{I}_i = \text{const}$ at $\mathbf{r} \in V_i$

From Equation 42 we have:

$$\Delta T_a(\mathbf{r}) = \sum_{i=1}^{N} \mathbf{I}_i \int_{V_i} [W(\mathbf{r},\mathbf{r}_0)] t \, dV_i = \sum_{i=1}^{N} I_i^x A_i^x(\mathbf{r}) + I_i^y A_i^y(\mathbf{r}) + I_i^z A_i^z(\mathbf{r}) \qquad (43)$$

where the following integrals are expressed as A_i^x, A_i^y, A_i^z:

$$A_i^x = \int_{V_i} (\lambda w_{xx} + \mu w_{xy} + \nu w_{xz}) \, dV_i$$

$$A_i^y = \int_{V_i} (\lambda w_{xy} + \mu w_{yy} + \nu w_{yz}) \, dV_i$$

$$A_i^z = \int_{V_i} (\lambda w_{xz} + \mu w_{yz} + \nu w_{zz}) \, dV_i \qquad (44)$$

Considering the discrete and final character of the assemblage of observation points $\{\mathbf{r}_k\}$, $k = 1, ...M$, we obtain a system of M linear equations in $3N$ unknowns:

$$\Delta T_k = \sum_{i=1}^{N} A_{ik}^x I_i^x + A_{ik}^y I_i^y + A_{ik}^z I_i^z \qquad (45a)$$

Here, we have used the notations $\Delta T_k = \Delta T(\mathbf{r}_k)$, $A_{ik}^{x[y,z]} = A_i^{x[y,z]}(\mathbf{r}_k)$. To compensate for the probable influence of a linear constituent in the measured field upon the observations, additional parameters, namely b, c, d, corresponding to the $M \times (3N+3)$ model with account of the linear trend, may be introduced into Equations 45a

$$\Delta T_k = \sum_{i=1}^{N} A_{ik}^x I_i^x + A_{ik}^y I_i^y + A_{ik}^z I_i^z + bX_k + cY_k + d \qquad (45b)$$

where X_k, Y_k are horizontal coordinates of the observation points.

System 45 reflects the discrete formulation of the problem: from M readings of the field we approximate the unknown magnetization as a piecewise-constant vector function given at the dissection of the volume $\{V_i\}_1^N$.

If we assume that the magnetization vector retains constant direction in the volume V, System 45 is simplified:

$$\Delta T_k = \sum A_{ik} J_i + \text{trend} \qquad (46)$$

where $A_{ik} = A_{ik}^x \cos D \cos I + A_{ik}^y \sin D \cos I + A_{ik}^z \sin I$, with D and I being the given declination and inclination vectors of magnetization. This $M \times N$ (or $M \times (N+3)$) system of equations corresponds to the model with scalar magnetization.

Under the assumption of homogeneous magnetization: $(I_i^x, I_i^y, I_i^z) = (I^x, I^y, I^z)$ for all i, System 45 is converted into

$$\Delta T_k = I^x A_{Sk}^x + I^y A_{Sk}^y + I^z A_{Sk}^z + \text{trend} \qquad (47)$$

where $A_{Sk}^x = \sum A_{ik}^x, A_{Sk}^y = \sum A_{ik}^y, A_{Sk}^z = \sum A_{ik}^z$.

System 47 corresponds to a classical formulation of the problem within the framework of a homogeneous magnetization hypothesis.

Let us consider one more principal, particular case, when the relative distribution of physical properties, i.e., magnetic intensity $|\mathbf{I(r)}|$ within the dissection of volume $\{V_i\}$, is known but we do not know its direction, though it is believed to be similar for all bodies of the dissected volume. Such a situation may arise, for instance, if at the previous stage of interpretation variations of magnetization intensity were determined with a given direction using System 46, and at the present stage, when these intensity variations are fixed, interpretator would like to refine the primary direction of magnetization. In this case we obtain a linear system similar to System 47:

$$\Delta T_k = l A_{Sk}^x + m A_{Sk}^y + n A_{Sk}^z + \text{trend} \qquad (48)$$

relative $l = \cos D \cos I$, $m = \sin D \cos I$, $n = \sin I$, where

$$A_{Sk}^x = \sum A_{ik}^x |\mathbf{I}_i|, A_{Sk}^y = \sum A_{ik}^y |\mathbf{I}_i|, A_{Sk}^z = \sum A_{ik}^z |\mathbf{I}_i|$$

A principal aspect in each discrete formulation of an inverse problem is the fact that the method of dissection of $\{V_i\}$ is an heuristic procedure which may affect considerably the properties of the obtained solutions. One of the possible ways to avoid the multivaluedness, originating in the course of the procedure, is reparameterization of the model by combining the blocks — dissected originally with undue details — by analyzing the parameters of the regularized solutions.

As soon as dissection $\{V_i\}$ is given, all properties of the model are determined by the parameters of linear systems, such as Systems 45 to 48, and may be studied by means of linear algebra. In this case, problems of uniqueness of a solution and its stability to the errors in the input data are solved easily. A stochastic approach to these systems allows use of the procedures of linear regressive analysis to describe the statistical properties of the model.

Summarizing, we note that Equations 45 correspond to a general discrete formulation of the problem, i.e., detecting vector piecewise-constant magnetization. Equations 46 express a geological-structural inverse problem when distribution of physical properties inside a seamount is detected with a known direction of the magnetization vector. Systems 47 and 48 are inverse paleomagnetic problems when the direction of magnetization is unknown whereas distribution of physical parameters is assumed to be either homogeneous or known.

B. SOLUTION OF A DIRECT MAGNETOMETRY PROBLEM FOR SEAMOUNTS

In the course of the iterational simulations of geological-geophysical models, there arises the need to perform numerous computations of the anomalous field of the current model corresponding to the given stage of interpretation. Therefore, the procedure for solution of direct problems affects greatly that of the inverse ones. The more accurate and efficient are the algorithms of the direct problem solution, the more adequate is the model which can be simulated for an acceptable time period. Due to the latest achievements in the solution of direct gravimetric and magnetometric problems (Strakhov et al., 1986), it became possible to estimate in a short time and with high accuracy anomalous effects of complex geometrical bodies when their surface and volume are given in different ways.

When investigating seamount magnetism the choice of the solution of the direct problem affects greatly the choice of the model for the inverse problem. The constrained and approximate character of the approach by Vacquier (1962) and Talwani (1965) was reflected in the simple formulation of the inverse problem for the case of homogeneous magnetization. The appearance of more "geological" algorithms (Plouff, 1976) resulted in attempts to simulate more complex and comprehensive models (Kodama et al., 1978; Sager et al., 1982).

Traditionally, there are two principal techniques of presenting information about the geometry of an anomalous source, i.e., the seamount topography.

The first technique was developed by Vacquier (1962) and involves approximation of the seamount volume by rectangular prisms. Depths to the top and basement of each prism and its position in the xy-plane are given. Russian scientists such as Kaminsky and Simovsky (1976) use the same technique for specifying and approximating the topography. The latter may be performed automatically if we specify the depth on some rectangular grid (irregular, in general) on the xy-plane. If the horizontal dimensions of the prisms coincide with the grid size, the depths to the upper and lower bounds are taken directly from the initial massives; if not, they can be estimated by interpolation. This approach allows layer-by-layer description of the seamount volume, and, rather complex geometrical structures can be simulated if we give each cell of the grid the order number of the body whose elementary volume corresponds to that cell. The obvious demerit of the approach is its low accuracy in the topography approximation (zero order), which has a particular effect in the near-surface parts of the seamount body where the right angles of the prisms cause notable marginal effects in the model field. This demands an increase in the number of elementary volumes to achieve the acceptable properties of the approximation.

The second approach to topography specification is to describe it by a system of broken lines corresponding to some definite depth contour lines. In this case, a complete coincidence with the cartographic image of the topography can be achieved. This approximation is quite natural and easy to grasp. The approach is convenient and efficient when calculating surface integrals by Talwani's technique with the supposition of homogeneous magnetization over all the volume. Its accuracy depends upon the number of levels and may be rather high. In the latest modification by Plouff, it became possible to dissect seamounts in the vertical direction. Pronounced expressions were obtained for the integrals which Talwani calculated from the quadratures.

Now we show how the results Strakhov et al. (1986) were used to compile the optimally accurate and quick-acting algorithms for the solution of the direct magnetometry problem for seamounts.

Suppose that on the (irregular) rectangular grid in the *xy*-plane we have depths to the surfaces $h_1(x, y)$ and $h_2(x, y)$ which mark the top and the bottom of the seamount body (or its part, i.e., a layer). We call the elementary body the one whose projection onto the *xy*-plane coincides with the rectangular figure $[x_i, x_{i+1}] \times [y_j, y_{j+1}]$, and the depths to the upper and lower bounds are determined by h_1 and h_2 in the intersections of the grid. We shall assume that the magnetization of each elementary body is constant. Then, their magnetic effects can be found from the integration of the corresponding functions over the surface which contains four vertical bounds along with the top and the bottom ones. Integrals over four lateral bounds are calculated directly by Strakhov's formula for homogeneous polygonal figures. The top and bottom bounds are not usually flat, since the four arbitrary points of the space are not necessarily in one plane. In the first order approximation we change these quadrangle irregular figures into four flat triangles to which Strakhov's formula can be applied.

We consider, for instance, the case of the top bound. Calculating from four angular points the average depth h_{ij}^1 to the top edge, we ascribe its value to the point with co-ordinates

$$(x_i^1, y_j^1) = \left((x_i + x_{i+1})/2, \ (y_j + y_{j+1})/2 \right)$$

Four new triangular bounds are formed, approximating the upper surface of the elementary body, from the bounds connecting the point (x_i^1, y_j^1, h_{ij}^1) with four angular summits of the top bound. The same procedure is applied to the other top and bottom irregular surfaces of elementary bodies. As a result we obtain an approximation of a seamount topography shown in Figure 10. The proposed technique allows us to use most completely the information about the topography from grid massives. Its advantages compared to that of the prisms is obvious from Figure 10 g, c.

We note that in the assumption of the magnetization homogeneity of a seamount body, it is not necessary to perform integration over the inner vertical bounds of the elementary bodies; thus, it is possible to reduce considerably the computations for this individual case.

A similar procedure can be used when the topography information is given by depth contour lines. In this case we can just approximate the elementary body surfaces by the triangular bounds based on the system of broken lines describing depth contour lines. The simplest way is to organize the input information as follows:

1. Adjacent contour lines should meet one of the following requirements:
   - to have equal number of bounds
   - one of the contour lines has only one point
   - one of these lines has twice as many bounds as the other
2. The numbering of summits of adjacent contour lines allows us to find an algorithm to unite the summits in triangular bounds and, thus, obtain a surface which best agrees with the geometry of a seamount.

An example of such a topography approximation is presented in Figure 10h. Figure 10f shows comparison Plouff's technique and illustrates the advantages of the former in its accuracy and sophisticated reflection of the topography details.

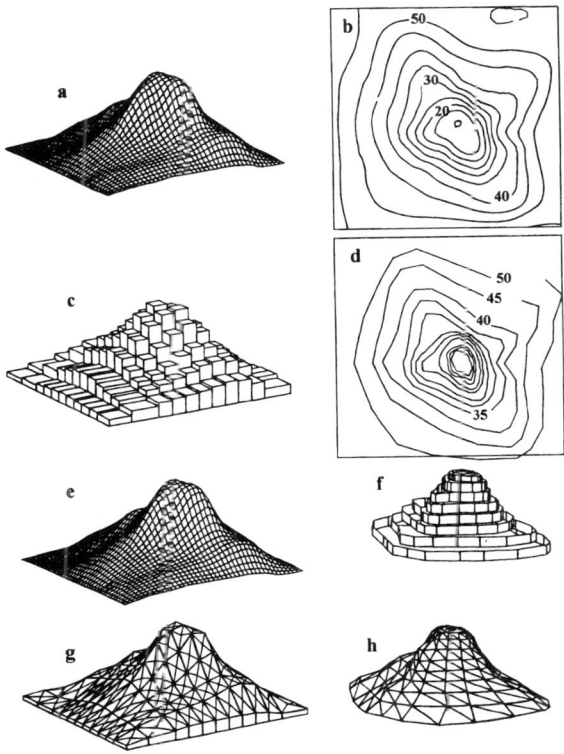

Figure 10 Methods of topographical approximation and disintegration in direct problem solution: (a, b, e) isometric configuration and seamount topography isobaths (initial data); (c) approximation according to Vacquier; (d) approximation according to Talwani; (f) approximation according to Plouff; (g, h) approximations used in the present work.

The model of a seamount magnetic structure is simulated by the technique of generalized linear inversion, which was described in detail previously. It is convenient to divide it into two parts for large complex seamount constructions.

First, to perform a more sophisticated dissection of a seamount, a direct problem was to be solved and a factor matrix for a system of equations describing the most detailed model was to be formed and written to the external storage device. A considerable reduction in calculations is achieved by such a procedure because no direct problem was to be solved when selecting various ways of regularization, reparameterization, and amalgamation of the model.

Singular-value decomposition of the model matrix permits us to analyze its stability and to investigate the uniqueness of the solution. The regularized solution is constructed with account taken of the field errors, *a priori* knowledge of the magnetic properties of rocks, and constraints on the variance of parameters. The analysis of the parameter resolution, character of the solution, and reparameterization of the model results in amalgamation, merging, recognition or exclusion of poorly resolved blocks, and in the simulation of a stable model.

C. DETERMINING THE DIRECTION OF MAGNETIZATION FROM THE MAGNETIC FIELD SPECTRUM AND TRANSFORMATIONS

When solving paleomagnetic problems, great attention is paid to those determinations of magnetization vector direction which are almost independent of the geological model selected.

We consider the method for determining the magnetic moment components

$$M_x = \int_V I_x \, dV, \quad M_y = \int_V I_y \, dV, \quad M_z = \int_V I_z \, dV$$

of the anomalous body with some volume V in the lower half-space from the two-dimensional spectrum of its magnetic field. The knowledge of these moments allows us, in turn, to find an integral (averaged by volume) magnetization direction vector of the body with angles

$$D = \arctan(M_y / M_x), \quad I = \mathrm{acrtan}[M_z / (M_x^2 + M_y^2)^{1/2}]$$

The given approach is a generalization for the three-dimensional case by O. A. Soloviev (1962). It is supposed that the dimensions of a rectangular figure in the plane where the values of anomalous field are given are too large to reconstruct by means of a numerical analysis of the field spectral behavior in the low-frequency range.

We use the spectral representation of a magnetic field on the plane (Gunn, 1975)

$$\Delta T(u,v) = \frac{2\pi}{s}[i(\lambda u + \mu v) + v s] \int_V [i(I^x u + I^y v) + I^z s] \exp[-i(\xi u + \eta v) - \zeta s] \, dV \quad (49)$$

From Equation 49 we obtain easily the expressions for the spectra of fields ΔX, ΔY, ΔZ:

$$\Delta X(u,v) = \frac{2\pi}{s} iu \int_V \left[i(I^x u + I^y v) + I^z s\right] \exp\left[-i(\xi u + \eta v)\zeta s\right] dV \quad (50a)$$

$$\Delta Y(u,v) = \frac{2\pi}{s} iv \int_V \left[i(I^x u + I^y v) + I^z s\right] \exp\left[-i(\xi u + \eta v) - \zeta s\right] dV \quad (50b)$$

$$\Delta Z(u,v) = 2\pi \int_V \left[i(I^x u + I^y v) + I^z s\right] \exp\left[-i(\xi u + \eta v) - \zeta s\right] dV \quad (50c)$$

Completing the differentiation with respect to u and v of spectra ΔX, ΔY, ΔZ in Equations 50 and calculating derivatives at $u = v = 0$, we obtain:

$$\Delta X'_u(0,0) = 2\pi \left(-\int_V I_x \, dV + i \int_V I_z \, dV\right) \quad (51a)$$

$$\Delta X'_v(0,0) = 0 \tag{51b}$$

$$\Delta Y'_u(0,0) = 0 \tag{52a}$$

$$\Delta Y'_v(0,0) = 2\pi\left(-\int_V I_y\, dV + i\int_V I_z\, dV\right) \tag{52b}$$

$$\Delta Z'_u(0,0) = 2\pi\left(\int_V I_z\, dV + i\int_V I_x\, dV\right) \tag{53a}$$

$$\Delta Z'_v(0,0) = 2\pi\left(\int_V I_z\, dV + i\int_V I_y\, dV\right) \tag{53b}$$

From these formulas we obtain the expressions:

$$2\pi M_x = \mathrm{Im}\,\Delta Z'_u(0,0) = -\mathrm{Re}\,\Delta X'_u(0,0) \tag{54a}$$

$$2\pi M_y = \mathrm{Im}\,\Delta Z'_v(0,0) = -\mathrm{Re}\,\Delta Y'_v(0,0) \tag{54b}$$

$$2\pi M_z = \mathrm{Re}\,\Delta Z'_u(0,0) = \mathrm{Re}\,\Delta Z'_v(0,0) = \mathrm{Im}\,\Delta X'_u(0,0) = \mathrm{Im}\,\Delta Y'_v(0,0) \tag{54c}$$

Now, we show how it is possible to obtain the same results with spectrum of the field ΔT_a without its recalculation in $\Delta X, \Delta Y, \Delta Z$. It is enough to remember that in a spectral domain, the Fourier images of fields ΔT_a and, for instance, ΔZ, are bound by the expression (Lourenso and Morrison, 1973):

$$\Delta Z(u,v) = \frac{s}{i(\lambda u + \mu v) + vs}\Delta T(u,v) \tag{55a}$$

The factor before $\Delta T(u, v)$ is a frequency response of the field transformation filter: ΔT_a into ΔZ. The relation of spectra $\Delta X(u, v)$ and $\Delta Y(u, v)$ to $\Delta T(u, v)$ is the same:

$$\Delta X(u,v) = \frac{iu}{i(\lambda u + \mu v) + vs}\Delta T(u,v) \tag{55b}$$

$$\Delta Y(u,v) = \frac{iv}{i(\lambda u + \mu v) + vs}\Delta T(u,v) \tag{55c}$$

Differentiating Equation 55a we obtain:

$$\Delta Z'_u(0,0) = \frac{1}{\nu + i\lambda} \Delta T'_u(0,0) \tag{56a}$$

$$\Delta Z'_v(0,0) = \frac{1}{\nu + i\mu} \Delta T'_v(0,0) \tag{56b}$$

In a similar manner, we obtain other formulas of this type:

$$\Delta X'_u(0,0) = \frac{i}{\nu + i\lambda} \Delta T'_u(0,0) \tag{57}$$

$$\Delta Y'_v(0,0) = \frac{i}{\nu + i\mu} \Delta T'_v(0,0) \tag{58}$$

Separating the real and imaginary parts in Equations 56 to 58, we obtain the final expressions equivalent to Equation 54:

$$2\pi M_z = \frac{\nu}{\lambda^2 + \nu^2} \operatorname{Re} \Delta T'_u(0,0) + \frac{\lambda}{\lambda^2 + \nu^2} \operatorname{Im} \Delta T'_u(0,0) \tag{59a}$$

$$2\pi M_z = \frac{\nu}{\mu^2 + \nu^2} \operatorname{Re} \Delta T'_v(0,0) + \frac{\mu}{\mu^2 + \nu^2} \operatorname{Im} \Delta T'_v(0,0) \tag{59b}$$

$$2\pi M_x = \frac{\nu}{\lambda^2 + \nu^2} \operatorname{Im} \Delta T'_u(0,0) - \frac{\lambda}{\lambda^2 + \nu^2} \operatorname{Re} \Delta T'_u(0,0) \tag{60}$$

$$2\pi M_y = \frac{\nu}{\mu^2 + \nu^2} \operatorname{Im} \Delta T'_v(0,0) - \frac{\mu}{\mu^2 + \nu^2} \operatorname{Re} \Delta T'_v(0,0) \tag{61}$$

In the course of practical realization and using the interpolation procedure, we deal with the magnetic field given at the intersections of a rectangular grid $L_x \times L_y$, at $L_x = M\Delta x$, $L_y = N\Delta y$, where Δx, Δy is the sampling interval. A two-dimensional discrete Fourier transform (DFT) is applied to such a massive:

$$F(k,l) = \sum_{m=0}^{M-1} \sum_{n=0}^{N-1} f(m,n) \exp\left[-2\pi i \left(km\Delta x / L_x + nl\Delta y / L_y\right)\right]$$

$$k = 0, 1, \ldots, M-1; \quad l = 0, 1, \ldots, N-1. \tag{62}$$

It is convenient to calculate a two-dimensional DFT by the algorithm of a one-dimensional fast Fourier transform (FFT). A discrete approach to the problem has some peculiar consequences that should be considered when calculating with formulas given in the present section.

First, the lowest frequencies $2\pi\Delta x / L_x$ and $2\pi\Delta y / L_y$ for which spectral values are calculated correspond to the harmonics whose wavelength coincides with the grid dimensions. For magnetic potential derivatives, the integral ratio is satisfied (Helbig, 1963):

$$\int_{-\infty}^{\infty}\int_{-\infty}^{\infty} \frac{dU}{dt}\, dx\, dy = 0$$

Therefore, in the discrete case, when dimensions of the grid are sufficient for an adequate description of the anomaly, the approximate equalities are to be observed:

$$\sum\sum \Delta T_{mn} \approx \sum\sum \Delta X_{mn} \approx \sum\sum \Delta Y_{mn} \approx \sum\sum \Delta Z_{mn} \approx 0 \qquad (63)$$

To calculate the spectral derivatives at the zero point, it is necessary to have at least two reliable spectral values along each axis near the reference point in the frequency plane. Expressions 63 give spectral values at zero, because the sum of readings corresponds to a constant component in the transformed massive, i.e., the harmonic with zero frequency. Therefore, to estimate spectral derivatives at zero, it is possible to use their values at $u = 2\pi\Delta x / L_x$, $v = 2\pi\Delta y / L_y$. However, it should be kept in mind that procedures of extrapolation, smoothening, and padding with zeroes as the massive margins, used in numerical spectral analysis, distort most of the lower-frequency part of the spectrum. From the experience, it follows that best results can be achieved when estimating spectral derivatives by harmonics with wavelengths corresponding to the main period of the anomaly (which have maximum amplitude and signal-to-noise ratio).

Second, in contrast to a continuous transformation, a DFT in the form of Equation 62 is not symmetric with respect to zero. It is not important in the discrete case because a DFT requires the recurrence of the function beyond the rectangle $L_x \times L_y$. However, in practice, digitization is chosen in such a way that the most typical parts of the anomaly are located in the center of the massive. This corresponds to a cyclic shift of the readings by one half the period, i.e., multiplication of the spectrum by $e^{-i\pi(k+l)} = (-1)^{k+l}$. Spectral readings, the index sum of which are odd, are multiplied by (-1). This property of spectra obtained through DFT should be taken into account when performing calculations according to Equations 54, 59 to 61.

Now, we show how it is possible to calculate, in practice, the components of a magnetic moment of an anomalous object from the transformed fields avoiding calculation difficulties caused by numerical differentiation of the spectrum.

Linear transformations allow us to find fields ΔX, ΔY, ΔZ from the anomaly field ΔT_a when conditions are favorable (excluding polar and near-equatorial parts). When such transformations are performed it is easy to calculate the magnetic moment components if we take the known integral ratio:

$$2\pi M_z = \int_{-\infty}^{\infty}\int_{-\infty}^{\infty} x\, \Delta X\, dx\, dy \int_{-\infty}^{\infty}\int_{-\infty}^{\infty} y\, \Delta Y\, dx\, dy \qquad (64a)$$

$$2\pi M_x = \int\int_{-\infty}^{\infty} x\,\Delta Z\,dx\,dy \qquad (64b)$$

$$2\pi M_y = \int\int_{-\infty}^{\infty} y\,\Delta Z\,dx\,dy \qquad (64c)$$

In the discrete case, moments are calculated approximately from:

$$2\pi M_z = c\sum\sum (m-m_0)\Delta X_{mn} = c\sum\sum (n-n_0)\Delta Y_{mn} \qquad (65a)$$

$$2\pi M_x = c\sum\sum (m-m_0)\Delta Z_{mn} \qquad (65b)$$

$$2\pi M_y = c\sum\sum (n-n_0)\Delta Z_{mn} \qquad (65c)$$

where c is a constant depending on the choice of physical units of measurements and scale. Indices m_0 and n_0 determine the center of the data massive and are found from the discrete analogs of the integral ratios for magnetic field components:

$$\int\int_{-\infty}^{\infty} y\Delta X\,dx\,dy = 0; \quad \int\int_{-\infty}^{\infty} x\Delta Y\,dx\,dy = 0$$

The procedure of estimating the magnetic moments of a body by integrating the transformant of field ΔT_a is more stable than numerical differentiation of the field spectra, though it is more effort-consuming. We should consider that expressions such as Equations 65 are also sensitive to the procedures of extrapolation, padding with zeroes, and smoothening of the massive's margins. Therefore, the initial field behavior should be analyzed thoroughly prior to transformation and integration: the most representative part of the anomaly should be placed, if possible, in the center of the massive and Conditions 63 should be observed before and after the preliminary data processing (smoothening of margins, padding with zeroes).

The above method of determining the direction of magnetization, using Fourier transforms, ensures quite reliable results if the anomaly is ascribed to a considerably localized object and is traced within a large segment, with the margins of the anomaly extending to the normal field. However, situations occur whereby the field has a complex interfering character due to the occurrence of abundant sources. In this case, interpretation in the terms of an individual object is meaningless, and it is expedient to perform the analysis for the entire assemblage of sources within the framework of a statistical approach (Le Mouel et al., 1972; Spector and Grant, 1970).

Now, let us consider once again the spectral approach to the magnetic field given in Equation 49. Let us assume that only the scalar value, i.e., magnetic intensity, is random, whereas the direction remains regular. Such a supposition does not seem to be too restrictive because of the following qualitative formulations. At present, it is widely assumed that the distribution of directions of magnetization intensity in volcanic rocks is well approximated by the Fisher distribution over the sphere. As was shown by Cox (1964), the Fisher distribution is practically equivalent to the distribution that arises when the assemblage of random vectors of the same length (but with a random direction distributed regularly over the sphere) is added to a regular vector. This assemblage of random vectors does not seem to affect considerably the magnetic anomaly, except for some abnormal cases. Its input in the field spectrum will be close to that of the "white noise".

Therefore, at the given stage of the problem investigation, we can assume models with only a random magnetization intensity. Assuming that, we use Equation 56 instead of Equation 49. Its form will be:

$$\Delta T(u,v) = 2\pi \mathcal{D}(\lambda, \mu, \nu, l, m, n, u, v) J\left(|\mathbf{I}|, \zeta; u, v\right) \tag{66}$$

Here, we have used the notations
$\mathcal{D}(\lambda,\mu,\nu,l,m,n,u,v) = [i\,(\lambda u + \mu v) + \nu s][i(lu + mv) + ns]/s$ a nonrandom factor related to a magnetization direction and external field.

$$J(|\mathbf{I}|,\zeta;u,v) = \int_0^\infty f(u,v,\zeta) \exp(-\zeta s)\, d\zeta$$

is a random component related to the magnetization intensity distribution. If we recalculate field ΔT_a into a pseudo-gravitational anomaly (integrating it along (λ, μ, ν)), the expression for $\mathcal{D}$ will be simplified: $\mathcal{D} = \mathcal{D}(l,m,n,u,v) = [i(lu+mv)+ns]/s$.

The energy spectrum of such a field is

$$\bar{E}(u,v) = 4\pi^2 \mathcal{D}\mathcal{D}^* <JJ^*> \tag{67}$$

where the notation $<\ >$ means averaging over the assemblage. The factor $\mathcal{D}\mathcal{D}^* = (l \cos \varphi + m \sin \varphi)^2 + n^2$ determines the effect of magnetization direction upon the energy spectrum, whereas angle $\varphi = \arctan(v/u)$ is an angle in the spectral plane. The factor $\mathcal{D}\mathcal{D}^*$ does not depend upon s.

Remembering that $l = \cos D \cos I$, and performing simple trigonometric transformations, we obtain the expression for $\mathcal{D}\mathcal{D}^*$ as:

$$\mathcal{D}\mathcal{D}^*(\varphi) = \cos^2 I \cos^2(\varphi - D) + \sin^2 I = 1 - \cos^2 I \sin^2(\varphi - D) \tag{68}$$

The maximum of the expression corresponds to $\varphi = D + \pi$, whereas its minimum will be observed at $\varphi = D + \pi/2$ and has a numerical value equal to $\sin^2 I$.

If we know the spectrum of magnetization intensity distribution $<JJ^*>$, and extract it from Equation 67, we can easily determine declination and inclination (D and I), by calculating at various angles φ the value of the field spectrum component remaining. When spectrum $<JJ^*>$ is unknown (this is common in practice) additional suppositions should be formulated about it. For instance, we can suppose that the magnetization is isotropic in the xy-plane. Then, its spectrum must have the property of circular symmetry. In the latter case, the maximum and minimum of Equation 68 should be searched for among the field spectral values calculated at the same s (along the circumference with the center at zero on a spectral plane). In this case, $\sin^2 I$ is determined from the minimum-to-maximum ratio.

The correlation function related to the energy spectrum by a pair of Fourier transforms also may serve as an indicator of the magnetization direction. For the case of isotropic fields it also will have circular symmetry. Therefore, performing a series of transformations like "reduction to the pole" with various angles of magnetization direction, it is possible to find such values for which the correlation function is symmetric. This approach has greater visuality than estimates from the energy spectrum, though it is more labor-consuming in the calculational aspect.

The present work used a simpler and more efficient approach to check the circular symmetry of a correlation function than that used in the original work by Le Mouel et al. (1972). Values of a two-dimensional correlation function are considered as a distribution of some density (masses) on the xy-plane. Asymmetry of such a system is characterized by the ratio between the lengths of the two main axes of inertia $e = I_1/I_2$, where I_1 is the length of the greater semi-axis, and I_2 is that of a smaller one. I_1 and I_2 are calculated according to a standard method through the secondary moments of the correlation function with respect to zero on the xy-plane. The direction of the greater semi-axis of the inertia ellipse indicates the predominating strike of the anomalies.

The purpose of a series of transformations of the "reduction to the pole" type is to obtain parameter e as close to 1 as possible. Values of the guiding cosines l, m, n of such transformations will ensure the desired magnetization direction.

When calculating the correlation function by DFT of the energy spectrum obtained from the initial field DFT, the function acquires a periodical character and is distorted along the margins of the plot. To exclude the effect of the least reliable values upon the estimate of the inertia ellipse, we consider only the values of the correlation function which exceed a given threshold, for instance, 0.3 or 0.5 of its maximum.

D. USE OF A TRANSFER FUNCTION (ADMITTANCE) BETWEEN THE TOPOGRAPHY AND MAGNETIC FIELD

In the general case, the magnetic anomaly spectrum depends both on the magnetization properties and on the form of their object, in particular on the volume of a seamount, which "modulates" that magnetization. Application of Equation 68 to estimate magnetization direction proves most useful if magnetization is isotropic and topography (object's shape) is expressed poorly and is also isotropic. At present, we have no reliable data to reject the hypothesis on the isotropic character of magnetization inside most seamounts, but the situation is different in the case of the topography. Most seamounts have typical topographic features far from symmetric and isotropic. Therefore, it seems important to be able to reduce the effect of the topography upon the estimate of magnetization from the magnetic field spectrum.

The transfer function (or admittance) between the topography and magnetic anomaly can be assumed as a spectral function that is free of the topography effect to a considerable extent. This procedure has been used widely for at least 10 years for the gravimetric determination of the mechanism of isostatic compensation. The proposed application of that procedure in our case allows us to find reliable directions of magnetization from the phase of the admittance between the topography and magnetic anomaly ΔZ. In its turn, it is not difficult to recalculate the spectrum of anomaly ΔT_a into that of ΔZ and to use the latter in the calculation of the admittance. We note that the phase of the admittance coincides numerically with the phase of the mutual spectrum between the topography and field ΔZ.

The optimal estimate of the admittance is found from the expression (Bendat and Piresol, 1983):

$$R(u,v) = \frac{<E_{HZ}(u,v)>}{<E_{HH}(u,v)>} \tag{69}$$

where $<E_{HZ}>$ is an estimate of a mutual spectrum of the topography and a magnetic field (input and output signals) and $<E_{HH}>$ is the estimate of the energy spectrum of the topography (input signal).

For linear systems, the admittance indicates the distribution of the gain coefficient in the frequency domain and phase shifts when the signal passes through the system. Comparison of estimates (Equation 69) and theoretical functions is used to investigate numerous physical and other phenomena and to simulate models and predict the response of the systems to input effects.

Unfortunately, the relation between the source shape and magnetic field is not linear, except in the cases when the object of magnetization has a simple geometrical structure like a horizontal layer, rectangular prism, etc. Therefore, the contribution of the topography and magnetization to the magnetic anomaly spectrum is difficult to distinguish without any additional suppositions on their peculiar characteristics.

The approximate linear relation between the topography and magnetic anomaly may be expressed as follows. Let us assume that magnetization allows the expression in the form of

$$\mathbf{I}(\mathbf{r}) = \mathbf{I}_0 + \mathbf{I}_N(\mathbf{r}) \tag{70}$$

where $\mathbf{I}_N(\mathbf{r})$ is a random component (noise) and $\mathbf{I}_0 = j_0\, \mathbf{J}_0$ is an average value. The topography (upper edge) is assumed to be described by a finite function $H(x,y)$, whereas the lower edge coincides with the plane $z = 0$. Let us also assume that anomaly ΔT measured at the level $z = z$ is recalculated for ΔZ.

Using the Parker expansion (Parker, 1972) for the magnetic field spectrum and leaving only a linear member there, we have:

$$\mathcal{F}(\Delta Z) = e^{-sz}[i(lu+mv)+ns]J_0\,\mathcal{F}(H) + \text{noise} \tag{71}$$

Here, all members of the transformation not considered assumed to be noise, the effect of a random magnetization component, and other inaccuracies of the simulation. It follows from Equation 71 that the factor $J_0 e^{-sz}[i(lu+mv)+ns]$ plays the role of admittance in such an approximately linear system. Since $J_0 e^{-sz}$ is a real number, the phase of the transfer function is determined from $i(lu+mv)+ns$ and is equal to

$\arctan\left(\dfrac{l}{n}\cos\varphi + \dfrac{m}{n}\sin\varphi\right)$, where $\varphi = \arctan\dfrac{v}{u}$ is an angle on a spectral plane.

Changing from the guiding cosines l, m, n to D and I angles, we obtain the expression for the admittance:

$$\Phi(u, v) = \arctan [\tan I_0 \cos (\varphi - D)] \qquad (72)$$

where the angle $\pi/2 - I$ (co-inclination) is expressed through I_0. The obtained expression describes the admittance phase as a theoretical function of the magnetization direction. It is clear that the phase depends only upon the angle φ of the spectral plane and is free of s. Maximum $\Phi(\varphi)$ corresponds to angle $\varphi = D$ and equals to I_0. Points where $\Phi(\varphi)$ passes through zero correspond to angles $\varphi = D + \pi/2 - I$.

Let us obtain an experimental estimate $\hat{R}(u, v)$ and its calculated phase $\hat{\Phi}(u, v)$ which, due to inaccuracy of simulation and noises can be a function of not only the angle φ. For better statistical robustness we shall change from phase $\hat{\Phi}(u, v)=\Phi(\varphi, s)$ to function $\Phi(\varphi)$ which is determined as a result of $\Phi(\varphi, s)$ averaging by s along rays with an angle φ starting at point (0, 0) on a spectral plane:

$$\Phi(\varphi) = \frac{1}{S} \int_0^S \Phi(\varphi, s) \, ds \qquad (73)$$

The averaged phase obtained in such a way will be used as an approximation of Equation 72 when estimating D and I.

Thus, the procedures considered in this Chapter for processing and interpretation ensure the solution of the principal problems of marine magnetometry. Significant geological results have been obtained using these approaches (see Chapters 4 and 5). So, we can hope that their systematic application in various geodynamic situations may provide in the near future qualitative new results on some fundamental problems of the World Ocean bottom structure and evolution.

Chapter 3

Mid-Oceanic Ridges and Deep Oceanic Basins: AMF Structure

V. Yu. Glebovsky, S. P. Maschenkov, A. M. Gorodnitsky, I. I. Belyaev,
A. M. Filin, S. V. Mercuriev, N. A. Sochevanova, S. V. Lukyanov,
G. M. Valyashko, E. A. Popov, and K. V. Popov

CONTENTS

I. The Canary-Bahamas Geotraverse (The Northern Part
of the Central Atlantic) 67
II. Basins of the Central and Southern Atlantic 81
   A. Cape Verde and Brazil Basins 81
   B. The Cape Basin 85
III. The Spreading Features of the Northwestern Indian Ocean
in Late Cenozoic 88
   A. The Arabian Basin 88
   B. The Somali Basin 92
   C. The Carlsberg Ridge 99
IV. The Central Basin of the Indian Ocean 107
V. The Tadjura Ridge 114
VI. The Juan de Fuca Ridge 123
VII. The Northwestern Pacific Basin and the Shatskiy Rise 133

I. THE CANARY-BAHAMAS GEOTRAVERSE (THE NORTHERN PART OF THE CENTRAL ATLANTIC)

Since 1987, "Sevmorgeologia" (under the auspices of the Russian Committee for Geology and the Earth's Interior) has been carrying out a medium-scale complex geological-geophysical research project in the area of a low-rate spreading in the northern part of the Central Atlantic, in particular at the Canary-Bahamas geotraverse (CBGT). The area under study is a 700-km-wide band between 23 and 29° N extending from North America to the African coast (Figure 1). The program of research is a part of the Russian program of oceanic geotraverses elaborated for the mapping of 11 transoceanic bands characterizing practically all tectonic regimes and morphostructures typical of the World Ocean (Maschenkov and Pogrebitsky, 1992). The principal goals of the research are to efficiently form a real-time database of precise geological and geophysical information, to study the structure of the oceanic lithosphere, and to explore potential ore fields. As compared with the on-land situation, data derived from medium-scale geophysical regular surveys at sea are scarce, if any. That explains the different choice of the technique applied to oceanic and on-land geotraverse research (Monger, 1986). Generalized data of the regular medium-scale survey on both the oceanic and on-land geotraverses may be used as a basis for global tectonic concepts on the lithosphere evolution.

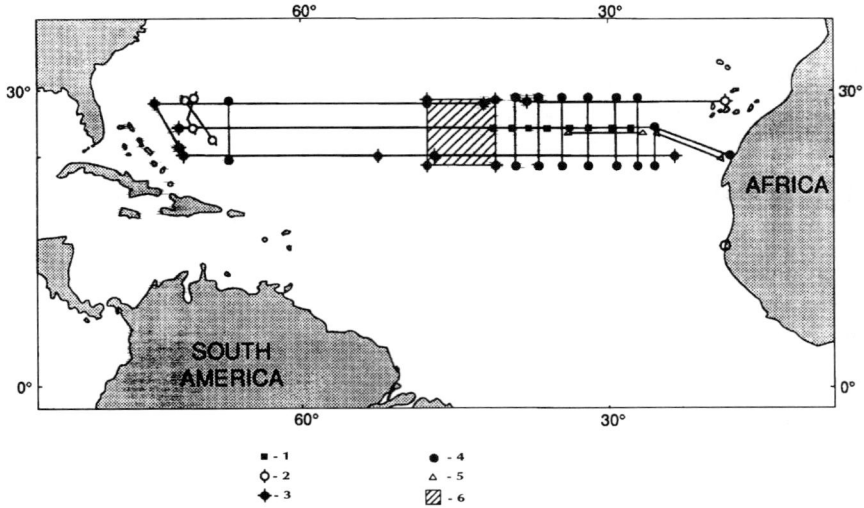

Figure 1 State of knowledge of the Canary-Bahamas geotraverse: (1) geological sampling, heat flow measurement; (2) two-vessel variant of seismic reflection; (3) gravimagnetic survey; (4) seismic reflection; (5) seismic refraction; (6) detailed survey.

An individual oceanic geotraverse is a unified system of observations, utilizing the following types of surveys (Figure 1):

- A grid geomorphological, gravity, and magnetic survey in the geotraverse band
- A regular grid of seismic (reflected waves) and seismoacoustic profiles, supplemented by two-vessel experimental seismic works
- Deep seismic sounding by refracted waves along the geotraverse axis, accompanied by heat flow measurements and geological sampling
- Detailed geological and geophysical studies within test areas located in the places that have the best prospects for ore field exploration

By now, the regular grid gravimagnetic survey within the CBGT has been completed in its central part over the Mid-Atlantic Ridge zone and adjacent areas of more ancient oceanic crust. The entire eastern part of the band from the African coast to 48° W is covered by a regular grid of seismic, bathymetric, gravimetric, and magnetometric profile observations.

Some preliminary results obtained in the central part of the CBGT were published in Bocharova et al. (1991), Maschenkov et al. (1993), and Maschenkov and Pogrebitsky (1992).

The main target of the magnetic survey is the study of temporal and spatial changes in the oceanic spreading regime and their correlation with the structural inhomogeneities of the oceanic lithosphere revealed by other techniques.

Within the geotraverse, the magnetic field of the northern part of the Central Atlantic has been sufficiently examined. Some areas of the CBGT were studied intensively not long ago, within the framework of RIDGE and SARA programs (Patriat et al., 1990). The central part of the geotraverse was selected as a place where the Natural Oceanic Laboratory was set to collect new information that would enrich our knowledge on the crust structure and evolution in the region (Tucholke et al., 1991).

Nevertheless, the majority of the magnetic surveys by Russia and other countries were mainly performed either to study particular regional and detailed problems or as works accompanying other oceanic research (for instance, together with a hydrographic sounding by the Russian program). Lack of a uniform geological aim for the hydromagnetic surveys performed in the area has led to considerable irregularity in the study of the region and to qualitative inhomogeneity of the obtained data. As a result, abundant magnetometric information on the magnetic field of the CBGT could not be used for a detailed geological analysis of the oceanic structure and evolution in the geotraverse band.

Magnetometric study within the framework of the Russian program on oceanic geotraverses (RPOG) was carried out everywhere along the regular grid of profiles. The planning of that grid was based on the position of the old tracks, with the gaps of the previous survey covered and the entire magnetic information within the geotraverse reliably fitted and generalized (Figure 2). A computer database was established for all types of geotraverse information. The old analog records, plots, and maps were digitized and reworked using a new computer technique and software. A coherent digitized database is used to plot the maps and provide geological interpretation. It is supplemented constantly by fresh survey data.

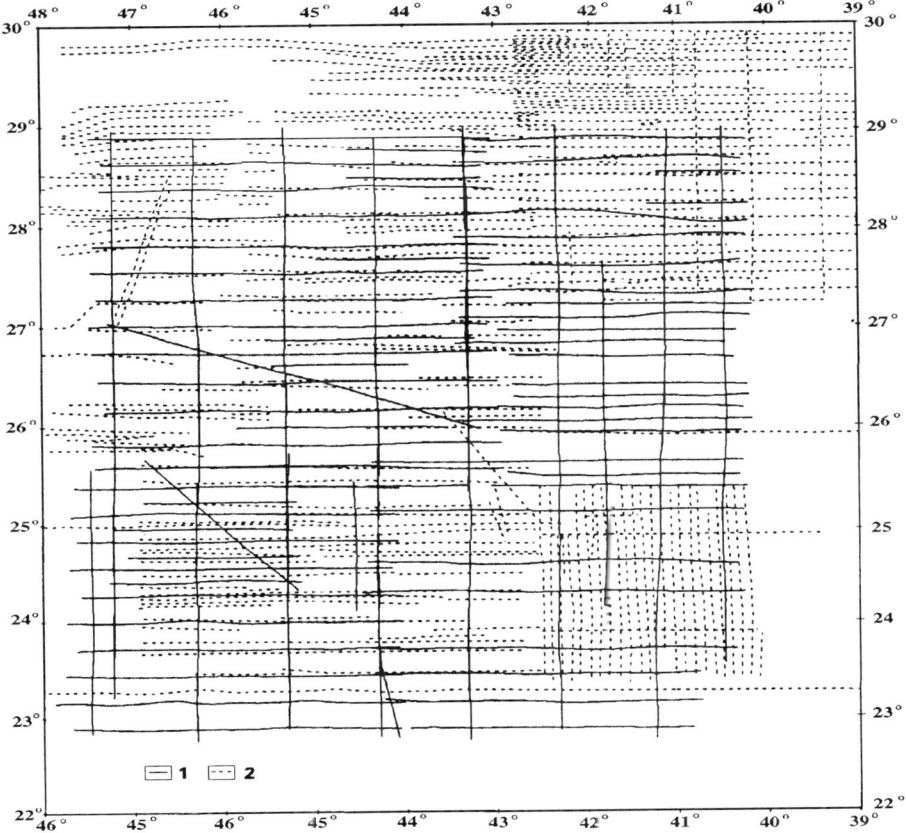

Figure 2 Tracks of magnetic survey: (1) our own bearing tracks; (2) ordinary (routine) tracks.

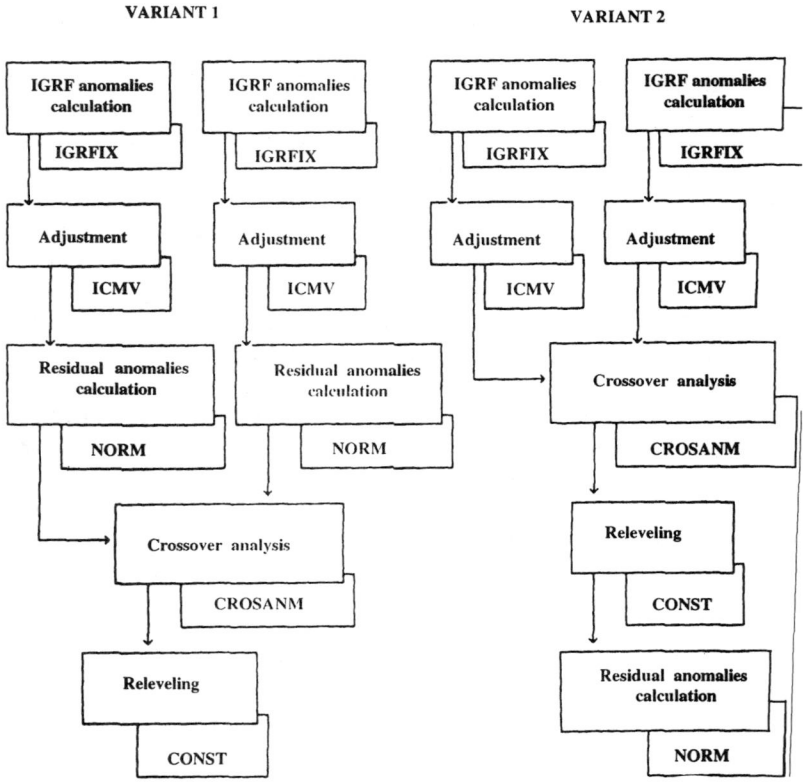

Figure 3 Compilation scheme for composite AMF chart.

A unification of heterogeneous magnetometric information for each individual survey has been realized in general as software for IBM-compatible PCs. Being universal, the scheme includes the following (see Figure 3):

- Data editing
- Identification of interpolated (between the fixes) coordinates and corresponding values of magnetic field
- Calculation of coordinates of transect points and residuals in these points
- Iterational fitting of the survey
- Calculation of IGRF (International Geomagnetic Reference Field) normal magnetic field
- Calculation of anomalous component against the IGRF field
- Calculation of the internal (original) field of reference (OGRF)
- Calculation of the residual anomalous magnetic field (AMF)

The practice of the processing of real data sets on the CBGT showed that the plan and number of operations may vary from survey to survey. That depends upon the magnetic field features and technique of observations, and also upon the quality, form of presentation, and storage of the primary information.

When processing and generalizing results of a long-term survey, the chief problem is rooted in the choice of an optimal model of a normal magnetic field and in the external and internal fitting.

Two principal approaches were used to bring the observed values of the magnetic field to a unified reference level:

- The normal magnetic field has been estimated from the analytical concept of the field and with the account of satellite data (models IGRF, DGRF, MAGSAT, etc.) (IAGA, 1986; Peddie, 1982; Verhoef and Macnab, 1989)
- The original reference field has been estimated from the observed values of magnetic field with use of the graphical or standard mathematical techniques of averaging (Technical recommendations ..., 1990)

Attempts were made to use direct analytical models of the normal field to infer the anomalous component from the magnetic data observed within the geotraverse. However, that caused a notable discrepancy in the level of reference of individual magnetic maps, apparently caused by an intensive secular anomaly in the region (Parkinson, 1986).

The internal level of reference of the OGRF was estimated by original formalized techniques of smoothening (filtration) of the primary magnetic data. Magnetic lineations are known to have distinct amplitude-frequency characteristics within vast areas, which vary gradually along the trend of the basin opening (Kolesova, 1985). A specific character of oceanic magnetic anomalies allows for the usage of known algorithms to resolve a field into components. Magnetic anomalies of a certain frequency range can be recognized within individual oceanic regions when the observed values of the total magnetic induction are being filtered.

The estimation technique of the internal reference field (OGRF) comprised a sliding weight averaging of the measured total magnetic induction over the area and approximation of the averaged values by a parabolic surface. As a result of practical estimates, parameters of an averaging filter were obtained that agree well with those inferred by Kolesova (1985) for the Atlantic magnetic field.

The summary maps of the AMF compiled against the OGRF field usually do not bear any information on the extended regional magnetic anomalies and, in fact, are maps of remnant anomalies ΔT_{res}. If compared to the real maps of ΔT_a, compiled by analytical models of the normal field, they have a number of advantages. They allow us to

- Compare and correlate long-term data presented exclusively as maps of the AMF ΔT_a and ΔT_{res}
- Plot magnetic field graphs in a large vertical scale and, thus, trace the fine structure of magnetic field
- Improve resolution of magnetic maps for the recognition of both magnetic lineations and tectonic features of the oceanic bottom structure

Therefore, maps of residual magnetic anomalies are preferable for geological interpretation.

Magnetic information on the CBGT is presented by two data types, namely, our own regular data and previous irregular sets. When planning the magnetic survey, we used the principle of partial overlapping of the grid and completing study; this made it possible to generalize the materials obtained earlier and to include them in the coherent database.

Our own geotraverse grid of magnetic survey was performed by research vessels *Geolog D. Nalivkin, Zapolyarnyi,* and *Akademik A. Karpinskiy* in 1988, 1989, and 1990, respectively.

The internal iterational fitting allowed us to reduce the root-mean-square RMS error in each of the three surveys, from 24-26 nT to 5-14nT. Such RMS error is only one half or even one third of those we had in the previous survey. So, we could use our own grid of magnetic survey as a reference for the fitting with the previous routine survey.

The problem of an external fitting has a poorly eleborated theoretical basis (Technical recommendations ..., 1990); however, its solution is essential for the formation of a coherent database.

Two ways of fitting reference surveys were tried (see Figure 2):

1. An independent processing of each survey in agreement with a full scheme of unification (mentioned above) and mutual fitting of residual magnetic fields estimated with respect to the level of original magnetic field.
2. An independent processing of each survey like that done in the first approach but up to the estimation of the IGRF AMF, mutual fitting and formation of a common database of the IGRF ΔT_a, calculation of original normal field and corresponding anomalous component over the entire area under study.

The second way obviously has a number of advantages, since

- The residual of the external fitting has acquired a distinct physical sense, that is, a real secular change ignored in the model of the IGRF normal field
- Fitting of isolated profiles, calculated according to the model similar to that of the grid survey, has become easier
- Smoothening (by moving average) of the entire data block, after its fitting by IGRF anomalies, is considerably more reliable than that performed after each survey, because the data density for the calculation of the OGRF normal field becomes higher; a great area is covered, increasing the accuracy of averaging and preventing the possible disagreement at the boundaries of individual survey areas

Because of these considerations, we used the second method of external fitting to achieve agreement between reference surveys.

The IGRF model of the normal field has been simulated automatically; this required precise time information at each point of observation. The survey performed with the minimum time gap may be fitted best. If the time interval between the surveys is great, it seems preferable to use the first method of fitting due to umprecise accounts of the secular changes in the analytical models of the normal field.

Internal fitting of the routine magnetic survey, performed in the geotraverse band in 1988, does not seem possible due to extremely rare grid (if any) of the cross tracks. Therefore, the processing procedure was limited by a listing, selection, and editing of the data in the interactive regime or by plotting of working charts of the ΔT_a graphs.

The choice of the internal fitting technique depends upon the available precise information on the operating time for each survey. For most of the old magnetic maps, we know only the time of the beginning and termination of the expedition, which does not allow an accurate calculation of the IGRF normal field and the corresponding anomalous component. In such cases, the AMF was estimated against the original level of reference and then was adjusted to the level of reference in the geotraverse observations. The described procedure was performed after the residual matrices in the

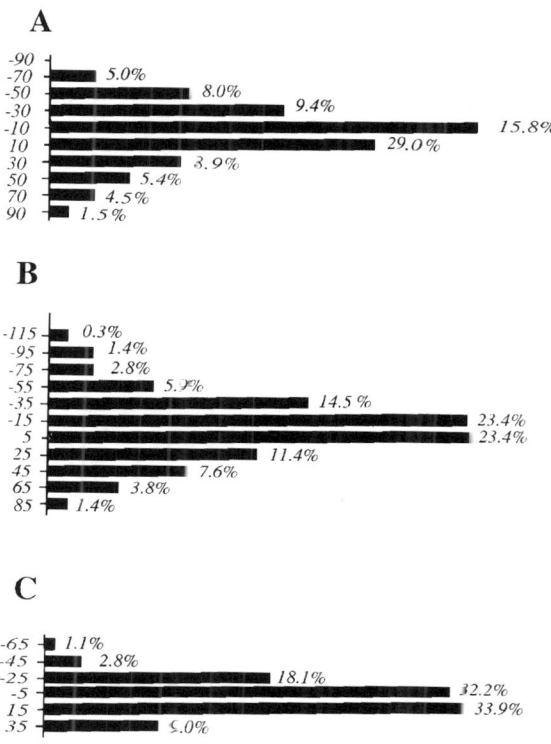

Figure 4 Residual distribution when fitting old magnetic survey data: (A) survey 1974;(B) survey 1978; (C) survey 1989. Residuals are in nT.

transect points of reference and routine magnetic profiles were analyzed. Figure 4 illustrates the resulting histograms of residual distribution.

Residuals of the external fitting of various parts of the analyzed region varied from 15 to 20 nT.

The integrated maps of AMF ΔT_a were compiled automatically on the coherent database formed in the process of compilation, unification and generalization of the entire accumulated information for the geotraverse band. The density of the summary grid of field survey (the tracks being spaced by 3 to 10 km and, in the central part of the area, by 5 km) allowed a compilation of the conditional maps, the scale being 1:1,000,000. Figure 5 shows a reduced copy of the chart of the ΔT_a graphs of the AMF. Then, all magnetic data were gridded by computer according to the minimum curvature technique (Briggs, 1974). The grid dimensions were 5 by 5 km and corresponded to the summary density of the survey grid. Gridding results are presented in Figure 6 as a map of magnetic field ΔT_a zero.

In addition, a map of AMF isolines was compiled by hand to trace precisely the axes of magnetic lineations and fracture zones, as well as to plot the map of isochrons and tectonic elements.

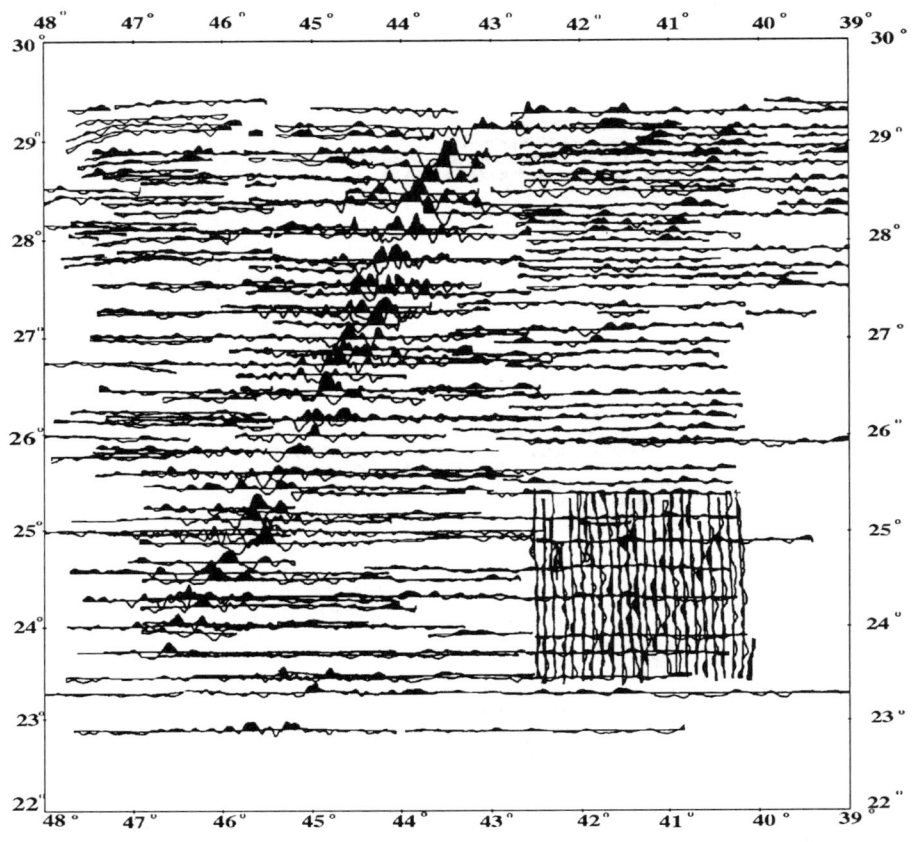

Figure 5 AMF chart, graphs of ΔT_a. Filled areas: positives.

As follows from the magnetic data analysis, magnetic lineations of the northeastern trend of about 80° prevail in the magnetic field of the area under study. Most intensive anomalies are confined to the Mid-Atlantic Ridge crest. They are pronounced against the neighboring magnetic lineations of both signs. Large transform faults are presented by sublatitudinal elongated narrow bands of negative or low magnetic field. They change the morphology of the axial anomaly, violate the general subsequence of magnetic lineations, and divide the region of study into four areas (large spreading cells) (Roest, 1984).

Interpretation of AMF includes its traditional geochronological analysis with the compilation of the isochron map (i.e., of the identified magnetic lineation axes).When identifying the axes we used the standard model of the active magnetic layer, compiled in agreement with the scale of geomagnetic reversals (La Breque et al., 1977) for two fixed spreading rates, namely, 1.5 and 2.0 cm per year.

The field of the model has been computed by the Merkuriev program (St. Petersburg division of IZMIRAN) with different degrees of smoothening according to the technique by Tissau and Patriat (1980). The technique considers that, in fact, some transition zones do exist between the bodies of the normal and reversed magnetization, located in the upper part of the oceanic basement. The width of these zones is given by a coefficient of smoothening R, and model fields were computed at various R values (1, 0.6, 0.4).

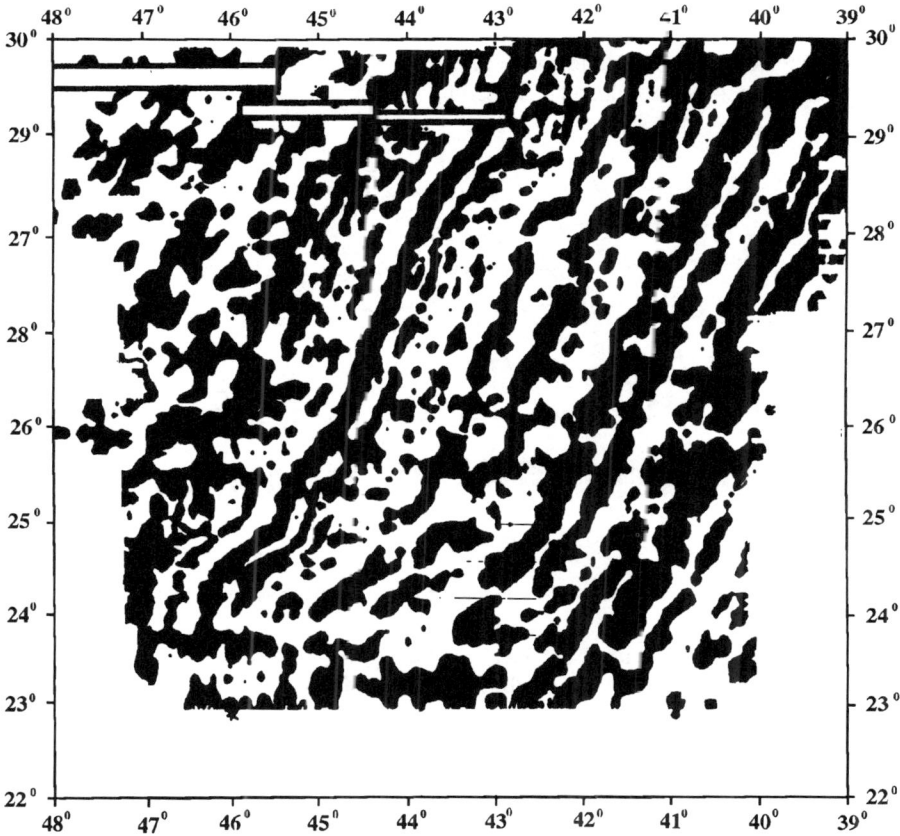

Figure 6 AMF chart plotted from gridding; ΔT_a isolines (spaced every 50 nT).

Close spacing of survey tracks and also good correlation of the observed and simulated graphs of the magnetic field allow us to identify almost all anomalies of the Lamont sequence located within the study area (Figure 7). Despite distinct interprofile correlation of typical anomalies (Figure 5), identification has been done profile by profile, so that we can learn in detail the kinematics of the region opening.

Figure 8 shows, besides the principal axes of magnetic lineations, the largest fracture zones within the CBGT. The northern and southern zones are large transform faults, namely Atlantis and Kane. These transform faults are expressed distinctly by the displacement of the Mid-Atlantic Ridge axis and are traced along the displacements of magnetic lineations (Klitgord and Schouten, 1986). The rest of the analyzed area seems to have no faults with remarkable displacements, but some faults located north of the Kane fracture zone do. Most of the sea area located between the Kane and Atlantis faults is characterized by magnetic lineations with slight displacements that appear, disappear, and migrate in turn (Maschenkov et al., in press).

The region is characterized by the occurrence of V-shaped structures discovered at the Recent Ridge, located north of the Kane and south of the Atlantis fracture zone (Figure 8). They were first discussed in the works of Muller and Roest (1992), Rona (1976), Rona and Gray (1980), and Sloan and Patriat (in press). These structures are placed symmetrically with respect to the mid-oceanic ridge crest but their strike differs considerably from the basin opening trend (transform fault strike). No simple mechanism has yet been elaborated capable of explaining the observed opposing strike of V-shaped structures. We have only hypothetical suppositions as to their dependence upon the absolute motion of the lithosphere plate assemblage (Bocharova et al., 1991) and migration of the segments along the spreading axis (Klitgord and Schouten, 1986).

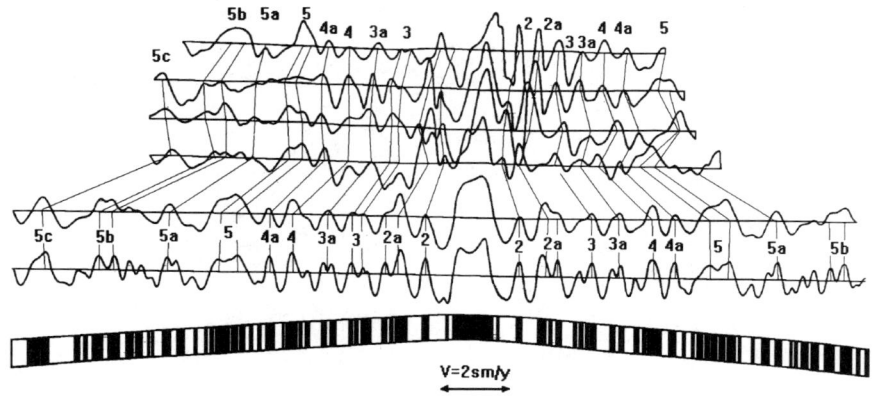

Figure 7 Results of identification according to La Brecque magnetic geochronological scale.

A scheme of magnetic lineation axes (Figure 8) was used to detect the details in the rate variations of oceanic floor spreading. We assume that the spreading rate is an average horizontal derivative of the floor age in the opening direction. The spreading trend was assumed to be in agreement with the Kane fracture zone strike. Table 1 shows the results of the spreading-rate determination. When analyzing the obtained rate characteristics, we should consider two things. First, spreading rates south of the Kane fault were determined approximately due to the less dense grid of survey and uncertain identification in the region because of the distorting effect of the transform fault. Second, the obtained rates may differ from the real linear spreading rates because the seafloor spreading trend may disagree in some places with the strike of the Kane fracture zone. Meantime, the relative character of the rate variations in time has been determined reliably well for the entire area, analyzed irrespective of possible errors in the estimated linear rates of spreading.

The following conclusions follow from the analysis of the spreading rates (Table 1):

1. The rifting rates on the opposite sides of the mid-oceanic ridge axis, between the Kane and Atlantis faults, differ slightly. They are notable only for the Oligocene, that is, 25 to 42 million years ago.
2. A slight discrepancy (0.1 to 0.3 cm per year) in the estimated spreading rates has been noted in some segments along the ridge axis. It is not regular and may be ascribed not only to the spreading processes (in this or that spreading cell) but also to the errors in identification. First, as the magnetic anomaly axis was drawn from the ΔT_a extrema, it may be shifted off the center of the magnetic source in the case of a nonvertical magnetization. Second, since magnetic sources are extended at a great distance from the ocean surface, we cannot account for their echelon structure in our estimations. More precise estimations of the crust age and spreading rates were obtained from the magnetic data recalculated for vertical magnetization (Maschenkov et al., in press).

3. There exists a certain regularity in the spreading rate changes in time. The rate was maximum at the initial stage of the area formation. In the Maestrichtian-Pleistocene (67 to 56 million years ago), it was about 1.8 cm per year. Further on, it decreased gradually and reached its minimum (about 1.2 cm per year) by the end of the Miocene. Then it started to increase, and by now, it is about 1.4 cm per year.
4. South of the Kane fracture zone, the spreading kinematics shows its asymmetry. If the latter is not caused by the possible errors of the identification mentioned above, there should be genetic difference in the process of a seafloor formation on opposite sides of this large transform discontinuity.
5. In the general plan, the results of the performed identification coincide well with the published data (Klitgord and Schouten, 1986; Roest, 1984; Rona, 1980), but are more detailed. The detected symmetry in the position of anomalies of the Lamont sequence, at least from 1 to 13, disagrees with the conclusion (Aplonov and Popov, 1991) on the remarkable asymmetry in the seafloor spreading in the analyzed region. The varying kinematic parameters along the strike of the Mid-Atlantic Ridge verify the statement on the occurrence of at least four spreading cells in the CBGT band, which are separated by large transform faults and V-shaped discontinuities (Maschenkov and Pogrebitsky, 1992).

Table 1 **Spreading linear rate V determined approximately**

No. anomalies	2	2A	3	3A	4	4A	5
Age, millions of years	1.7	2.9	4.3	5.4	6.5	7.6	9
V_{av}, cm per year	1.2	1.5	1.35	1.45	1.5	1.45	1.55
Interval V, cm per year	0.8-1.65	1.0-1.6	–	1.1-1.75	1.3-1.8	1.25-1.8	1.3-1.75
No. anomalies	5A	5B	5C	5D	5E	6	6A
Age, millions of years	11.2	14.5	16.2	17.4	18.5	19.5	21
V_{av}, cm per year	1.5	1.4	1.4	1.4	–	–	–
Interval V, cm per year	1.4-1.75	1.4-1.45	–	–	1.35-1.4	–	–

New techniques of survey, processing, and interpretation, applied to the study of various levels of segmentation of the spreading centers, allowed us to go further in the understanding of tectonic, magmatic, and hydrothermal processes that occur in the newly formed oceanic crust. This study acquires a principal meaning due to its relation with the metallogeny of the ocean. Since the CBGT is a complex program of multi-parametrical geophysical mapping, the usage of the entire set of data provides for a unique opportunity to trace the correlation between the revealed heterogeneity of the spreading process (from the interpretation of the AMF) and other geophysical characteristics.

We considered the problem of the Mid-Atlantic Ridge segmentation using the bathymetric, gravimetric, and magnetometric data. Unfortunately, the reflection by CDP method performed in the MAR vicinity cannot be used because, so far, the survey and processing technique within the standard CDP method does not allow the stratification of a consolidated crust directly under the ridge due to the distorting effect of wave noise.

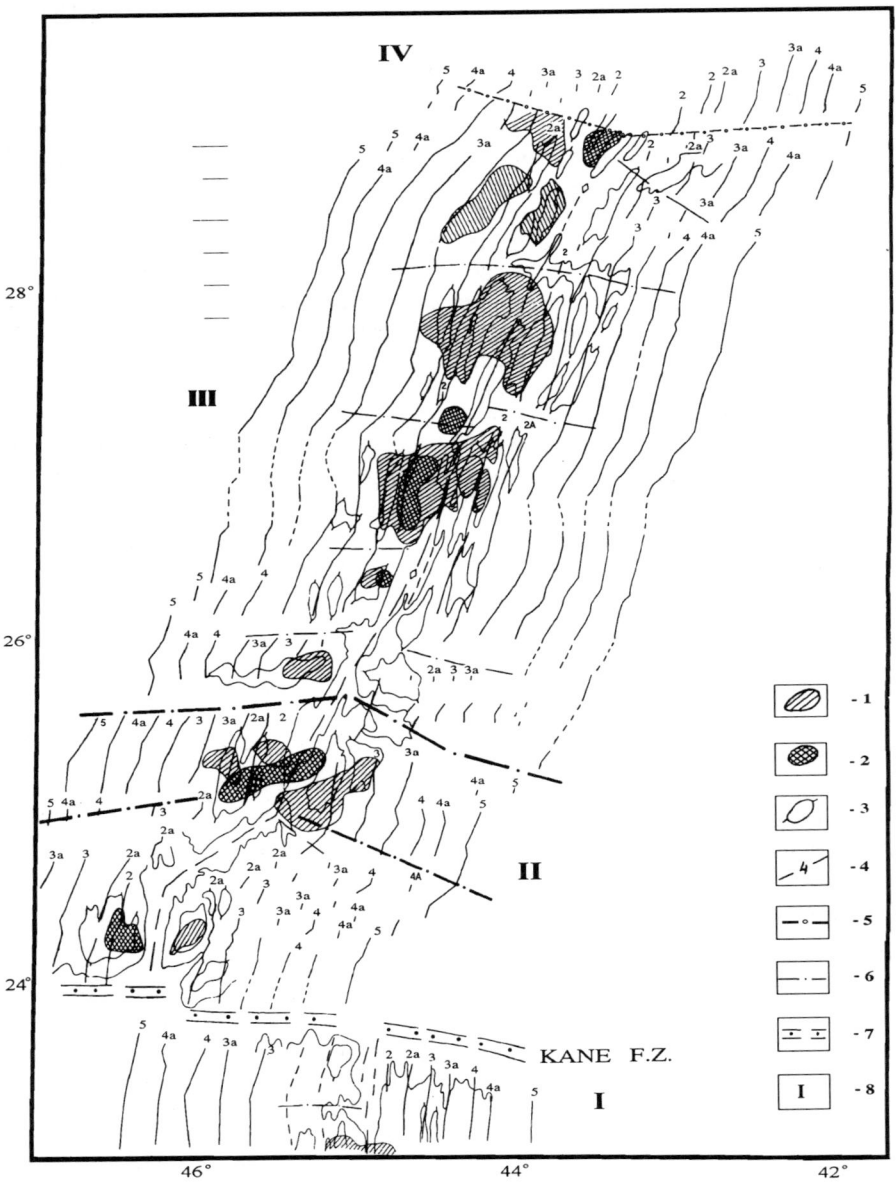

Figure 8 Scheme of tectonic and structural elements of the seafloor: (1) intensive negative mantle anomalies; (2) intensive negative residuals; (3) isobaths bounding ridge crest zones; (4) axes of identified magnetic lineations, with dashed lines corresponding to conventional correlation; (5) V-shaped structures; (6) nontransform faults; (7) transform fault zone; (8) mean-scale Earth segments, geophysical characteristics in Table 2.

When performing the segmentation of the MAR, information was presented in two ways. The analyzed geophysical parameters, such as the depth to the bottom, the AMF, Bouguer mantle anomalies, residual gravity anomalies from the results of a cross-spectral analysis of bathymetric and gravimetric information, mapped in a uniform scale, allow us to consider their structure in the integrated form. A compilation of the most resistant characteristics typical of the rift is presented as a complex interpretation scheme (Figure 8). The rift structure and its principal deformations can be described with considerable certainty by a combination of various techniques.

The information can be also presented for analysis as a profile editing of the considered geophysical characteristics along the rift axis and a compilation of a synthesized profile describing the behavior of a given parameter in this or that direction.

The combinatory analysis of the data allowed us to distinguish four large blocks within the Mid-Atlantic Ridge (Table 2), namely

- The block south of the Kane fault with anomalous values of absolutely all parameters
- The block between the Kane fault and 26° N (a V-shaped structure), characterized by bending of the rift valley and numerous displacements of the magnetic lineation axes
- The block between two V-shaped structures (26 to 28°30 N), characterized by best coincidence in the strikes of the geomorphological lineaments and magnetic lineations, and also by typical negative Bouguer anomalies of the "bull-eye" type
- The block between the V-shaped structure of 28°30 N and the Atlantis fault similar to the second block by many parameters

The four blocks described can be divided into smaller segments from the zoning by various parameters. Their position is mapped well by multibeam echo-sounding and gravimetric survey (Lin et al., 1990).

Simultaneous analysis of the data allows us to consider V-shaped structures bounding the segments as a reflection of a three-dimensional character of the spreading process. The regions hosting such structures are characterized by a considerable fragmentation of the lithosphere that is verified by changes in the effective thickness of the plastic plate estimated from the results of a cross-spectral analysis of the gravity field and bottom topography.

Arising from the existing theoretical concepts on the processes of a partial melting and magma migration under the axis of a low-rate mid-oceanic ridge and from numerical simulations (Morgan, 1991), the rift segmentation may be explained in two ways:

1. Segmentation is believed to be a direct reflection of a mantle upwelling and melting. This model is conventionally called a "thermal" scenario.
2. Segmentation is assumed as a sequence of the lithosphere extension, in particular, in the regions where the mantle upwelling and crust formation are tilted (maximum tension tensor) toward the spreading trend. This model is conventionally called a "geomechanical" scenario. This second scenario seems to correspond to the case of a V-shaped structure formation, whereas realization of the first scenario is possible in segments similar to the block, we distinguished between two structures of this type (block III in Figure 8).

The ratio between the Bouguer mantle anomalies (describing thermal inhomogeneity) and residual gravity anomalies (describing as well the changes of elastic parameters of the lithosphere) is precisely the information that allows identification of this or that type of segmentation. Knowing this ratio and using supplementary geomorphological and magnetometric information, we may suppose that both scenarios of segmentation act between the Kane and Atlantis faults.

Table 2 **Principal tectonic blocks of the Mid-Atlantic Ridge in the CBGT band**

Number on map of complex interpretation	Geographic position	Expressed in geophysical parameters			
		Bathymetry	Axial magnetic lineation	Δg_{MB}	Δg_{res}
I	South of the Cane fault	W-shaped valley with large neo-volcanic uplift	Wide low-amplitude	No intensive local minima	Negative anomaly of regular structure
II	The Cane fault, 28° N	V-shaped valley with rare bending	Wide, one-peak high-amplitude	Series of local minima	Intensive elongated negative anomaly Bifurcation of axial zone of low values
III	26-28°5 N	Continuous valley with small central neo-volcanic uplift	Wide, high-amplitude, complicated	"Bull-eye" anomalies	Series of negative anomalies
IV	Atlantis fault, 28°5 N	Displacement of valley with W-shaped profile	Wide, one-peak, high-amplitude	Series of local minima	Intensive local anomaly Bifurcation of axial zone of low values

So, the geomechanical scenario seems to prevail in blocks II and IV, expressed in the series of "bull-eye" anomalies and considerably intensive residual gravity anomalies against the discontinuities and V-shaped structures. On the other hand, the thermal scenario seems to prevail in block III, though in some segments a mixed version may occur. It is interesting to note that these genetic differences in the formation of a new oceanic lithosphere occur against the average stability of spreading rates inferred from a geochronological analysis of the AMF (see above).

The results presented in this chapter describe only some new data on AMF and other geophysical characteristics obtained in routine survey within the CBGT program. Expeditions and geological interpretation of magnetometric and other information on the geotraverse are not completed. We hope that our data, in combination with results of other oceanic programs, will contribute greatly to the international study of ocean geology.

II. BASINS OF THE CENTRAL AND SOUTHERN ATLANTIC
A. CAPE VERDE AND BRAZIL BASINS

Despite the considerable scope of knowledge on the Atlantic AMF, the geomagnetic situation in the deep basins of the Central Atlantic, namely, in the Cape Verde and Brazil Basins, has not been studied properly yet. That may be explained by the proximity of these basins to the equator, which may cause certain problems in their geomagnetic study, and also by extremely complex magnetic anomalies. So, magnetic lineations have not been recognized practically over the entire near-equatorial zone of the Atlantic. Meantime, the magnetic field pattern and its relation with the structure and tectonic evolution of the oceanic crust seem to be of principal concern due to the tectonic complexity of the region, abundant transform faults, and areas of volcanic activity, and due to the absence of any reliable geochronological information.

Two study areas in the Central Atlantic, namely, the Cape Verde and Brazil Basins, were chosen for detailed geomagnetic survey during the 31st cruise of the R/V *Dmitry Mendeleev* (Figure 9). The northern part of the study area in the Cape Verde Basin is bounded by the Cape Verde transform fault, which follows the 15° N parallel and is called the "15° fault" in some papers. South of the study area there is a system of Vema transform faults. In the longitudinal direction, the study area is located between the eastern slope of the Mid-Atlantic Ridge, on the west, and the uplift of the Cape Verde Islands, on the east. According to the data of the route geomagnetic survey within the Cape Verde Basin, the study area is located northwest of magnetic lineation 34 (about 80 million years), between anomalies 34 and 33 (72 to 76 million years). The age of the bottom within the study area may presumably be estimated as 76 to 80 million years (Campanian-Maestrichtian). Magnetic lineation 34 is shifted along the Cape Verde transform fault for about 80 to 100 miles. At the same time, we should note that according to the map of the axes of paleomagnetic lineations compiled for the World Ocean (Karasik and Sochevanova, 1981), anomalies 21 and 25 are not shifted along the Cape Verde fault at the eastern flank of the Mid-Atlantic Ridge. It should be also noted that the position of lineation 34 southeast of the study area, determined from a unit route of the 14th cruise of the R/V *Glomar Challenger*, needs thorough verification. If there is no displacement of lineation 34 south of the Cape Verde transform fault, the bottom age within the study area should be over 80 million years (Santonian, Coniacian).

From the geomorphological survey data, bottom depth within the study area varies from 4500 to 6200 m. In its northern part, there is a pronounced sublatitudinal depression, related probably to the Cape Verde transform fault. In the rest of the area, located at the southern side of the Cape Verde fault, individual uplifts and depressions can be traced mainly by the sublatitudinal trend.

A geomagnetic survey was performed alongside the geomorphological survey, with gravity measurements and CSP along the grid of the sublatitudinal and submeridional profiles spaced 20 miles apart on average. Besides, geomagnetic measurements were performed along the supplementary diagonal profiles and along the deep seismic sounding profile. Since the area of detailed geomagnetic survey was located near the equator, and thus, related variations of geomagnetic field may occur, numerous measurements of magnetic field were carried out along the reference profile following the line of deep seismic sounding. No considerable variations of magnetic field were detected there.

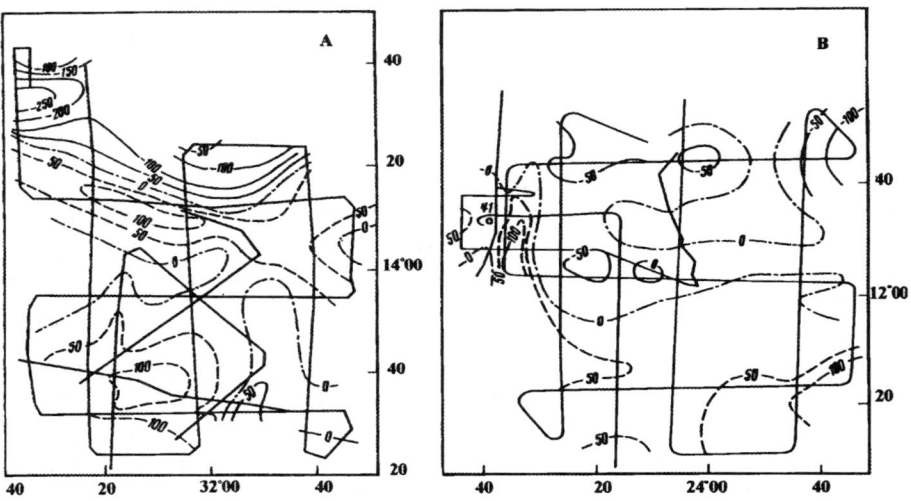

Figure 9 Isodynam map of AMF for the detailed magnetic survey areas in the (A) Cape Verde and (B) Brazil Basins. Isodynam interval is 50 nT.

An RMS of the detailed geomagnetic survey at the study area showed that it does not exceed 15 nT. From the results of the detailed geomagnetic survey, a map of the AMF ΔT_a has been compiled (Figure 9A). The analysis of the map showed that within the study area, the magnetic anomaly field is rugged. Total anomalous amplitude within the study area reaches 400 nT (from −250 to 150 nT). Distinct principal sublatitudinal and submeridional tectonic elements typical of the Cape Verde Basin were revealed in the anomalous field structure of the study area, where no signs of magnetic lineations are known. The northern part of the area is complicated by a margin of the transform fault and is characterized by negative magnetic field anomalies. The study area is located approximately 1200 km from the Mid-Atlantic Ridge axis, and therefore the intensity of magnetic anomalies caused by the basalt layer of the oceanic crust is relatively low (Nazarova, 1981). Meanwhile, if we assume the absence of magnetic field inversion within the area under study, its structure would be determined first by the topography and magnetization intensity of the basalt layer top. The fact that the sediments within the study area are not thick may also support the presence of such relations. When we compare the map of the AMF with the bathymetry of the area and CSP data, it is clear that the zone of negative magnetic anomalies is confined to the transform fault depression bounded by normal fault scarps. The zone of maximum horizontal gradient of the anomalous field is attributed to the southern slope of the fault. This step of the topography seems to be related to the fault manifested in magnetic anomalies. In the northeastern part of the area, the zone of negative anomalies is cut by a northeast lineation. The general trend of the zone (about 110°) corresponds approximately to the Cape Verde fault strike. Maximum thickness of sediments is confined to the graben section. The southern part of the graben with the crest 4600 m in altitude is characterized by a change of the magnetic field sign and positive values of ΔT_a. A narrow belt of positive anomalies ΔT_a with an amplitude over 100 nT is registered over the valley at the southern slope of the fault.

South of the Cape Verde fault zone, the bottom topography of which is slightly rugged, the AMF of the rest of the study area has a relatively smooth character. Positive anomalies with amplitudes a bit higher than 100 nT predominate there. At the same time, despite poor intensity of anomalies, northeast lineaments with the 35 to 45° azimuth are distinctly traced in that part of the study area. Together with the sublatitudinal lineaments, they seem to determine the main structural plan of the AMF. They are verified well by the CSP data and correlate with the discontinuities expressed by the steps of the acoustic basement topography and cut-in of the sublatitudinal structural elements. These lineaments, which are orthogonal with respect to the sublatitudinal field lineaments and are connected with the transform fault, seem to be confined to normal fault fissures, which formed a basalt ridge topography at the Mid-Atlantic Ridge slopes. Such ridges are typical of the Atlantic Ocean, where the spreading rate is not high.

In the southern part of the study area, a maximum positive anomaly (ΔT_a is over 150 nT) is confined to the depression, whereas the uplifts are characterized by field values close to zero. As has been noted earlier, the most probable longitudinal position of the study area is west of anomaly 34, which corresponds to the epoch of reverse magnetic polarity (early Campanian, 76 to 79 million years). That can partially explain the inverse correlation between the anomalies ΔT_a and bottom topography. In this connection, there arises a question on the origin of the negative magnetic anomaly attributed to the depression in the fault valley in the northwestern part of the study area. There, we have a direct correlation between the sign of the anomaly and acoustic basement topography. If these anomalies result from the basalt layer, we may suppose a considerable displacement of the northern and southern flanks along the fault valley that would correspond to the epochs of different polarity in that case. However, according to the map of magnetic lineations for the Atlantic, the northern flank of the Cape Verde fault is older than anomaly 34 in this area and corresponds to the late Cretaceous time with a direct magnetic polarity (80 to 112 million years). Thus, we may suppose that the negative anomaly in the fault depression is related to the uplift of the oceanic deep-crust layer composed of gabbro and serpentinite. This is verified partly by the finding of serpentinite in the cores of manganese nodules uplifted from the bottom at that part of the fault.

Magnetic simulation was performed to study the origin of active magnetic layers (Belyaev et al., 1989b) using experimental data taken by magnetic survey within the study area.

In accordance with the maps of magnetic inclination, average magnetic inclination was assumed as 26°. The depth to the active magnetic layer base was assumed as 10 km. The average depth to the acoustic basement composed of basalt was taken as 5 to 5.5 km. This corresponds with the thickness of an active magnetic layer (5 km) and thus includes the entire crust from DSS data for consideration.

Magnetic simulation was performed for profiles 1 and 7, crossing the latitudinal depression in the acoustic basement along the meridian.

As can be seen from the simulations for profile 7, the effective magnetization for the magnetic layer in the depression is two to three times higher than that in the northern part of the profile. In the case of profile 1, effective magnetization of rocks in the depression is also considerably higher (five to ten times), than that in the southern part of the fault. If we start the simulation assuming thickness of an active magnetic layer to be 0.5 km, we should suppose even greater contrast in magnetization of adjacent blocks, which would contradict the data on magnetization of oceanic basalt and serpentinite obtained from deep sea drilling (Nazarova, 1987).

The obtained results provide evidence for the possible relation of magnetic field anomalies over the fault in the northwestern part of the study area and rocks from the deep layers of the oceanic crust.

An area for detailed geomagnetic survey of the Brazil Basin was taken with the following coordinates: 11°25 to 12°30 S and 23°35 to 24°40 W. Geomagnetic studies within that area were performed for the first time. In that latitudinal interval, magnetic lineations were recognized only far to the east, in the Mid-Atlantic Ridge zone. There, in the vicinity of the Ascension Islands, lineations 3 and 5 were recognized. In the mentioned longitudinal interval, the system of magnetic lineations was distinguished only south of the transform fault following approximately 15°S. If we assume that displacement of magnetic lineations along that fault continues, we may suppose that the study area is located approximately in the region of lineations 29 and 30 (65 to 67 million years).

Geomagnetic survey within the study area was performed along the grid of sublatitudinal and submeridional profiles spaced 20 miles apart, on average.

As follows from the map (Figure 9B) of the Brazil Basin study area, the magnetic field there has a smooth topography. Anomalous amplitudes are less than 100 nT as a rule, and the horizontal gradient of the field is small. In the southern part of the study area, positive values of the field predominate, and in the northern part, they change for negative anomalies. Against that background, a zone of positive magnetic anomalies with amplitude up to 130 nT and high horizontal gradients is pronounced in the northwestern part of the area. A positive anomaly with an amplitude reaching 200 nT is also registered in the southeastern part of the area under study. Besides, sublatitudinal and submeridional lineations are clearly traced against a smoothed field.

Comparison of the materials of geomagnetic survey with the sounding and CSP data provide evidence of a direct correlation, in the first approximation, between the magnetic field anomalies and topography of the acoustic basement resting at a depth of 5 to 6 km; the anomalies are shifted slightly, probably due to the proximity of the study area to the equator. In the northern part of the study area, a seamount has been recognized at profiles 1 and 3. A small magnetic anomaly (to –78 nT) is registered over that seamount. An elevation of the seamount over the floor is 1.3 km. Dredging results show that the mount is of volcanic origin.

Two orthogonal systems of lineations, sublatitudinal and submeridional, being in a subordinated relation, are traced at the isodynamic map of the magnetic field. If two principal sublatitudinal lineations, those following isolines 0 and 50 nT, are compared to the sounding and CSP data, it is possible to suppose that they seem to be related with disjunctive dislocations parallel to the system of transform faults in the Brazil Basin. Judging from the magnetic lineation displacements, transform faults bounding the analyzed lithosphere block are spaced north of the study area, between 5 and 10° S, and south of it, in the region of 13° S. Submeridional lineations of the field seem to be related to the basalt basement ridges bounded by the system of gentle normal faults and parallel to the Mid-Atlantic Ridge axis.

In relation to these lineations, it seems interesting to study the positive magnetic lineations in the northwestern and southeastern parts of the area confined to the sides of submeridional depressions. The latter seem to be of tectonic origin. From the comparison of the sounding data, positive magnetic lineations with amplitudes up to 100 nT are confined there to the depression slope filled by sediments but not to the rises of basalt basement. One may suppose that submeridional ridges are formed under the influence of gentle normal faults that leads to the sliding down of individual blocks of the oceanic crust toward the rift zone at small nodule plane angles. The result is that deep serpentinite layers causing magnetic field anomalies may be uplifted to the bottom surface at the flanks of the originating depressions. That is verified partly by the results of the dredging carried out in the northwestern part of the study area at the flank of the depression under consideration. Samples of serpentinite and gabbro were taken during the dredging. Magnetic simulation was performed by the "Interpretation" program (the technique has been described earlier) to study the origin of an active magnetic layer along profile 8; the latter is a latitudinal transection across the depression filled by sediments. Magnetic inclination for the model was assumed to be $-20°$. The depth to the basement of an active magnetic layer was assumed in the model as 10 km with the depth to its top being 5 to 5.5 km, according to CSP data. The simulation results (Gorodnitsky et al., 1990c) show that magnetization of an active magnetic layer within the part of the depression represented by blocks 12 to 16 and in its eastern part, blocks 12 to 14 in particular, are, respectively, two and seven times as much as the effective magnetization within its other parts.

In the light of this, we shall stress the anomalous character of the Earth's crust section obtained in the result of the deep seismic sounding in the depression of the northwestern part of the study area. Both geological and seismic data verify the supposition that the observed anomalies have a serpentinite origin. Meantime, a system of negative anomalies, one of which is related to a sea volcano, is traced along the northern frame of the study area. These anomalies seem to trace the sublatitudinal fault zone along which volcanic constructions evolve.

Results of the detailed area geomagnetic survey in the Cape Verde and Brazil Basins allow us to determine some general structural features of the AMF probably typical of the deep basins of the Central Atlantic:

1. The AMF usually does not have submeridional magnetic lineations, therefore, the field structure in these areas does directly reflect the specific features of geological and tectonic structure of oceanic crust.
2. The structure of the AMF is determined mainly by the topography of the acoustic basement composed by basalt and also by two systems of submeridional and sublatitudinal lineaments orthogonal with respect to each other. The first, sublatitudinal system is related to the developed system of transform faults dividing the Central Atlantic lithosphere into separate blocks. The second, submeridional system is related to the ridges of the Mid-Atlantic Ridge separated by gentle normal faults (angles no more than 15°) and leveled by isostasy.
3. Intensive magnetic anomalies confined to the deep layers of the oceanic crust, uplifted to the surface and probably composed of serpentinite, play a leading role in the anomalous field structure of the Cape Verde and Brazil Basins. These anomalies are usually confined to the flanks of the depressions formed by submeridional normal faults. The origin of these anomalies requires further additional investigation. However, it is possible to suppose already now that the deep rocks of the oceanic crust may be uplifted to the surface due to displacement of blocks of the crust along gentle normal faults at the depression. That aspect is principal when analyzing the origin of magnetic anomalies in the deep regions of the World Ocean.

B. THE CAPE BASIN

The Cape Basin, located in the southeastern Atlantic and bounded by the Walvis Ridge, from the north, and by a system of the Falkland-Agulhas faults, from the south, belongs to one of the deepest basins of the World Ocean. The bottom depth there reaches 5500 m. Analysis of geomagnetic survey results obtained in the Cape Basin and on the continental slope of South Africa allows us to recognize there a system of the Cenozoic and Mesozoic magnetic lineations, from the rift anomaly to anomaly M-12 (128 million years).

Detailed mapping of the Mesozoic lineations in the Cape and Argentine Basins and also in the zone of a continental rise at the transition to the African shelf together with the gravity and tectonic materials were used for the Late Mesozoic reconstructions of the Southern Atlantic.

The Cape Basin itself and magnetic lineations recognized in it, their fine structure, intraplate volcanoes, and dislocations, are poorly investigated in the geomagnetic aspect.

Geomagnetic survey was performed on the 31st cruise of R/V *Dmitry Mendeleev*. It followed the route from Montevideo to the Cape Agulhas crossing the Mid-Atlantic Ridge and the Cape Basin approximately along 36° S. An area survey was also fulfilled at the segment of the seamount in the Cape Basin. The route of geomagnetic survey in the Cape Basin crosses the undeformed lithosphere block between two transform faults, a bit south of the route followed by the R/V *Glomar Challenger* on its 39th cruise. It was located across the trend of magnetic lineations. The results of the route geomagnetic survey in the Cape Basin are presented as graphs ΔT_a at the integrated geophysical profile (Gorodnitsky et al., 1990c).

The Cenozoic magnetic lineations are registered distinctly east of the Walvis Ridge, in the western part of the Cape Basin. From a comparison of the map of the axes of paleomagnetic anomalies with the magnetic lineations, determined from the data of the earlier geomagnetic survey performed in the Southern Atlantic, they are identified as anomalies 24 to 33 (57 to 79 million years). It is clear from the comparison of the magnetic field anomalies, recorded along the route, with the topography of the acoustic basement, which has a complex character and tectonic origin within the Cape Basin, that the fine structure of magnetic lineations is determined mainly by the topography of the basalt basement and disjunctive displacements.

Gradient magnetic survey with the proton differential magnetometer DPM-1 was also attempted alongside the measurements of geomagnetic intensity in the Cape Basin. Sensors of the device were towed behind the vessel; the distance from the vessel to the nearest sensor was 200 m, and distance between them was 200 m. As the result of the survey, gradient characteristics of the Cenozoic magnetic lineations were obtained, and boundaries of the zones of direct and reverse polarity were specified. It allowed revision of the spreading model for the Southern Atlantic and provided the basis for the magnetic simulation by the Montevideo-Agulhas geotraverse. From the estimates, the average rate of the Cape Basin semi-opening during the interval of 79 to 57 million years was about 2.2 cm per year. A detailed area survey was performed along the segment of magnetic lineation 34 located in the Cape Basin.

The area of detailed geomagnetic survey at the magnetic lineation 34 segment is bounded by coordinates 35°10 to 36°25 S and 02°00 to 03°40 E. Geomagnetic survey within the study area was performed together with the sounding, gravimetric observations and CSP along the grid of sublatitudinal and submeridional profiles spaced 10 to 20 miles apart, on average.

At the same time, as follows from analysis of the compiled map (Figure 10), the principal anomalous zone within the study area is the zone of magnetic lineation 34 (about 80 million years, Santonian). The amplitude of the anomaly within the area is over 360 nT. The anomaly 34 zone crosses the central part of the area. Its general trend is about 340°. In the western part of the area, the second magnetic lineation with a positive amplitude of about 25 nT is traced parallel to anomaly 34. It is precisely these elongated magnetic lineations that determine the structure of the AMF within the study area.

The map of isodynams shows distinctly that besides the general northwestern trend of the principal lineations of the magnetic field, there are declinations of isodynams orthogonal to the first system trend and that seem to be related to disjunctive displacements parallel to transform faults. In the Cape Basin, that system of lineations is not so pronounced as in the Central Atlantic basins, where the systems running parallel to transform faults form their individual group of magnetic anomalies. Meantime, in the northeastern part of the area, at the segment with coordinates 35°30 S and 03°00 E, a rupture of isoline 100 nT and its closure are recorded. It goes further to the southwest along the rupture and closure of isoline 200 nT in the zone of anomaly 34 maximum, and further southwestward along the typical declination of isolines to the closure of isoline 50 nT. The trend of this zone is about 55°. Parallel lineations are registered in the northwestern corner and southern part of the study area.

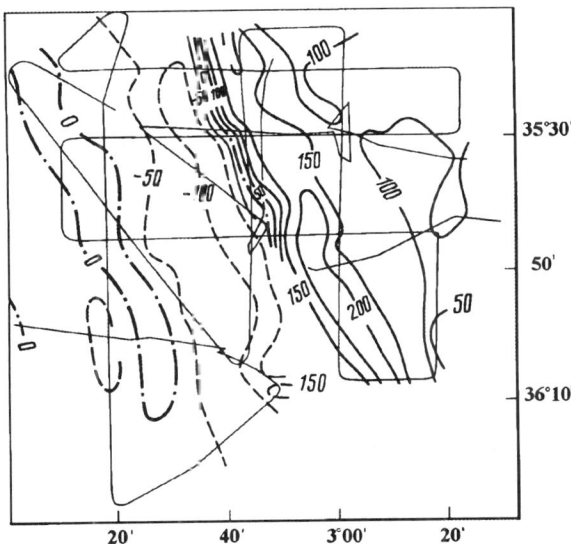

Figure 10 Isodynam map of AMF for the detailed geomagnetic survey area in the Cape Basin along lineation 34 segment. Isodynam interval is 50 nT.

Comparison of graphs ΔT_a with the CSP data provides evidence of the relation between the anomalies ΔT_a with the topography of the basalt acoustic basement, lying at a depth of about 5.5 km, against the background of the two mentioned lineations of the AMF.

At the segment of anomaly 34, besides the area geomagnetic survey, a geomagnetic survey over the seamount was carried out in a small scope. The seamount, described by coordinates 36°30 S and 08°03 E, is confined to the northwest ridge. Its minimum depth is 2796 m, and the summit is elevated over the bottom at 2 km.

The data of the geomagnetic survey have shed light on the structure of the AMF over the seamount indicating its volcanic origin. A positive magnetic anomaly with an amplitude of over 670 nT is registered over the seamount summit.

A local isometric character of anomaly ΔT_a over the seamount allows its first approximation by a magnetic dipole to estimate the parameters of its magnetization

(Gorodnitsky et al., 1990c). It is possible to suppose that the African plate has shifted northward since the time of the seamount formation. Magnetic moment of the dipole was ~1.7×10^{11} Am^2. The seamount volume is ~50 km^3; the cone has a basement diameter of 10 km and an altitude of 2 km. The estimated effective magnetization of the seamount rocks is about 2 A/m.

III. THE SPREADING FEATURES OF THE NORTHWESTERN INDIAN OCEAN IN LATE CENOZOIC

A. THE ARABIAN BASIN

The Arabian Basin is one of the most magnetometrically explored regions of the northwestern part of the Indian Ocean. It is bounded on the northwest by the continental margin of the Arabian Peninsula, the north by the Meerey Ridge, the northeast by the continental margin of the Indostan Peninsula, the east by the Maldivsky Ridge, and the southwest by the Carlsberg Ridge (Figure 11).

In the Arabian Basin beyond the magnetic quiet zone the high-intensity anomalies were discovered by Matthews (1966) and Taylor (1968). In 1966, Bowin and Vogt discovered the symmetrical sequence of anomalies in the eastern part of the Somali Basin. The trending of these anomalies in the Arabian Basin is N90° E and N120° E in the Somali Basin, and it essentially differs from the trending of the Carlsberg Ridge N130° E to N140° E. Having first identified the specific anomalies number 25 and 26, McKenzie and Sclater (1971) identified the sequence of anomalies of Paleocene age (from 23 to 28) in both basins. The spreading rate for this period relatively to the Carlsberg protoridge with the latitude trending is 6.5 cm per year, calculated with the Heirtzler et al. (1968) time scale.

The geophysical and geological data collected during the 23nd cruise of the D/V *Glomar Challenger* (Init. Rep. DSDP, 23, 1974) made it possible to improve the motion of the Arabian Basin floor structure and its age. Using the results published by McKenzie and Sclater in 1971 as well as bathymetric, seismic, and magnetic profiles obtained by D/V *Glomar Challenger, Vima, Conrad,* and *R. Whitmarsh* marked out new faults — the Rudra and the Shiva — situated to the southeast of the Anachita fault, which was discovered earlier, and then identified anomalies with numbers 18 to 24 (Whitmarsh, 1974). In many respects this identification was defined by the micropaleontological age (46 million years) of the oldest sediments at the basalt contact, exposed by the 221st DSDP site (Init. Rep. DSDP, 23 ,1974).

A new step in the investigation of the Arabian Basin is connected with the systematic magnetic studies in the given region (Karasik et al., 1986a, 1988).

In the Arabian Basin the system of low-amplitude anomalies with NW-SE trending, arranged to the northeastern side of the Carlsberg Ridge, is replaced by the system of intensive (with a middle range of 200 to 400 nT and a maximum range to 600 nT), mostly long-wave magnetic anomalies with latitude trending, partly displaced on 30 to 80 km in the submeridion direction. The basin is cut into the separate spreading cells by the earlier known faults Anachita, Brahma, Shiva, and the newly discovered faults, Ganesha and the others. The magnetic profiles within each cell usually are characterized by a high degree of similarity. And the profiles from the different cells might differ a little in width, amplitude, and skew of anomalies, and also in distance between them. To the east of 65° E and to the south of 17° N some complications appear in the field structure: a number of anomalies disappear, the distance between the anomalies changes irregularly, and a single chain of anomalies seems to have skew trending. The sudden rise of intensity up to 750 nT is observed over the Fedyenskiye Ridge — a big uplift of basement with NE-SW trending, which was recognized from seismic data (Babenko et al., 1980). It is reasonable to connect this rise with the approach of the sources to the observation surface. However, it is essential that the overwhelming majority of the anomalies in this region keeps the almost strict latitude trending, which is typical for the whole Arabian Basin. The anomalies of the same trending are surely observed over the Fedyenskiye Ridge, thereby testifying that the latter represents an uplifted block of the ocean crust. To the north of 17° N the structure becomes more regular.

The most complex structure of magnetic anomaly is observed in the transition zone from the northeastern outlying area of the ridge to the deep-water part of the Arabian

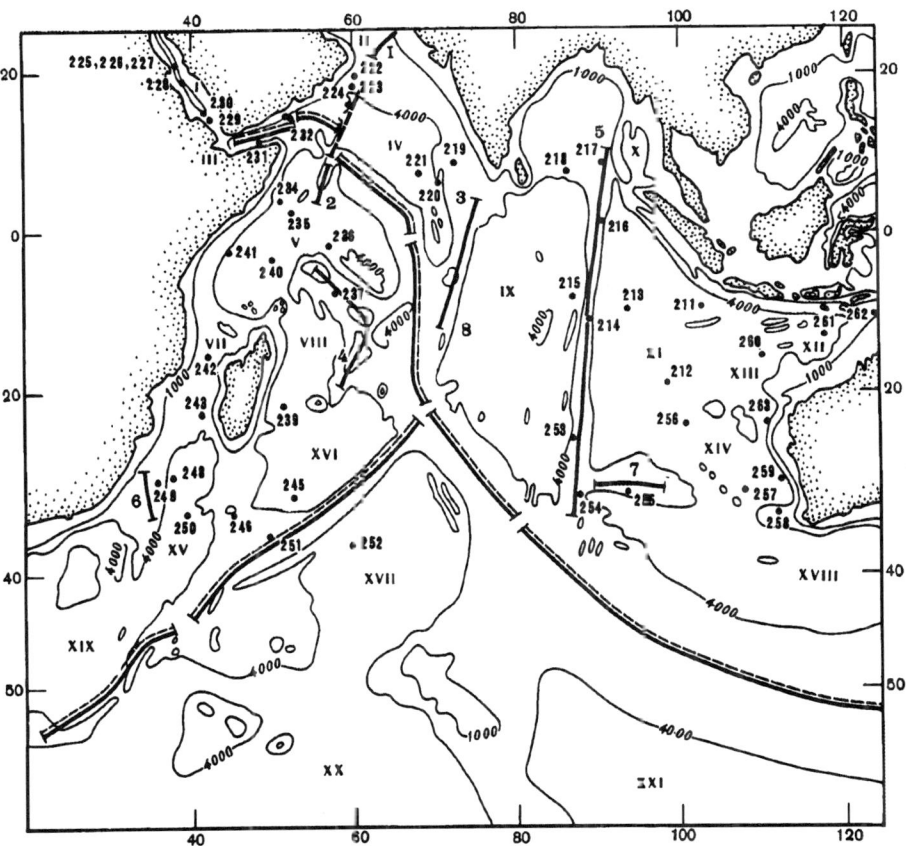

Figure 11 Simplified bathymetric map (isobaths 1000 and 4000 m) and principal structures of the Indian Ocean. Basins: (I) Red Sea; (II) Oman; (III) Aden; (IV) Arabian; (V) Somalian; (VI) Amiran; (VII) Comoro; (VIII) Mascarene; (IX) Central; (X) Andaman; (XI) Varton; (XII) Argo; (XIII) QV; (XIV) Perth; (XV) Mozambique; (XVI) Madagascar; (XVII) Crozet; (XVIII) South Australian; (XIX) Agulhas; (XX) African; (XXI) Antarctic. Ridges and swells: (1) Murray R.; (2) Chain R.; (3) Chagos-Laccadive R.; (4) Mascarene R.; (5) Ninetyeast R.; (6) Mozambique R.; (7) West Australian R.

Basin. Here we can see a system of short-wave anomalies with low intensity within 100 to 200 nT having NW-SE trending, which transform into the anomalies of the Arabian Basin with sublatitude trending.

The structural peculiarity of magnetic anomaly of the Arabian Basin is the presence of two discordant systems of linear magnetic anomalies, shown on the lineation map (Figure 12). Here, the south sequence cuts down the north one at an angle near to 40°, and this corresponds with fast spreading with respect to the latitude axis of the proto-ridge and with slow rotation with respect to the modern axis of the Carlsberg Ridge. The identification of the anomalies on the northern slope of the Carlsberg Ridge gives the sequence of anomalies with the same trending: from anomaly 5, and sometimes younger, to anomalies 7 to 8, and sometimes 9. Thus, we succeed in separating out the region of ocean crust development, which appears on the modern spreading axis during the last 26 to 28 million years. Though the dating of the beginning of the last spreading episode has no pretensions to be new, for the first time it is made on reliable magnetometric data with high accuracy.

The tracing of transverse disarrangements along the shifts of the anomalous structure leads to separation of the fault system, characterized by shifts of both signs with amplitudes up to 40 to 45 km.

The selection of the observed profiles by inversion-spreading model of the layer with magnetic activity within the Arabian Basin gives, with the greatest degree of confidence, the axes of anomalies 26, 25, 24, 20, and 13; due to this, it is possible to correlate the overwhelming majority of anomalies (Figure 13) and make the map of their axes. For the first time in the Arabian Basin the anomalies 10 to 18 were identified, situated in the "deaf zone", where it was previously impossible even to correlate anomalies. On the north of the basin, anomalies 28 and 29 were traced out, as were two even older anomalies, which are hypothetically identified with anomalies 30 and 31. Thus, the ocean crust on the axis of the Carlsberg protoridge has been estimated to have formed from 70 (?) to 67 until 30 million years ago, in other words during a period of about 40 million years.

Turning to the north sequence, we note that its northern part is older than anomaly 24 and is characterized by an inhomogeneous field. If on the west we can see a distinct display of linear anomalies 25 and 26, which are followed by anomalies 27 to 29, being settled out positively enough on the several profiles, and maybe anomalies number 30 to 31, then to the east from 65° E, the normal sequence is broken: the dropping out of separate anomalies and irregular change of distance between them are observed. The structural breaking of the field is characterized by the presence of discordant elements toward the main structures of latitude trending. Dating of these events is hindered and not positive enough because the short, unbroken intercepts of the profiles are not representative and do not provide strong diagnosis.

Linear rates of spreading are rather unstably distributed in time and space. The earliest period of rotation, distinguished hypothetically before epoch A-29, had a rate about 30 km per million years. Then came the long period of high-rate spreading, when the rates were about 56 km per million years. During epoch A-23, the rate fell to 25 km per million years, which naturally is connected with the first contact between India and South Asia in the early Eocene. Next, slowing to 16 km per million years occurred during epoch A-20; this coincided with the time of essential reconstruction of plate geometry and kinematics to the east of the triple junction of the Indian, Antarctic, and African plates due to the collision of the Indian and Asian continental blocks. Finally during epochs 9 to 7 the rate of spreading fell again, to 10 to 13 km per million years, whereas its direction changed on 40° to the east.

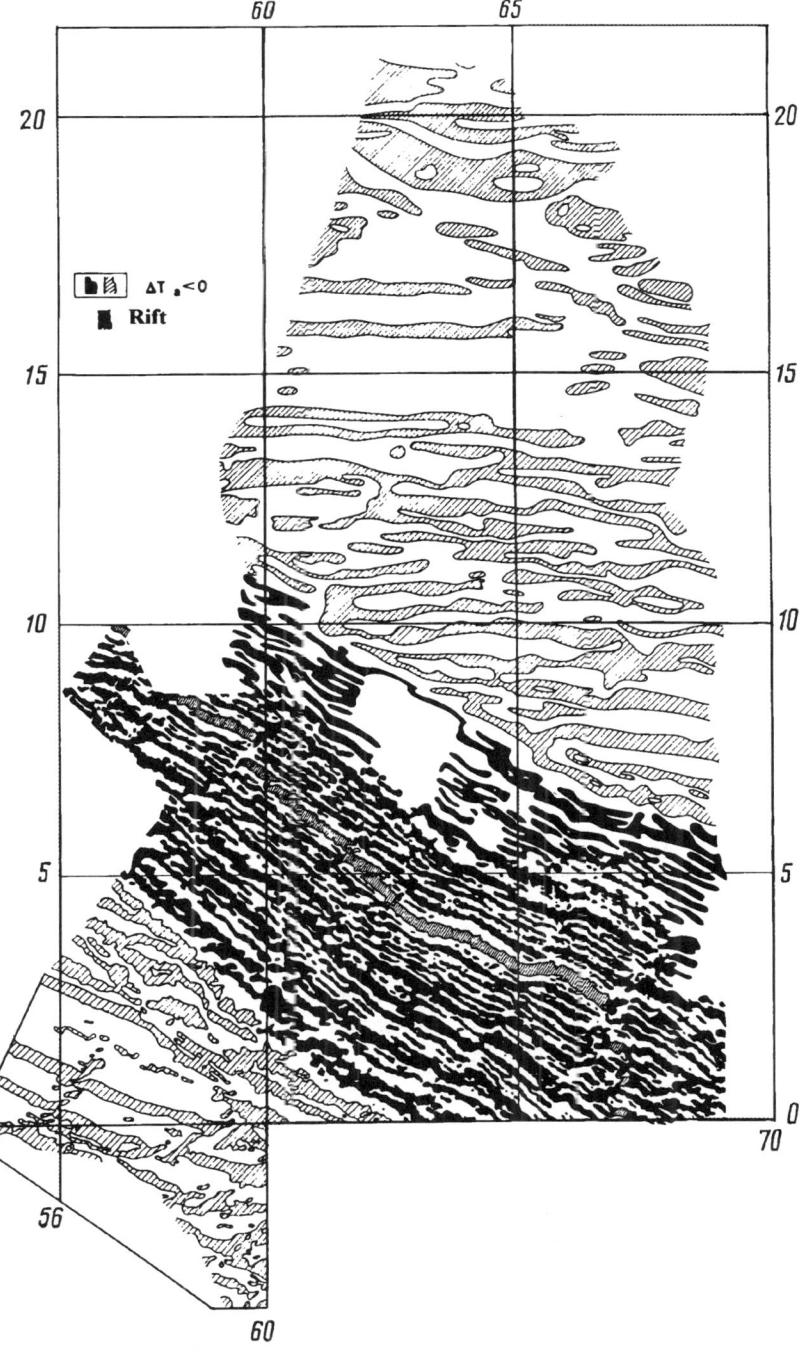

Figure 12 Chart of zero isolines of the Arabian plate AMF.

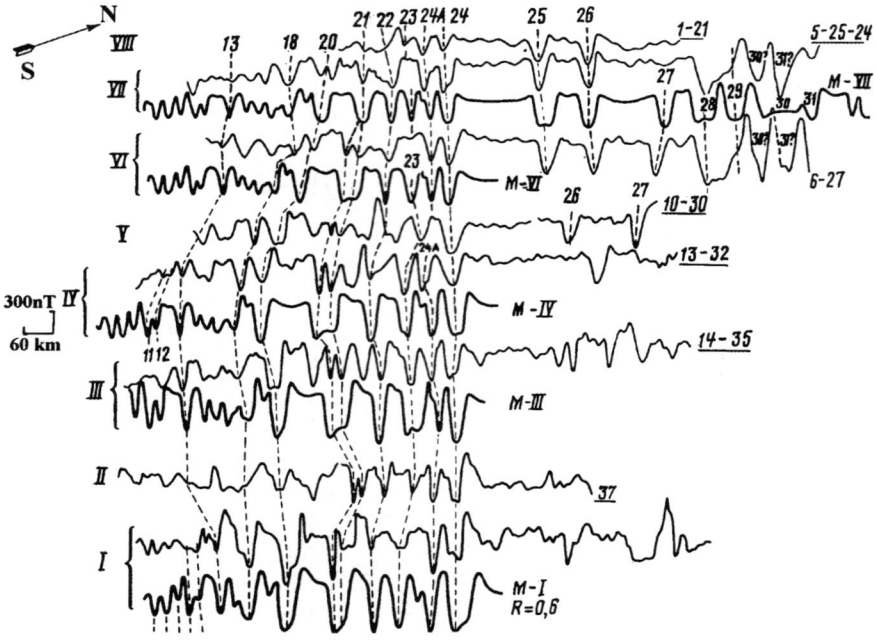

Figure 13 Typical profiles in the Arabian Sea from all spreading cells (I to VIII) and their geochronological interpretation. The dashed line shows anomaly correlation at the observed and simulated (M) profiles; anomalies are digitized in the Lamont system; P is a degree of smoothening of a simulated profile (Tisseau and Patriot, 1981).

Thus, the trend to the step-by-step slowing of the spreading rate of the Arabian Sea is observed, broken only with the initial periods of spreading.

The observed variation of spreading rates in various cells does not obey any distinct spatial regularities and cannot be explained in a step-by-step change of the linear rate of spreading as a result of removal from or approach to the pole of rotation. The data can indicate either widely asymmetrical spreading or jumps of its axis.

At the same time in cell 1 in the Arabian Basin, we discovered a short "stopping" of spreading in the interval 41.3 to 38.4 million years ago, which means a falling out of the part of the Indian plate with the square of about 330 × 55 km. We can scarcely explain this fact as a one-side spreading in favor of the Arabian plate; it is most likely that the axis jump took place, joining it with the part of the Indian plate. Similar effects evidently take place in the rest of the Arabian Basin, especially in its northeast zone, where a falling out of anomaly 25 and maybe others (Figure 14) is expected. Therefore we can assume that the main part in the formation of observed variations of the regular structure of anomalous axes belongs to unexposed jumps of spreading.

B. THE SOMALI BASIN

Until recently, the section of the Somali Basin extending between the Seychelles Bank and the Carlsberg Ridge was the least-known part of the northeastern Indian Ocean. In 1971, McKenzie and Sclater, using several "Owen" profiles (1962 to 1963), determined the trends of the anomalies and identified them with a sequence of Paleocene

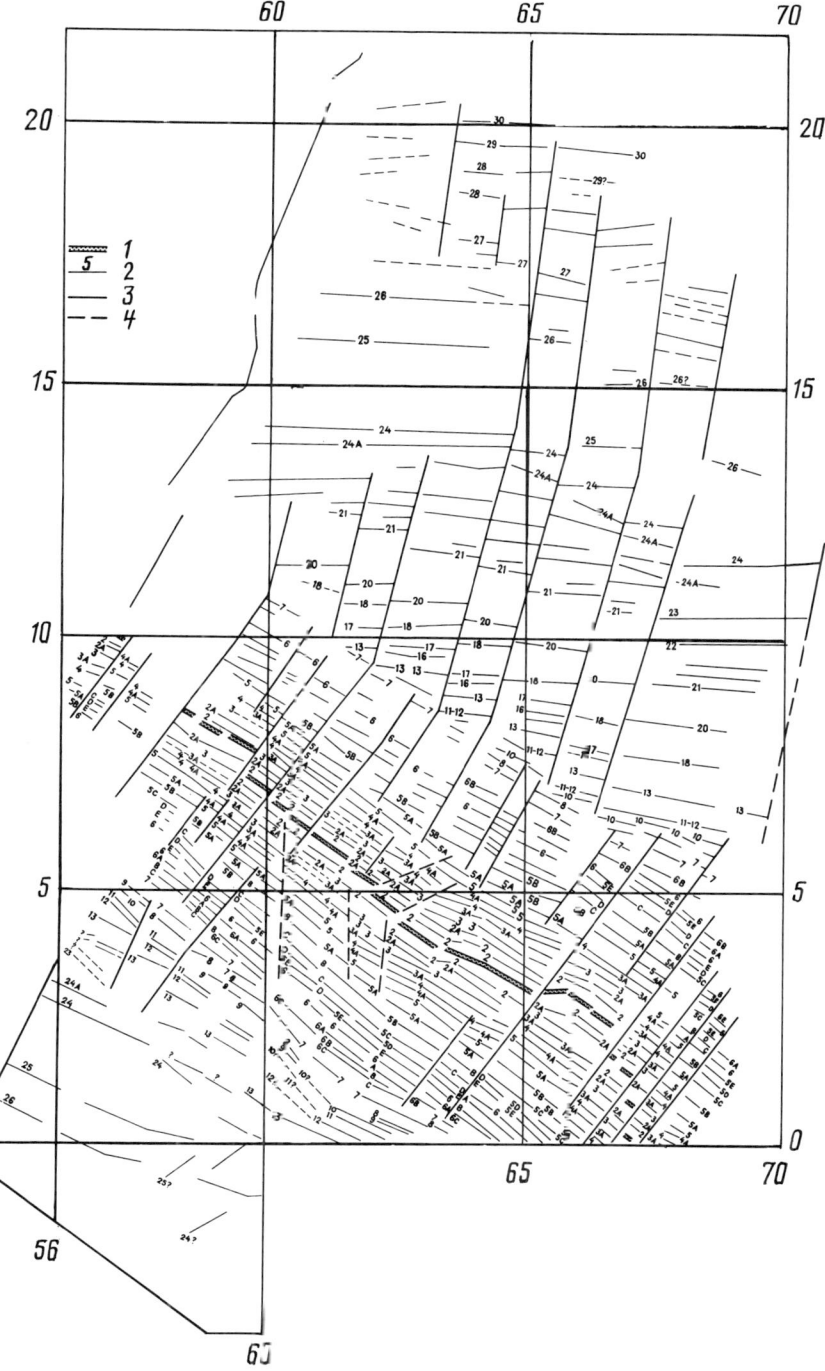

Figure 14 Chart of the paleomagnetic anomaly axes in the Arabian Sea: (1) rift anomaly; (2) anomaly axes and their numbers; (3) transform faults; (4) pseudo-faults.

anomalies (23 to 28, 60 to 70 million years). Since then, there was no essential revision of these views. The more detailed study in this part of the Somali Basin, made by R. Schlich in 1982, on the whole confirmed the results of McKenzie and Sclater's analysis. But it allowed refinement of the axes structure and definition of several faults that had not been found previously; it also provided identification by anomalies 6 and 7 by the only profile on the southeastern slope of the Carlsberg Ridge (Schlich, 1982). The history of the Somali Basin spreading of late 60 million years appeared to be unstudied. Besides, existing views about seafloor chronology contradicted deep sea drilling data. So the only site drilled by D/V *Glomar Challenger* in this part of the Somali Basin (N 236), gives the ocean crust age (by micropaleontological sediments) as about 57 to 58 million years whereas the "magnetic age", defined by the Heirtzler et al. (1968) time scale, is not less then 67 million years.

Thus, on the one hand, the magnetometric data are scarce, making it impossible to define linear magnetic anomalies younger than anomaly 23 (58 million years), while on the other hand, contradictions in the known data testified to the poor knowledge of the Somali Basin. And as a consequence there were rather rough ideas of seafloor structure and the history of its development.

In 1986 to 1988 we studied the Somali Basin by systematic magnetic mapping, to the southwest of the Carlsberg Ridge, in its northeastern part bounded to the NE-SW by the Carlsberg Ridge and Seychelles Bank and to the NW-SE by the Chain Ridge and the Medingley Plateau (Figure 11).

According to the map of the zero isolines (Figure 15), a typical ocean field presented by linear magnetic anomalies is on the sea region observed. The intensity and width of anomalies essentially vary, depending on the geomorpological provinces for which these anomalies are arranged.

Anomalies of the Somali Basin in its flat and deep-water part (Figure 15) are characterized by high regularity, lineation, and united (single, common) trending of anomalies. Here the high linear negative anomalies, with a range up to 600 nT and a width of 20 to 30 km, were developed. They are trending through the whole zone from the northwest to the southeast (115° NE). The maximum horizontal gradients reach the value of 30 to 40 nT/km. Intensive negative anomalies interchange with equal intensive positive anomalies with a length of 20 to 60 km and an amplitude of 250 to 300 nT. In the central part of this region (Figure 15) we can see a positive anomaly with a length about 200 km, which can be traced as a wide belt (band) in parallel to the main anomalies trend (115° NE). In its smooth part this wide positive anomaly is complicated by low-intensity short-period anomalies, well correlated with each profile. On the whole, positive anomalies predominate in the given part of the basin. This fact resulted from the structure of the Paleocene part of the geochronological scale, which is characterized by high/rapid spreading that formed the magnetoactive layer due for the observed anomalies.

The smooth floor of the Somali Basin is broken by numerous rocks, some of which have a height of 3 to 4 km. The main complication in the floor relief of the basin is the province of the submarine rocks (Figure 15, region 2) of volcanic origin, as it is to be likely, which begins near the southwestern bound of the equatorial mountain region and extends northeast to 57° E. In the magnetic field, submarine rocks are expressed by short-period intensive anomalies which destroy the linear field structure and cause isometric structures in isolines.

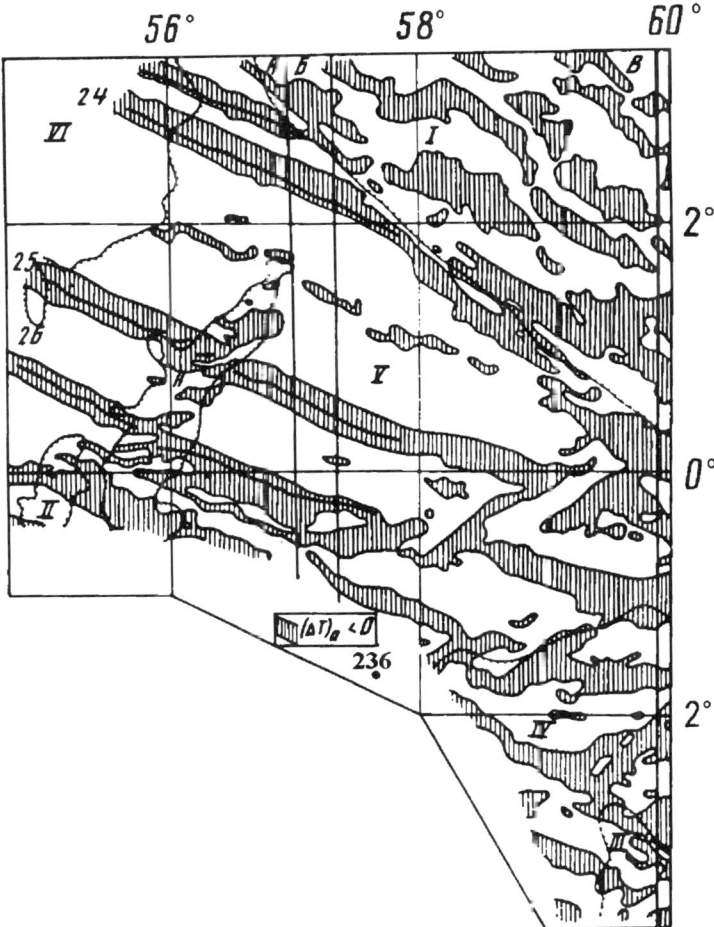

Figure 15 Fragment of the chart of zero isolines in the southeastern Somali Basin. The broken line shows the boundaries of geomorphological provinces (I) southwestern slope of the Carlsberg Ridge; (II) volcanic ridge; (III) periphery of the Medinclay rise; (IV) ridge-and-valley part of the Somali Basin; (V) plane part of the Somali Basin; (VI) deep part of the Somali Basin.

Trending of these anomalies, which coincides with the volcanic ridge trending, is found out to be perpendicular to (be normal to) the prevailing NW-SE trending of the linear magnetic anomalies of the Somali Basin. Thus, if we take into account the morphostructure of the volcanic ridge and its display in the magnetic field, we can consider this ridge as a motion track of the Somali plate over the hot-spot in the direction coinciding with the ridge trend.

In the southeastern part of the region next to the Medingley Plateau (Figure 15, regions III and IV) the floor relief is also complicated by the submarine rocks, which protrude over the foot more than 2.5 km, the height of which gradually decreases in the northern direction. This region of complicated relief is changed by the smooth floor of the Somali Basin. It is difficult to separate in the magnetic field these relief complications of the outlying area of the Medingley Plateau and of the region of hill relief of the southeastern part of the Somali Basin next to the plateau because of irregular anomalies structure progress in this part of the region.

The zone of high linear regular anomalies, which covers a considerable part of the region in the southwest, is followed in the north by the adjacent zone of accidental linear anomalies, which boundary agrees with the southwestern flange of the coincident Carlsberg Ridge, this zone has NW-SE trending in parallel with the ridge axis. Anomalies of this zone (Figure 15, region I) are characterized by breaking of anomaly lineation and its regularity. Low-intensity (amplitude 50 to 100 nT) short-period (10 to 30 km) anomalies are widely distributed in this zone, and their trending coincides with the NW-SE trending of the Carlsberg Ridge. Faults break the linear structure of the field into numerous spreading cells. Sometimes profiles from the different cells differ in amplitude, shape, and trending of anomalies, making it difficult to do thorough correlation and to identify anomalies. In the relief of this zone corresponding to the margin of the southwestern slope of the Carlsberg Ridge, the uplifts in the NW-SE direction and depressions separated the uplifts from each other, with step-by-step rising of the mean floor level in the northeastern direction (this fact corresponds with the general ridge slope relief).

In the structure of the AMF of the Somali Basin there is no discordant relation between the oldest and young sequences of anomalies, as is observed in the Arabian Basin. Both sequences of anomalies with various degree of lineation intensity and representative wavelengths have directions close (115 to 120° NE) to the ridge trend (135° NE). Such a relation of anomaly trends is in accordance with the kinematics of the Indian and African plates.

The main result of the geochronologic interpretation of the AMF of the Somali Basin is the distinction of the complete sequence of anomalies, from 23 (55 million years ago) to the axial anomaly. The anomalies from this sequence (Figure 16) are characterized by low amplitude (100 to 150 nT) and poor correlation, which can be explained by the presence of a great number of faults, connected with the change of spreading centers. Even so, for the first time, using marked anomalies 5, 6, 9, and 13, we were able to trace the complete sequence of anomalies all through the southwestern slope of the Carlsberg Ridge in the adjoining part of the Somali Basin (Figure 16).

Between the oldest anomalies, corresponding to the period of rapid spreading relative to the latitudinal axis of the Carlsberg protoridge, and the sequence of young anomalies there is a narrow zone (Figure 16) of low-correlated anomalies, corresponding to the period of nonstationary spreading. During this period the spreading rate decreased to 2.5 cm per year, and the axis of the Carlsberg protoridge was broken by numerous transformed faults, slightly shifting the anomalies of the same age for 10 to 20 km. As a result, a modern system of spreading centers was formed. Complex spreading geometry and kinematics of this period cause uncertain identification of anomalies. And for a complete analysis, it would be necessary to add a complementary sequence of anomalies from the Arabian Basin.

The results of our investigation have confirmed the sequence of high-intensity linear anomalies in the Somali Basin, identified as anomalies 23 to 27 (54 to 56 million years) on the results of some intersections by all researchers (McKenzie and Sclater, 1971; Schlich, 1982). But according to the zero isoline map shown in Figure 15, the linear structure of the field, corresponding to anomalies from the rapid-spreading period, changes essentially, moving farther along these anomalies in the southwestern direction. Anomalies with latitudinal and even NE-SW trending can be traced along with

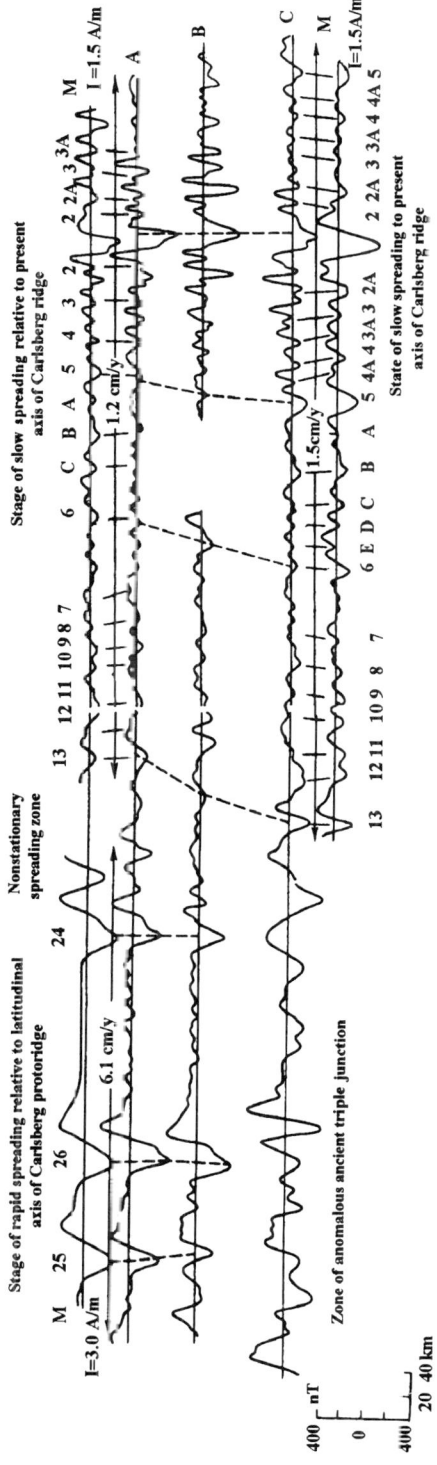

Figure 16 A correlation scheme of representative magnetic profiles A to C, aligned by the axis of rift anomaly, and their geochronological interpretation.

general NW-SE trending that causes occurrence of oblique lineaments that have a V-shape in the field structure. Anomalies 24, 25, and 26 can be easily traced through the profiles in the western part of the section (Figure 16, profiles A and B), but we have not found similar anomalies in the southeastern part of the Somali Basin.

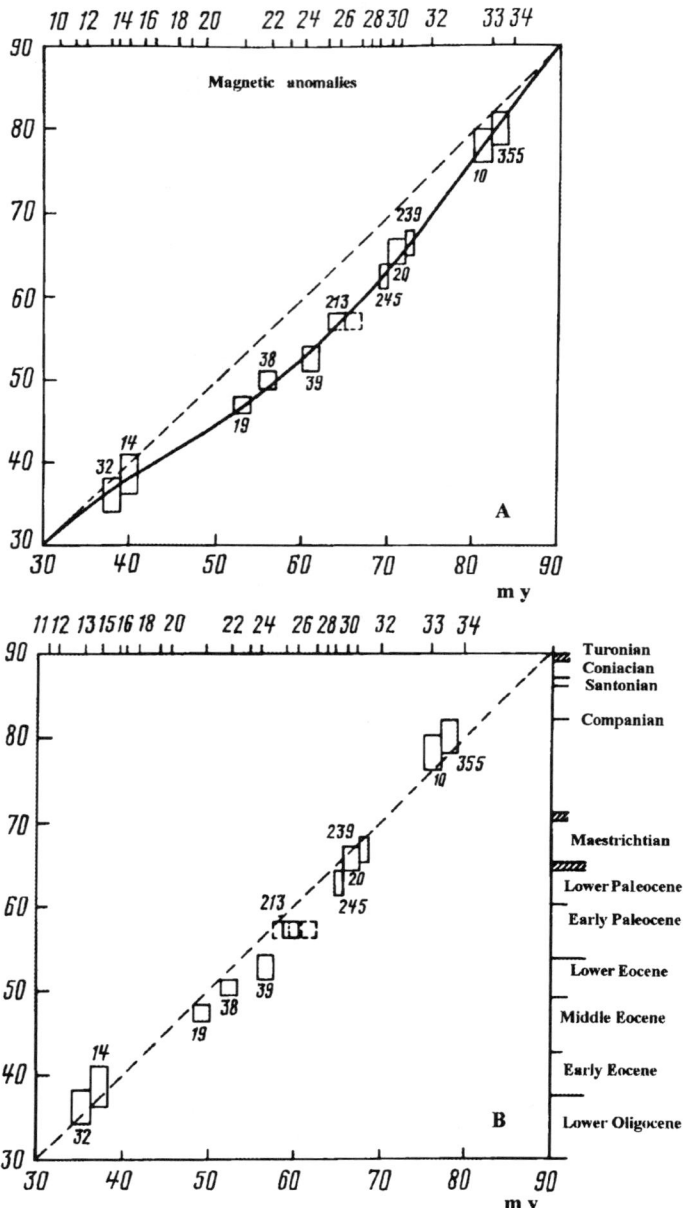

Figure 17 Comparison of magnetic and biostratigraphic ages in borehole 236 according to Schlich identification technique and according to our technique of fitting (dashed line) by (A) Heirtzler and (B) La Brecque scales.

This specific inhomogeneity in the field structure that we discovered indicates complex changing of spreading geometry in the southeastern part of the Somali Basin during the formation of anomalies 27 to 24 (55 to 64 million years ago) and provides a basis for a new view on some contradictions in geological and geophysical data. First, we speak about deep sea drilling data from site 236, according to which the age of the seafloor is about 57 to 58 million years, determined by the oldest sediments exceeding the basalt basis (Init. Rep. DSDP, 24, 1974). The age of the oceanic crust in the region of site 236, defined by magnetic anomalies identified by the Heirtzler et al. (1968) time scale, is not less then 67 million years, because the site is located to the south of the oldest identified anomaly 27 (65 million years). Schlich suggests solving this contradiction by age calibration of Heirtzler's inversion scale (Figure 17). The specific structure of the magnetic field anomaly with the V-shaped lineaments, in combination with the site 236 data, allows us to assume a triple junction of the Somali, Indian, and Antarctic plates, which probably existed during the formation of anomalies 27 to 24 epoch and migrated in the SW-NE direction. According to isotopic and geochronological data devived by analysis of young eruptive rocks of the Seychelles microcontinent (Dickin et al., 1986), the maximum of magmatic activity took place just in this period. This fact is one more argument in favor of the old triple junction.

C. THE CARLSBERG RIDGE

We have systematically investigated the Carlsberg Ridge and both its slopes, with adjoining parts of the Arabian and Somali Basins from the Owen fracture zone to the equator (more then 1500 km) (Karasik et al., 1986b; Merkuriev et al., 1988).

Due to the regular system of profiles, we have made more then 300 crossings of the Carlsberg Ridge. This gave us an opportunity to study in detail the rift zone of the ridge and the structure of its floor as well as to investigate the display of tectonic inhomogeneities in the structure of the AMF.

Morphologically the ridge is a system of linear uplifts and depressions separating them, which make provinces of rift rocks and flanks. The fracture zones shifting the ridge are represented by transform valleys and prefault ridges and have complex floor morphology that shows the changes of spreading geometry.

The longitudinal uplifts are traced along the axis of some transformed valleys; these uplifts divide the valleys into two parallel branches. The greatest depths of the ocean floor (to 5000 m) coincide with the points where the ridge meets transformed faults. The fracture zones break the structure of the Ridge into the system of spreading cells; the width in the central part of the ridge (63° to 64° E) is 150 to 200 km and decreases to 25 km in the southeastern and northwestern direction. These faults, especially the most significant of them, at about 67° E, with the lateral shifting amplitude along them being about 60 km (the Academic Vernadskieye fracture zone), are traced perfectly in the morphology of the floor characterized by the wide spectrum of the negative shapes of relief.

As an anomaly magnetic map (Figure 15) shows, the Carlsberg Ridge axis, throughout its extent, is marked with confidence by the intensive negative anomaly that has the greatest (within the ridge) amplitude (600 nT). The AMF of the ridge on the whole was formed by the system of mainly linear negative and positive anomalies which are parallel to each other, then trending coincides with the NW-SE (130° NE) trending of the ridge axis, its rift valley, and axial magnetic anomaly. The intensity of the anomalies within the ridge mainly decreases from its axis and rift rocks to the flanks. This process agrees with the generalized regional topographic profile of the ridge.

The high linear structure of the field, occurring on the Carlsberg Ridge and its flanks, is broken by numerous transformed faults, which divide the ridge into separate blocks of the same type. As a rule, the lineation of the field inside each block is very high. This is obtained by ensemble of the coherent profiles belonging to this block. The whole sequence of anomalies from one block is shifted toward the next block by the transverse fracture zones, which in the floor relief are adequate to the linear depression framed by prefault ridges. Shiftings along the faults in the northwestern and southeastern part of the Carlsberg Ridge are essentially different. This fact indicates inhomogeneity of the tectonic processes along its trending axis. In sum, in the investigated part of the Carlsberg Ridge we have distinguished about 20 transformed faults. Most of these are left-side faults and only in two cases in the central part of the ridge are there right-side faults.

On the whole, according to its structure, magnetic anomaly of the Carlsberg Ridge is typical ocean field. Local inhomogeneities in the AMF structure and in floor relief show the inhomogeneities of spreading at modern stage. The presence of numerous transverse faults indicates the intensive disjoint tectonics and block structure of the ridge.

The intensive negative magnetic anomaly that coincides with the axis of the ridge, in combination with the floor relief data, make it possible to judge the position and trending of the constructive boundary of the Indian and African plates, previously defined only by the earthquake epicenters and bathymetric data (Barazangi et al., 1969; McKenzie and Sclater, 1971). Taking into account the width and trending of the Ridge (130° NE), the sign of this anomaly shows that a direct magnetized body, formed during the epoch of Brunese direct polarity, is its origin. Linear magnetic anomalies, representing two complementary sequences with a symmetrical position about the Ridge axis, have appeared in the modern spreading centers and conform with ocean floor spreading during the last 26 to 28 million years.

The magnetic field of the southeastern part of the ridge is represented by the system of blocks of the same type, shifted relative to each other on fracture zones. High-linear easily defined anomalies (Figure 18) are developed within each block identified with spreading. We succeeded in identifying the complete sequence of anomalies from axial to anomalies 7 and 8 by the spreading model with a spreading rate of 1.2 to 1.6 cm per year. Due to identification and profile correlation throughout the ridge, a young sequence of anomalies was traced, and a map of the paleomagnetic anomalies axes was made (Figure 17).

The western part of the Carlsberg Ridge, to the east of the Owen fracture zone, represents a sequential system of spreading centers with elements of their overlapping. As a rule, the faults that divide the whole parts of the ridge in the floor relief are not traced in the transverse direction, and the shifting of anomalies of the same age is happening along oblique faults that are situated symmetrically about the axis of the ridge and that have a V-shaped form. The oblique faults in the structure of the magnetic field of the Carlsberg Ridge and their reflection in the relief of the floor as positive forms of the relief suggest that they are the result of the rift motion that migrates in the northwestern direction. The structure of the magnetic anomaly and relief of the western part of the Carlsberg Ridge floor indicate the complex rebuilding of the spreading centers that took place during the last stage of the northeastern Indian Ocean spreading.

To study the peculiarities of the rift genesis, we have investigated the traverses of the Carlsberg Ridge in its various parts that show all morphological signs peculiar to slow ridges. First, our attention was attracted by the axial part of the ridge with a well-defined V-shaped rift valley for which the modern tectonic and volcanic activity is arranged. Along the entire part of the ridge that we studied, we succeeded in tracing a rift valley with the relative depth of the bottom varying from 1500 to 3500 m. The valley width also varies from 5 to 20 km along the trend. On some isolated parts of the northwestern part of the ridge, the rift valley has a more complex structure, and its form becomes W-shaped. On several parts of the ridge, we managed to study in great detail the structural peculiarities of the W-shaped valley and to reveal the regularities of its change. On the given part of the ridge in its southern part, the rift valley has an asymmetric W-shaped form, and it is formed by two depressions separated by a small uplift, with the left depression deeper by 200 to 300 m than the right one. On the profiles with mid-position, the shape of the valley becomes symmetric with its left and right parts having a relative depth of 3500 m. Finally, moving farther to the north, the valley again takes an asymmetric form, but this time the right depression is deeper by 200 to 300 m than the left one.

Consequently, if we suppose that the maximum tectonic activity in the rift W-shaped valley, formed by two parallel depressions, concentrates in its deeper part, then on the given part of the ridge (1) the maximum of spreading component is arranged to its left part in the south, (2) in the middle the left and right parts of the valley are of the same activity, and (3) in the north of this part of the ridge the right part of the valley becomes more active. The changing of tectonic activity along the trend of the ridge is also represented in the magnetic field (Figure 19). If we follow the changing of the shape of the rift anomaly on the corresponding magnetic profiles, we can easily notice that this change depends directly on the relief changing of the rift valley.

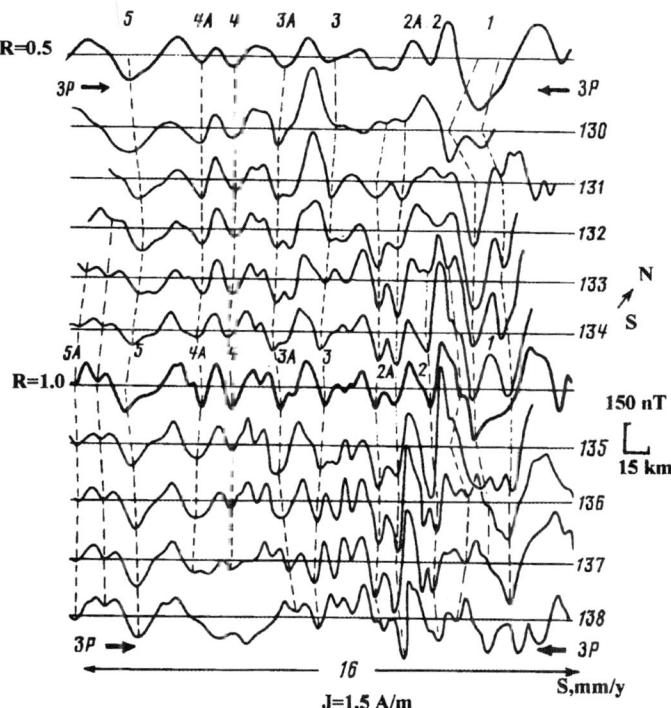

Figure 18 Typical magnetic profiles of one of the spreading cells on the southwestern slope of the Carlsberg Ridge, and their geochronological interpretation.

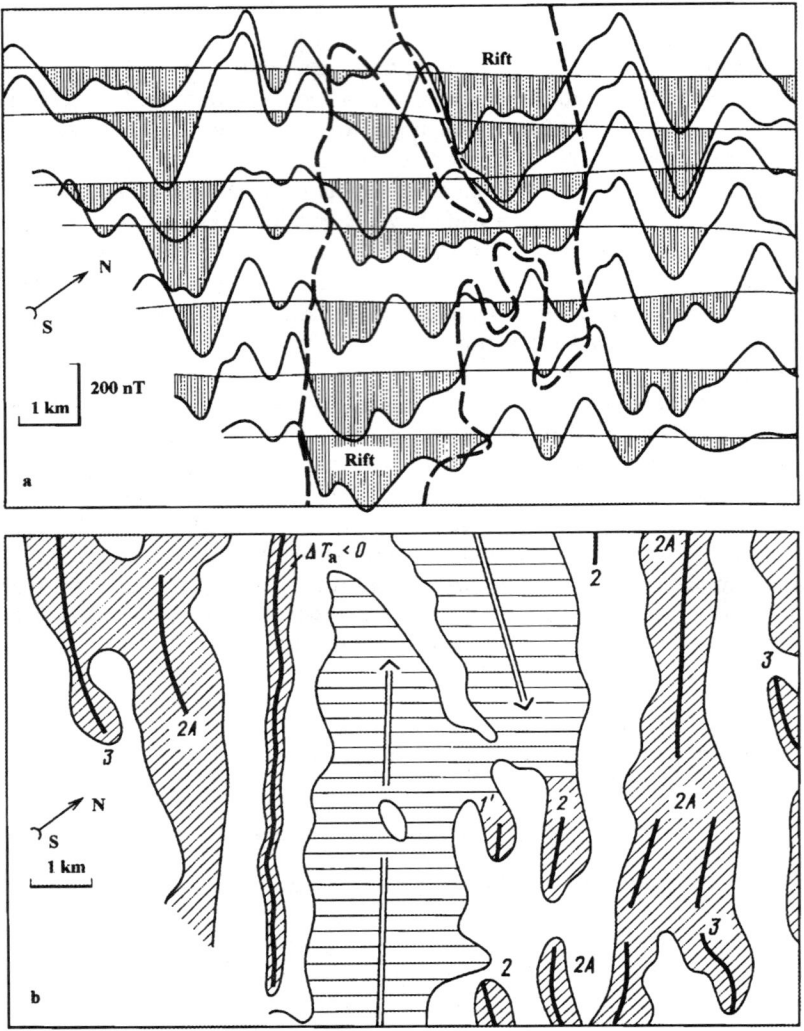

Figure 19 Overlapping of spreading centers at the Carlsberg Ridge near 60° E, and its reflection in bottom topography and AMF. (a) ΔT_a; (b) zero isolines; (c) topography plot.

It is reasonable to relate the changing of the rift anomaly shape along its trend to relief of the upper edge of the magnetic active layer and to inhomogeneous magnetization caused by asymmetric volcanic activity. The correlation between bathymetric and magnetic profiles makes it possible to reveal the following regularity: the minimum of magnetic anomaly coincides with the maximum depths within the rift valley. This fact shows that the greatest volcanic activity and, consequently, the biggest values of magnetization of basalt agree with a deeper branch of the W-shaped rift valley. Such overlap of spreading centers discovered on the "slow" Carlsberg Ridge, can be considered an analog of the overlap of spreading centers discovered on the "fast" rift of the Galapagos zone of spreading (Macdonald and Fox, 1983). This fact supports the genetic unity of the spreading process on "slow" and "fast" rifts and the versatility of the rift-genesis model for analysis of evolution of any regions of the World Ocean.

During the evolution of the Indian Ocean, stopping of activity on the old spreading axes with the simultaneous initiation of the new spreading centers took place repeatedly. This created the paleorift framed with the old symmetric sequences of linear magnetic anomalies. An example is the Mesozoic sequence of anomalies in the northeastern part of the Somali Basin and the near-western coast of Australia, the late Cretaceous Mascarene Basin anomalies, and anomalies in the northeastern part of the Indian Ocean with the youngest anomaly being 20 (about 46 million years ago) (Karasik and Sochevanova, 1981). Now more data appear, testified local jumps of the spreading axis, along with named higher examples of jumps of spreading axes connected with the global rearranging in the plate movement within Gondwana. As a rule the amplitude of these shifts does not exceed the first tens of kilometers, and the jumps themselves take place on the ridge axis within the spreading cell.

The jumps of spreading axis that were recorded in the magnetic anomaly field occurred in the studied part of the northwestern Indian Ocean during various stages of spreading about the Carlsberg Ridge. The possible jumps of spreading axis in the early stage of spreading can be indicated by the "drop-outs" of some single anomalies, observed in the northeastern part of the Arabian Basin, which we explain as the axis jump causing the addition of part of the Indian plate to the African plate. Then, during the rearranging of geometry of the spreading axis from its sublatitude trend to its modern NW-SE trend, the axis jump occurred as the result of the movement of the young rift to the northwest in the direction of the Aden Gulf. It seems likely that as a consequence of this process some fragments of the young part of the Indian plate spilled from it and added to the Somali plate. The lack of data on one sequence of anomalies often makes it possible to predict only an excess of anomalies or a lack of them in the complementary sequence; that is why the spreading axis jumps substantiated by sequences of anomalies are the most reliable. The overjumps, occurring on the modern stage of spreading and embodied in the floor relief, which has not changed yet, are of the most interest.

In the northwestern part of the Carlsberg Ridge we discovered a spreading jump, which was 2 to 3 million years old. On the correlation scheme of magnetic profiles crossing the axis of the ridge (Figure 20) we can see that the axial anomaly, which is clearly shown on all four profiles, has shift with an amplitude about 20 km. The older anomalies, number 3 and older (3A, 4, 4A, 5, 5A) can be traced on all profiles without any shift. Thus, there is the asymmetrical location of the anomalies on two southern profiles. According to this location, anomaly 2 follows the axial anomaly in the left sequence, whereas the right sequence has an excess of anomalies in the corresponding interval 0 to 25 million years. Consequently, the drop-out of anomalies 2A in the left sequence is compensated by the occurrence of an "extra" anomaly 2A on the part of the profile on the right of the ridge. The presence of two equal anomalies 2A on the same side of the axial anomaly testifies that the overjump of the spreading axis from one position into another (the new position is marked by a star on Figure 20) had taken place during the period 2.5 to 2.0 million years ago, which corresponds to the interval of the reversed polarity between anomalies 2A and 2. For some time after the overjump the "old" and "new" spreading axes were active simultaneously; then the activity continued spreading only in the new center. On the model profiles we show (1) symmetrical spreading during the epoch of anomalies 3 (4 million years) and 2A (2.5 million years) and (2) the jump of the spreading axis after anomaly 2A.

The overjump of the spreading axis, discovered as the result of inhomogeneity analysis of the structure of the magnetic anomaly field, was also embodied in the floor relief. We can clearly define a deep V-shaped rift valley with the rift rocks on both sides on bathymetric profiles G-29 and G-31 crossing the axis of the ridge relative to which the growth of the ocean floor took place within the steady-state spreading regime. The typical structure of the rift zone of slow ridges, observed on profiles G-29 and G-31, changes substantially on the non-steady state-spreading part. The non-steady state of spreading expressed by disruption of the linear magnetic anomaly order gave birth to complex floor relief. The rift valley has not already had its usual structure, and its position is more easily defined in the magnetic field by a marked axial anomaly. After the overjump of spreading axis, the new rift has not had time to make it self-evident in the floor relief, and the simultaneous activity in the old and new centers of spreading caused the shaping of positive relief forms on the part of the crust between them.

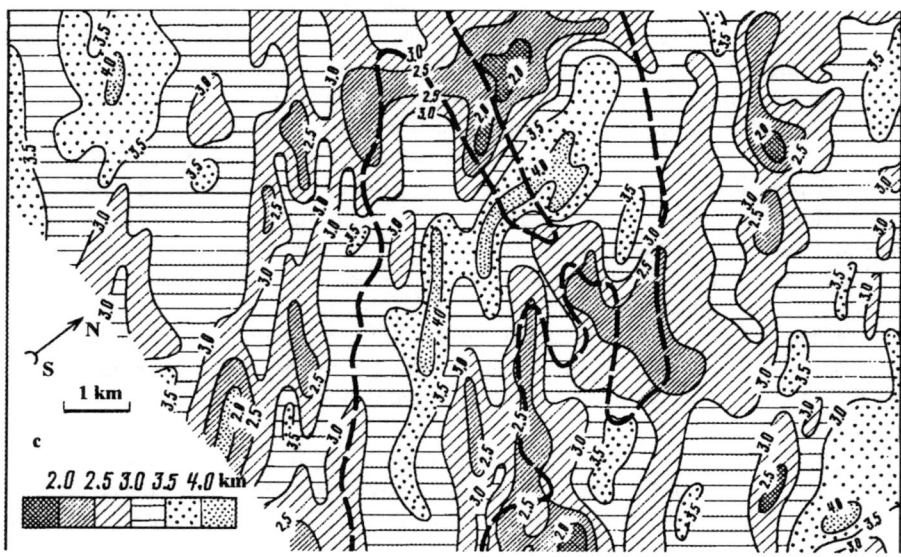

Figure 19 (continued). Overlapping of spreading centers at the Carlsberg Ridge near 60° E, and its reflection in bottom topography and AMF. (a) ΔT_a ; (b) zero isolines; (c) topography plot.

Analysis of magnetic anomaly fields and floor relief on the studied part of the ocean indicates that the overjump of spreading axis that occurred 2 to 2.5 million years ago in the northwestern part of the ridge was not the only one in the history of spreading of the ocean floor relative to the old and modern centers of spreading. As a result, the picture of spreading is asymmetrical on the isolated rift crossings along the drift line in both magnetic anomaly and the floor relief.

The revealed overjumps of spreading center within the stage of changing its old outlines into modern ones, and the relationship between the ages of old and modern sequences of linear magnetic anomalies testifies that the mechanism of the Carlsberg Ridge outline realignment differs significantly from the earlier models of geometric changing of spreading centers (Atwater and Mudie, 1973). The present views about the mechanism of these changes were formulated by them and others (Atwater, 1981; Hey, 1977, 1979; Hey et al., 1980) within the concept of propagating rift (PR).

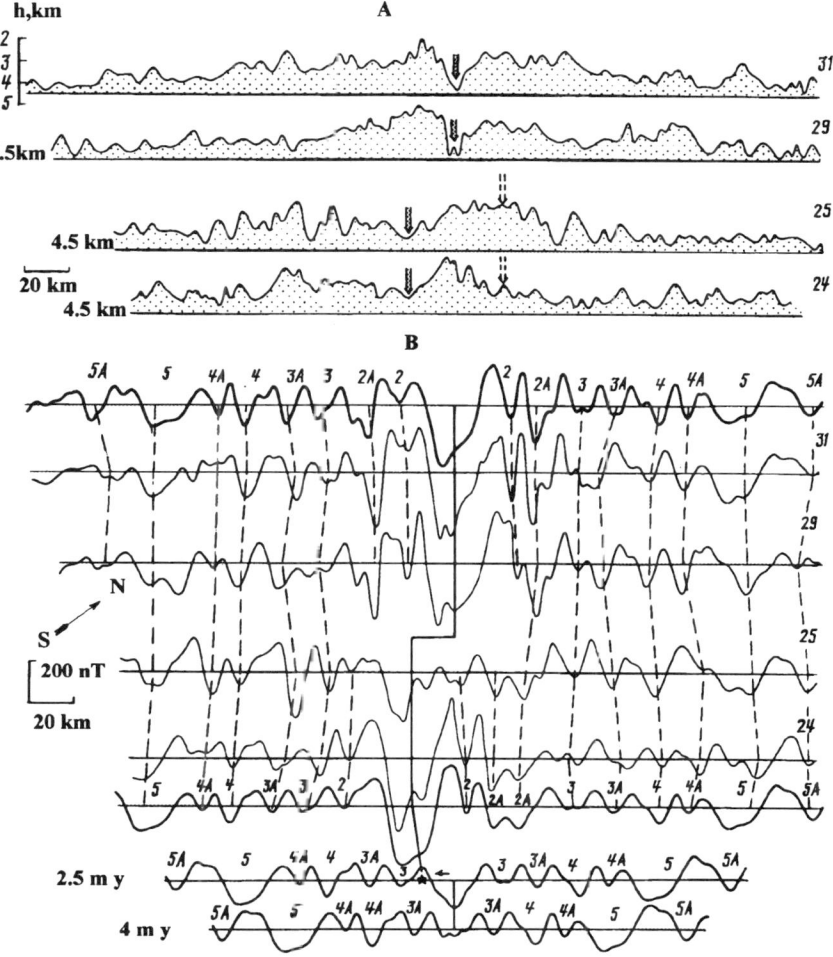

Figure 20 (a) Bottom topography and (b) AMF.

At present there is universally accepted view that the jumps of spreading center are the form of rift-propagating process (Hey et al., 1980). As a consequence of our investigations (Karasik et al., 1986, 1988), we proposed that during the shaping of the ocean crust of the Arabian Basin 40 to 25 million years ago, there was a substantial spreading non-steady state (Figure 21), expressed in the wide development of the rift-moving process, which determined the constant interchange of crust material between the Indian and African plates.

It is likely that the propagating of new rift also helped change the outlines of the Carlsberg protoridge to their present configuration. As the relationship between the oldest anomalies of the young system and the youngest anomalies of the old system indicates, the nonstop changeover is exhibited only in the extreme eastern part of the investigated area, where anomaly 9 follows anomaly 10 (Figure 21). Just to the west of the eastern most fracture zone on the ridge flank, anomaly 10 is followed by anomaly 8; farther to the west, anomaly 13 is followed by anomaly 7; and finally, at the extreme west of ridge flank between closely located anomalies 18 and 7, it has been impossible to distinguish any other anomalies (Figure 21). The easiest explanation of such a disposition of the axes is the progressive movement of the new rift into the lithosphere, previously shaped on the protoridge axis. As this takes place, azimuths of the old and new spreading centers differ so much that the new rift center invades only into the Indian plate, repelling the fragments of old lithosphere from it and connecting them to the African plate, where they should be searched for now.

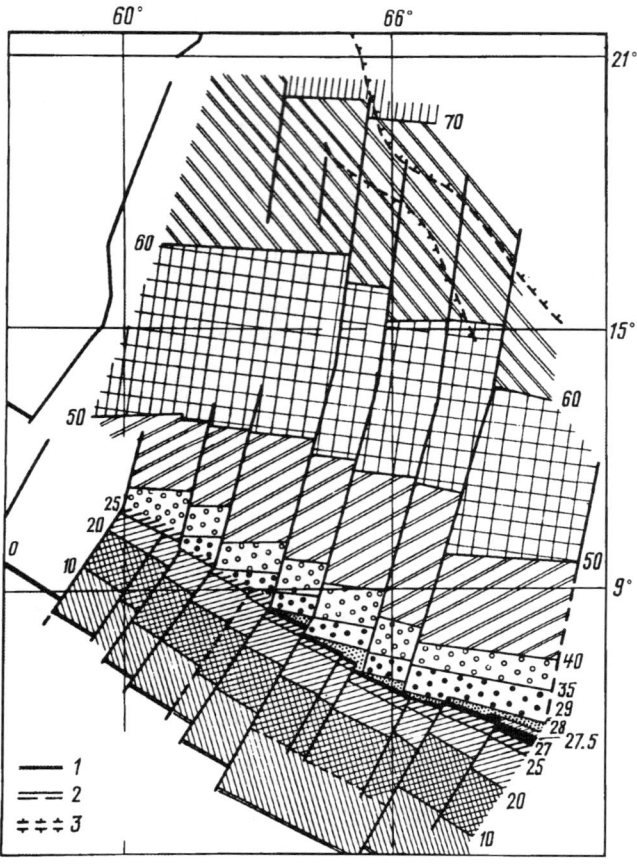

Figure 21 Isochron chart for the Arabian Sea: (1) axis anomaly; (2) fracture zone; (3) basement uplift.

This means that the above-described breaks of "normal" sequence of axes at the interfaces between two systems are caused not by the stop of spreading, which the previous investigators (McKenzie and Sclater, 1971; Norton and Sclater, 1979) assumed, but by discordant change from the protoridge outlines to the update outlines, which took place in the form of propagating of rift.

As our studies of anomalous magnetic structure and floor relief of the Carlsberg Ridge showed, less radical changes of spreading centers occur on the present stage of spreading (Mercuriev et al., 1988). According to the model, offered by Hey et al. (1980), the PR process brings into existence in the magnetic field structure the skewed lineations of V-shaped pairs of magnetic anomaly shifts, which were named pseudofaults. They spread skewed relatively spreading ridge and have orientation in the propagation direction. The V-shaped boundaries of these pseudo-faults represent the jumps of age and indicate the traces of the end motion of the moving rift and trace the motion of the plates relative to the end of the moving rift (Figure 21). In the floor relief the process of PR must also cause the occurrence of linear morphostructures with skewed orientation relative to rift tending. The disappearance of the rift causes the occurrence of some bathymetric trend in the floor relief. It represents graben structure (left rift), which extends skewed from the end of the dying center of spreading. The sequential system of spreading centers and pseudo-faults, marked by V-shaped structures in the magnetic field and the floor relief, which we discovered in the northwestern part of the Carlsberg Ridge, testified that in the present stage of spreading the isolated parts of the ridge axis have activity in both longitudinal and cross directions. Judging from the character of the shifts along the pseudo-faults and their orientation, the growing rift was moving in the northwestern direction spasmodically, with the rate close to the modern rate (1.2 to 1.5 cm per year).

Thus, the process of rift propagation played an important part in the formation of the magnetoactive layer and the structure of the AMF induced by it during the stage of spreading change of the ridge axis as well as during the present stage.

IV. THE CENTRAL BASIN OF THE INDIAN OCEAN

The Central Basin of the Indian Ocean occupies the oceanic part of the Australia-India plate and covers a considerable part of its floor. In the north-south direction, it extends from the Indostan Peninsula to the Mid-Indian Ridge crest, whereas in the west-east direction — from the submeridional Maldive Ridge to the parallel Ninety-East Ridge.

In geomagnetic terms, the Central Basin is studied poorly. By now, only basic geomagnetic survey has been carried out there (McKenzie and Sclater, 1971; Shreider and Trukhin, 1981). From the data analyses, several areas with different types of magnetic field anomalies can be distinguished there: discrepancy in the anomalies is from hundreds to thousands of nanoteslas over the most farthest basin south of the equator. The Afanasiy Nikitin Seamount is characterized by intensive magnetic anomaly, whereas the Lanka Ridge is not expressed in the magnetic field. North of the equator, northeastern elements are also observed against the background of meridional geomagnetic anomalies.

Two systems of magnetic lineations are distinguished in the Central Basin proper. The young system of magnetic lineations is related spatially and genetically to the modern spreading axes of the Arabian-Indian and Mid-Indian Ridges. In the southern part of the basin, anomalies of that group are identified up to 15; they have northwestern strike. Latitudinal magnetic lineations are recognized within the rest of the Central Basin; they are identified as anomalies 21 to 34, the age of which increases northward. The most ancient lineations in the Central Basin are shifted by a series of submeridional transform faults, namely Chagos, Indrani, Indira, and 86° E.

In 1976, a component geomagnetic survey was carried out during the 58th cruise of R/V *Vityaz'* (Shreider and Trukhin, 1981). The data processing resulted in estimating the inclination of the vector of rock magnetization and magnetic declination of its horizontal projection for meridional transections of magnetic lineation 30. Lineations were crossed in the longitudinal direction between 75 and 85° E and in the latitudinal interval between 9° S and the equator. The estimated magnetic inclination was 51° for one of the meridional profiles and 55° for another. The estimated paleolatitudes are 32 and 35° S, respectively; the azimuth of the opening axis was about 125°.

In 1984, in the course of the 31st cruise of R/V *Dmitry Mendeleev*, the detailed geomagnetic survey was carried out within two study areas in the Central Basin simultaneously, with gravimetric and geomorphologic survey, continuous seismic profiling, deep seismic sounding, and geological studies (Gorodnitsky et al., 1990c).

The bulk of the research was performed within study area V bounded by 3°10-5°10 S and 78°30-81°30 E. In the longitudinal direction, the area is located between the Indrani transform fault crossing its western margin and the Indira transform fault. That area of 2 by 2.5° was chosen for a detailed study of a fragile fold structure in the region of 4° S and 80° E detected in the previous cruises. It is a basement uplift 1.0 to 1.2 km high, bounded by faults and buried under the sediments 1.1 to 1.2 km thick. In the modern topography, it corresponds to a gentle high of over 500 m, with the slopes no more than 2. The high is complicated by gentle folds with altitudes of 30 to 70 m confined to the sublatitudinal depressions. North of the high, the basin bottom is uplifted at 50 to 200 m as compared to that in the southern direction (Neprochnov et al., 1988). From the west and east, the folded basement of the uplift is bounded by ancient dislocations of presedimentational origin. From the data of a continuous seismic profiling, that is verified by a gentle leaning of reflectors against the step-like fault scarps with altitudes of 500 to 700 m, along which the basement top is shifted. Dislocations, bounding the uplift from the north and south, are expressed on meridional profiles by a series of upthrust basement scarps with the altitudes reaching 500 m. In the sedimentary layer, they correspond to the asymmetric folds with steep limbs facing the depression.

Strikes of the principal tectonic elements are distinctly shown on the map of basement topography compiled by O. V. Levchenko (Neprochnov et al., 1988). A separate dislocated crust block, within which a folded uplift is sited, has a shape close to a regular parallelogram with the sides 100 and 140 km and bounded by subparallel faults, namely, ancient submeridional faults and young northeastern ones. Upthrusts and asymmetric folds have a sublatitudinal strike and are spaced in an echelon pattern along the principal northeastern dislocations.

An anomalous structure of the crust and upper mantle has been revealed within the study area from the results of Deep Seismic Sounding (DSS). Layers with velocities of 2.0 (sediments); 6.1; 6.9; and 7.6 km/s and corresponding thicknesses of 1.1; 1.3; and 4.2 km are detected there (Neprochnov et al., 1988). The thickness of the crust down to the Moho-boundary, characterized by a velocity of 7.6 km/s and possibly corresponding to the lower thick layer of the crust composed of serpentinite, is 6.6 km.

A normal section of the oceanic crust has been obtained within the study area for the basin part with an even bottom that has not experienced any considerable deformation. The section mentioned is composed by layers with the velocities of 2.0 (sediments); 5.6; and 6.6 km/s and corresponding thicknesses of 1.6; 1.6; and 4.6 km. The total thickness of the crust down to the Moho-boundary, characterized by seismic wave velocity of 8.2 km/s, is about 8 km.

Geomagnetic survey within study area V was performed along the grid of sublatitudinal and submeridional profiles spaced at 10 to 15 miles apart, on average (Figure 22), and also along the supplementary profiles during the DSS. Satellite navigation was used during the work. The total length of geomagnetic survey was 2260 miles. Geomagnetic survey was performed by a tow proton magnetometer MPM-5M. To better account for a head deviation effect at the cable of 300 m long, it was measured within an area with a normal magnetic field. A full span of a deviation curve within the study area is 7 nT. From the estimate, the RMS of the survey does not exceed 12 nT.

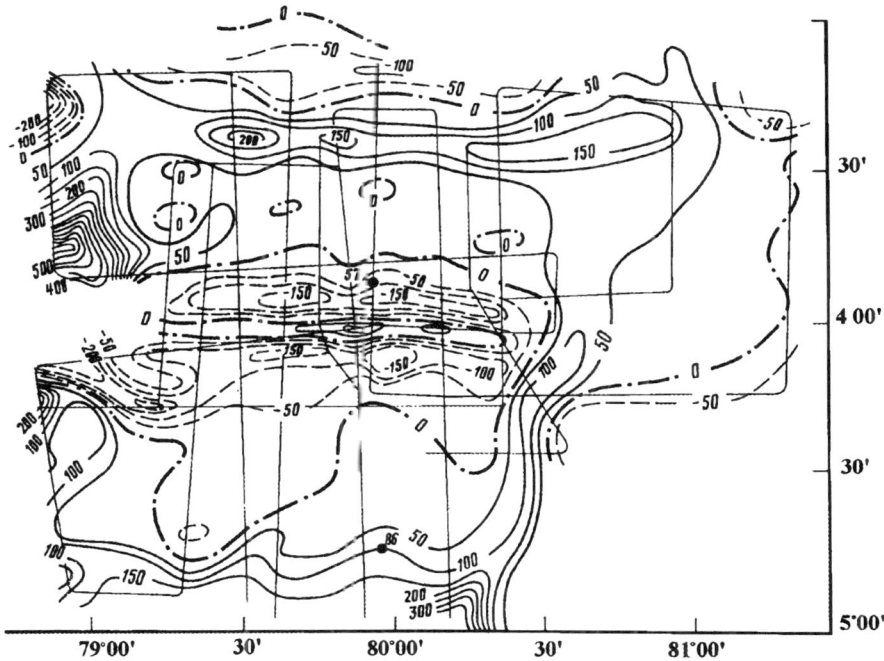

Figure 22 Isodynam map of anomalous magnetic field ΔT_a for the Central Basin of the Indian Ocean, study area V. Isodynams are drawn for each 50 nT. The map shows points of heat flow measurements and heat flow density in mW/m².

The anomalous field within study area V has a complex character (Figure 22). Its anomalies vary from 100 to 1000 nT and from 5 to 20 miles in the period. In the central part of the study area, that is, between 79°30 and 80°30 E, meridional profiles show distinct magnetic lineations, which can be identified as anomalies 32B and 32 (69 to 71 million years, early and middle Maestrichtian).

To specify the anomalies, an active magnetic layer has been simulated by a computer technique for which magnetic field graphs were used along the meridional profiles approaching the magnetic lineations across their strike. Spreading velocities were tried at various magnetic inclinations (Gorodnitsky et al., 1990c).

From the simulation results, it appears that magnetic lineation 32B (69.2 to 69.4 million years) was registered in the central part of the study area. In its northern part, anomaly 32 (69.6 to 71.0 million years) was also registered. An average spreading rate in the mentioned time interval is 4 cm per year. Anomaly 31 is located south of the study area. Therefore, we can suppose that a considerable increase of the spreading rate occurs between anomalies 32 and 31. That is verified indirectly by the results of geomagnetic survey performed between the transform fault of 86° and the East Indian Ridge. According to these results, the spreading rate in the lithosphere block between the mentioned structures within the segment of 32B-32, is 5.7 cm per year, and within the segment of anomalies 30 to 31, it increases to 12 cm per year. Meantime, an anomalous structure of the crust within the basement uplift at the study area, namely, the occurrence there of a thick layer with seismic wave velocity of 7.6 km/s, intensive dislocations of the sedimentary cover, and anomalous high heat flow (86 to 140 mW/m²), and also compressive stresses in that part of the Central Basin, allows an alternative explanation of a spatial convergence of magnetic lineations 32B and 32. It may result from the evolution of nappe-overthrust structures due to submeridional compression in the rear of the Indian plate and sliding of the crust blocks in the submeridional direction along the mechanically weakened serpentinite basement in accordance with a two-layer tectonic model by Lobkovsky (1988). In that case, a crust destruction may partially occur, which would lead to a convergence of the mentioned lineations. This problem requires further investigation.

Intensive submeridional anomalies with amplitudes to 900 nT are detected at the latitudinal profiles in the western and eastern parts of study area V. The most intensive of them can be traced distinctly in the longitudinal interval of 79°00 to 79°30 E and seems to be confined to the Indrani transform fault. As follows from the displacement of the anomalies along that fault (Figure 22), its western side is shifted for approximately 160 miles northward. Judging from the lineation 32B displacement, the lithosphere at the eastern side of the Indrani fault is 4 million years older than that of the western side. From the theoretical estimates (Gorodnitsky, 1985), the depth of the basement at the western side would be 0.7 km less than that at the eastern one; that is verified by the CSP data.

The second anomalous zone of a submeridional strike is recognized during the magnetic survey in the eastern part of the study area and extends approximately along 80°30 E. East of it, no sublatitudinal lineations are traced. Therefore, we may suppose that one more transform fault unknown earlier runs between the Indrani and Indira transform faults and crosses the Central Basin along 80°30 E.

Analysis of the isodynams ΔT_a map (Figure 22) shows that in the central part of the study area, lines of isodynams of 0 to 50 nT correspond well to the contour of the basement topography uplift recognized from the CSP data.

Two measurements of heat flow density were taken at study area V (Figure 22). The measured value, 57 mW/m^2 in the dome part of the fold uplift, agrees well with the theoretical estimate for the corresponding thickness of the cooling lithosphere (Gorodnitsky, 1985). An anomalous heat flow density of 86 mW/m^2 was obtained in the southern part of the uplift. An anomalous heat flow density of 140 mW/m^2 was registered in the northern boundary of the study area during the previous research. From the experience of the comprehensive geological-geophysical interpretation performed for the Azores-Gibraltar fracture zone in the North Atlantic (Verzhbitsky et al., 1989b), the occurrence of an anomalous heat flow in the area of nappe-overthrust structures may result from the dissipative heating in the compression zone.

Alongside the survey within the principal study area of the Central Basin, a site magnetic survey was carried out at study area VI, in its eastern part. Area VI was located between the Indira transform fault in the west and 86° E transform fault in the east and was bounded by coordinates 00°30 to 01°10 S and 83°40 to 84°40 E. Geomagnetic survey by MPM-5M magnetometer was carried out there alongside the sounding, gravity measurements, and CSP along the grid of SE profiles. The total length of geomagnetic survey within the area was about 650 miles. The RMS does not exceed ±12 nT. A schematic map of the magnetic field isodynams has been compiled as a result (Figure 23).

Judging by a 90-mile shift of magnetic lineations 31 and 32 along the Indira transform fault, located 2° west of study area VI, the latter should be spaced not far from anomaly 34 (80 million years old).

Within the study area, the bottom topography is even, with a mean depth of 4600 m. A northeast chain of seamounts is pronounced against that background, the seamounts are elevated over the bottom at a height of 1200 to 1400 m.

Within the study area, the magnetic field has a smooth character with small amplitudes and no sharp horizontal gradients (Figure 23). Local anomalies of the magnetic field with amplitudes of 250 nT are registered over the seamounts against those smooth fields. These anomalies have a dipole character with predominating negative values ΔT_a and seem to evidence a volcanic origin of the seamounts. That supposition has been verified by the dredging results: samples of alkaline trachybasalts were taken from the summit of one of the seamounts under study. The available number of profiles is not sufficient to delineate the anomalies for a representative choice of a geomagnetic model of seamounts. From the CSP data, the thickness of the conformed sediments at the rise of these seamounts reaches 2 km. So, the actual height of volcanic seamounts is no less than 3 km. According to the theoretical estimates relating the limiting height of oceanic volcanoes with the age of the bearing lithosphere, the great intensity of the AMF and great height of volcanic constructions suppose that the investigated volcanoes are superimposed over the lithosphere 15 to 20 million years old. That is verified well by the alkaline basalt differences sampled in the course of dredging. If we assume that the age of the bearing lithosphere is 80 to 85 million years, the age of the seamounts may be determined as 60 to 70 million years.

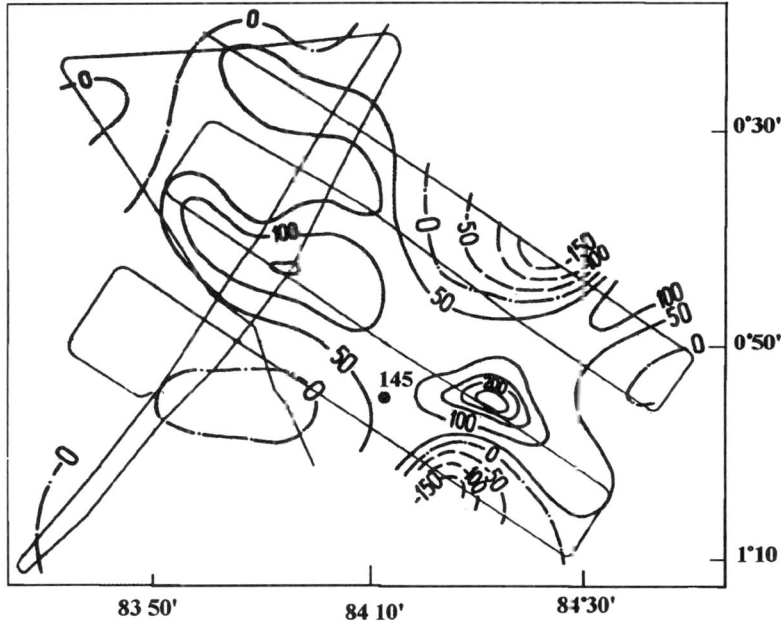

Figure 23 Isodynam map of anomalous magnetic field ΔT_a for the Central Basin of the Indian Ocean, study area VI. The map shows points of heat flow measurements and heat flow density in mW/m².

Due to the absence of detailed information on the seamount topography, a simplified simulation procedure by a dipole described above (Gorodnitsky et al., 1990c) was used for a paleomagnetic analysis of the geomagnetic survey results.

The obtained data allow us to conclude that the paleolatitude of the seamount at the stage of its formation was 31°. Thus, at the present equatorial position of the volcano, its meridional displacement is 30°, that is, about 3300 km, since the moment of its formation. If the age of the volcano corresponds to the late Maestrichtian-early Paleocene, the average rate of the Cenozoic meridional drift of the Indian-Australian plate was about 4.4 cm per year.

The dredging results of the seamount evidence of the subaerial character of a volcanic eruption. That allows us, using the data on the present-day depth to the mount summit ($h = 3200$ m), to estimate the time necessary for its subsidence together with the bearing lithosphere (Sorokhtin, 1974). According to formula, $h = 0.35 \sqrt{t}$. At $h = 3200$ m, we have $t = 81$ million years. Thus, both paleomagnetic analysis and theoretical estimate indicate that the seamount under investigation is an ancient paleovolcano of the late Cretaceous or early Paleocene age. Considering that the alkaline composition of volcanites and great height of the volcano evidence a younger age of the volcanic construction (less than 70 million years), we may suppose that the volcano or probably the entire volcanic chain subsided at a greater rate than the bearing lithosphere. That principal aspect may be explained by recent high tectonic activity of that region of the Central Basin, which is manifested as high seismic activity, occurrence of an anomalous high heat flow domain, and dislocations of the sedimentary cover.

It follows from the paleomagnetic analysis with the use of a magnetic dipole that the difference between the modern and ancient magnetic declinations for the volcano under investigation is 45°. That allows us to determine the northeast drift of the Indian-Australian plate during the Cenozoic time, which was complicated by a clockwise 45° rotation of the plate.

Geomagnetic survey with differential magnetometer DPM-2 was carried out within test areas and by traverses in the region of the intraplate dislocation during the 22nd cruise of R/V *Professor Shtokman*. The test area for geomagnetic survey was located between 2° S and 2° N, and between 82 and 84° E. Geomagnetic survey was performed along legs 1 and 2 as well.

A series of sublatitudinal magnetic anomalies from 21 to 34, shifted by a series of submeridional transform faults, namely Chagos, Indrani, Indira, and 86°, were identified in the analyzed part of the Central Basin. The Indira fault runs directly along the western slope of the Afanasiy Nikitin Rise. In that region, the age of the oceanic crust is determined by magnetic lineations 32 and 33B (71 to 77 million years, late Campanian-early Maestrichtian). The same age (78 million years) is detected from the deep drilling data in borehole 116 sited between the Indira and Indrani faults, 1° S and 81°24 E. Drilling results evidence the occurrence of intraplate deformations there. The differential magnetic survey performed within the study area revealed the intensive geomagnetic variations there, thus providing support for a gradient magnetic survey application in the near-equatorial part of the Indian Ocean.

A complex rugged topography of the AMF is clear from analysis of the charts plotted for the field. No signs of magnetic lineations were detected there. The discrepancy of magnetic anomalies that seem to have tectonic origin reach 450 nT there. In the northeastern part of the study area, intensive magnetic anomalies are confined to sea mounts and to the Indira fault zone. At the same time, within the analyzed study area, zones of northeast negative and positive anomalies are traced within the central and southern parts of the polygon, thus reflecting in plan the position of the fault systems with the 45° tilt toward the general meridional compression. The correlation between the magnetic field anomalies and crystalline basement topography is obvious. From the CSP data, a number of anomalies are confined to the basement depressions. One may suppose that these anomalies are confined to the fault zones filled by more magnetized matter and are segments along which intraplate motions occur. The crystalline basement uplift, recorded in the central part of the study area, also seems to be of tectonic origin and bounded by feathering faults. From geomagnetic data, the main fault is traced from profile 5 to profile 15. So, the preliminary qualitative analysis of the magnetic field anomalies allows the supposition that the latter are related mainly with the complex dislocated structure of the crystalline basement.

In the light of this, we should stress that from the DSS data, the total thickness of the oceanic crust in the study area is about 6 km (Neprochnov, 1988). Reflector II, characterized by velocities of 4.4 to 5.1 km/s, is 1.5 km thick, whereas the thickness of Reflector III (velocities of 6.2 to 6.8 km/s) varies from 3.5 to 6.7 km. The occurrence of anomalous high heat flow density (112 mW/m^2) in the southern part of the study area evidences the intensive intraplate dislocations within the area analyzed. One may suppose that in the environment of a relatively ancient oceanic crust of the late Cretaceous age and with no manifestations of recent volcanic activity, the heat flow anomalies there, as well as within study area VI of the 31st cruise of the R/V *Dmitry Mendeleev* (Belyaev et al., 1987b), may result from dissipative heating due to compression. This allows the supposition that the intraplate dislocations in the analyzed area of the Central Basin may also result from the formation of the nappe-thrust structures, according to the two-layer tectonic model (Lobkovsky, 1988).

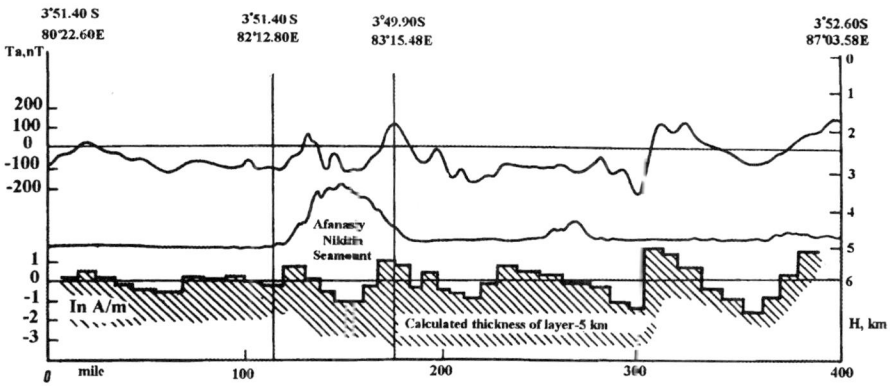

Figure 24 Geomagnetic profile across the Afanasiy Nikitin Seamount.

From analysis of the traverse survey (Figure 24), legs 1 and 2, transform faults are marked by typical anomalies of magnetic field. An intensive negative anomaly with an amplitude of 300 nT and width of 40 km was recorded over the 86° fault. A positive anomaly with an amplitude of 250 nT was recorded on the latitudinal traverse over the Indira fault bounding the Afanasiy Nikitin Seamount from the east. The second positive magnetic anomaly, with an amplitude of 160 nT, is confined to the rise of the western slope of that mountain. From the analysis of the magnetic field anomalies, a submeridional zone of magnetic anomalies with amplitude of over 200 nT may be delineated between 85°10 and 85°30 E. The comparison of these anomalies with those over the Indira and 86° transform faults made it possible to suppose one more deep fault there, probably of transform origin.

Effective magnetization of the active magnetic layer, the essential thickness of which was assumed as 5 km, was estimated along the mentioned legs. As was noted above, the thickness of Reflector II was determined from DSS data to be only 1.5 km.

The correlation of effective magnetization with actual I_n values of the measured magnetic parameters of the dredged samples may be achieved only when the thickness of the estimated active magnetic layer is doubled approximately, that is to 5 km.

During the 22nd cruise of the R/V *Professor Shtokman*, besides the traverse survey a detailed geomagnetic survey was carried out at the summit and on the rise of the western slope of the Afanasiy Nikitin Seamount proper. The summit part of the mountain is characterized by an intensive positive magnetic anomaly with an amplitude of 600 nT. There are practically no magnetic field anomalies on the rise of the western slope. Within both study areas, submersibles "MIR" uplifted basalt samples, the residual magnetization of which varied from 0.5 to 8 A/m, with the mean value being 2 A/m. Samples of picrite-basalt taken on dive 1/34 at Koenigsbergen factor Q are the most magnetized, that is, about 80. Trachybasalt was sampled at the very summit of the mountain. Value I shows the tendency to increase from the summit down the rise of the slope (from 1 to 2 A/m, near the summit, to 6 or 7 A/m, near the foot). Data from the traverse survey and estimation of effective magnetization and geomagnetic study of the Afanasiy Nikitin Seamount, as well as the survey of its summit in particular, provide evidence that the mountain is an intraplate basalt volcano of complex genesis. The age of the sediments adjacent to its base is dated as upper Cretaceous. At the same time, the mountain is characterized by the areas of direct and inverse magnetic polarity, with mean calculated I values about 1 A/m. One may suppose that the volcanic construction basement was set during the period of direct magnetic polarity in the late Cretaceous, whereas the later cycles of volcanic activity occurred during the Tertiary time.

V. THE TADJURA RIDGE

An advanced stage in the detailed geological-geophysical studies by operated submersibles (ROA) at the mid-oceanic ridges was started in the former U.S.S.R. in 1979 by the expedition to the Red Sea (IORAN), which acquired the name "PILAR". These expeditions have been regularly conducted since that time. In the course of these expeditions, spreading processes were studied in the Red Sea, at the Reykjanes Ridge, in the Tadjura Ridge, at the Juan de Fuca Ridge, in the vicinity of 26° N of the Mid-Atlantic Ridge, and in some other regions of the World Ocean (Monin et al., 1980; Zonenshain et al., 1981; Bogdanov et al., 1983; Monin et al., 1985; Valyashko et al., 1987a). Hydromagnetic study was an integral part of that research (Valyashko et al., 1985, 1987b, 1988; Shreider et al., 1990). As a result, it became possible to perform some reconstructions of the oceanic crust formation and of its main regularities. From analysis of magnetic properties of the rocks, attempts have been made to detect the correlation between magnetization in the active magnetic layer during its formation, and the AMF and tectonics of the study area.

A detailed geological-geophysical study of the axial part of the Tadjura Ridge was carried out over 2400 km^2 in 1983 to 1985 during the seventh and ninth cruises of the R/V *Akademik Mstislav Keldysh*. The Tadjura Ridge, located in the western part of the Gulf of Aden, has a sublatitudinal trend and extends to the African coast, where it comes onto the surface and acquires the name of the Azal Rift.

The complex geological-geophysical studies in the Gulf of Aden started during the period of the International Geophysical Year (Laughton, 1966; Laughton et al., 1970; Matthews, 1966). As a result, the strike of the Cheba Ridge and of transform faults crossing it at an angle of 40° was determined. Magnetic lineations to 5, inclusive, (10 million years) were recognized. A spreading rate equal to 2 cm per year was estimated.

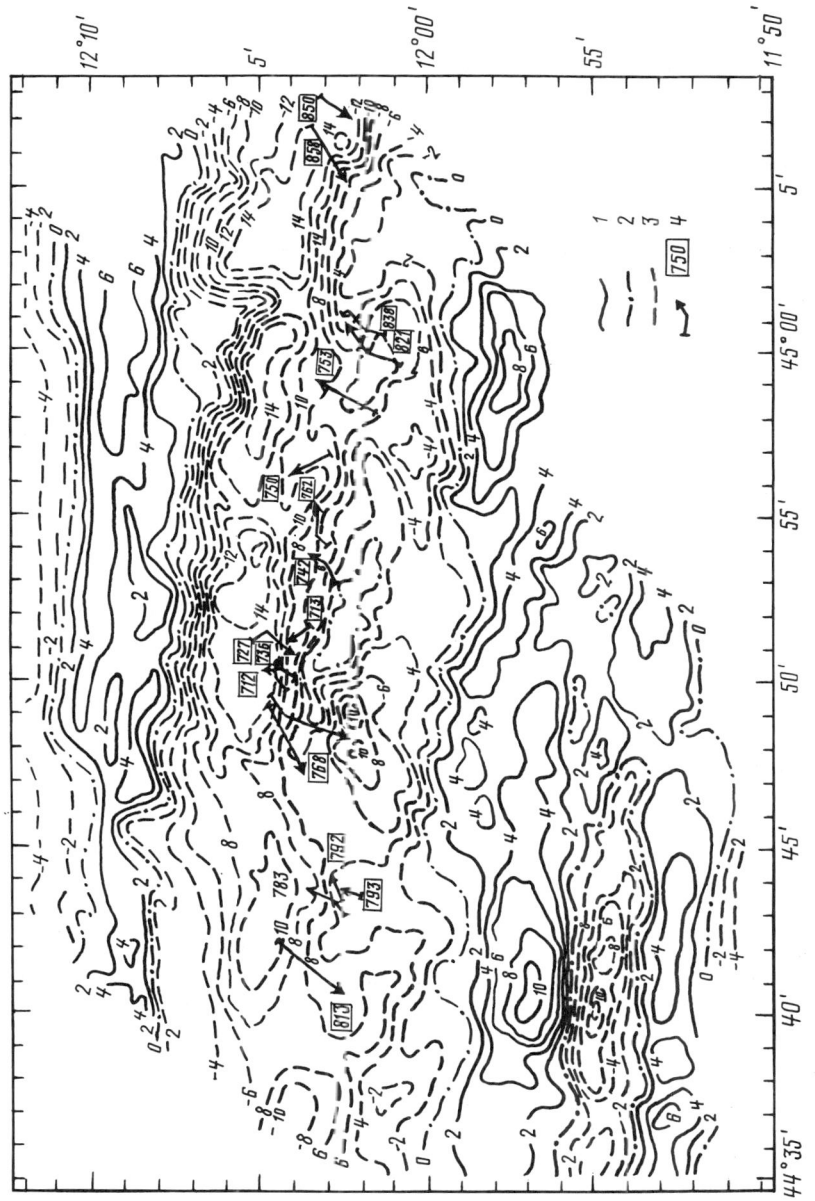

Figure 25 Isodynam map for AMF within the Tadjura study area. Isodynams are given in hundreds of nT. (1) Positive isodynams; (2) zero isodynam; (3) negative isodynams; (4) numbers and tracks of RAO "Pisces".

It has been assumed that the process of divergence between Africa and Arabia started about 20 million years ago. The final generalization of geophysical survey was made by Cochran (1981, 1982). Magnetic lineations to 2A (2.4 million years) in the north and to 3 (4.2 million years) in the south were mapped in the Tadjura Ridge.

The study area is located in the region of 45° E, in the area of the Tadjura Ridge and Cheba Ridge junction. Sixty hydromagnetic profiles spaced 0.5 to 1 km apart with the azimuth of 40° were performed there by magnetometer MPM-5M. The instrumental error was ±1 nT (Afonyashin et al., 1984). Navigation was realized by the Magnavox satellite system and by anchored buoys. The total navigational error did not exceed 130 m.

The AMF is characterized by a linear pattern. The map of the AMF is given in Figure 25. In the western part of the study area, the negative axial anomaly is 17 to 20 km wide. In its central part, it is divided into two, but further eastward it merges again in a single anomaly 15 km wide. The amplitude of the axial anomaly increases eastward from 1300 to 1500 to 1800 to 2000 nT. From the north and south, it is bounded by areas of inverse sign lineations altering with the distance from the axial anomaly. The amplitude of the peripheral anomalies also decreases regularly by one order: to 100 to 150 nT. In the topography, the area of the central anomaly corresponds to the most subsided area with a depth to 1400 m. There, an extrusive zone is recognized composed of volcanic constructions 50 to 100 m high and basement widths to 1.5 km. The inner rift is 4 to 5 km wide. To the north and south, the bottom is uplifted to a depth of 500 to 700 m, and it looks like a staircase made of listric faults, with their upper terraces being tilted 10° off the rift axis.

Comparison of the magnetic field anomalies and Tadjura bottom topography showed that they are related only in the regional plan. The axial trough corresponds to the central negative anomaly, whereas most large fracture zones are expressed both in the topography of the bottom and in magnetic field. The trends of magnetic lineations and of principal bottom features coincide. So, in the southern and northern parts of the study area, the trend azimuth of both is 278°, increasing to 310° in its central part. The degree of mutual correlation of magnetic field anomalies and its relations with the bottom topography is clearly seen from Figure 26.

Subparallel branches of the central anomaly are overlapped in the central part of the area at a distance greater than 20 km, whereas in the eastern part of the area, this distance is only 6 to 7 km. The amplitude of the anomalies in the southern branch is two to three times less than in the northern one. In the bottom topography, at profiles 12, 14, 16, 19, 21, and 24 (Figure 23), the axial trough also bifurcates. The southern depression subsides in the east-west direction from 1000 to 1390 m. The eastern part of the area is described by the same pattern. In the areas corresponding to the Shtokman and Keldysh faults, mutual correlation of anomalies is violated most distinctly.

Since intensity of the AMF is great there, it is easy to identify magnetic lineations. The central negative area is identified by a recent epoch, namely, Bruness. On both sides of it, a subchron Charamilio has been recognized, chrons 2 and 2a. In the southwestern part of the area, crust segments are 2.4 million years old. The northern boundary of the detailed survey coincides practically with the end of the subchron Charamilio, which is about 1 million years old. Further to the north, regional profiles covered the crust area of 2.8 million years old.

Amplitude of the axial anomaly should be of special concern. Sometimes it exceeds 2000 nT due to its proximity to the equator. Its value may be compared with that observed over the Reykjanes Ridge axial zone located in high latitudes. From the data by Vogt (1979), the Tadjura Ridge region belongs to so-called H zones, characterized by high concentration of ferrum oxidize and titanium in basalt. It seems interesting to compare magnetic properties of basalt in the Tadjura Ridge and the Red Sea. According to our materials, the mean value of the natural residual magnetization in basalt from the axial zone of the Red Sea is 35 A/m. Approximately the same value was determined for the basalt from the Tadjura (Table 3). Percentage concentrations of ferrum oxidize and titanium are, respectively, 10.99 and 1.14% in the Red Sea and 12.56 and 1.45% in the Tadjura Ridge (Al'mukhamedov et al., 1985). Average amplitudes of magnetic anomalies are 500 and 1800 nT, respectively. Thus, enrichment of basalts by ferromagnetite and different conditions for their crystallization (basalt plays the main role in the AMF formation) lead to a three- to four-fold increase of magnetic intensity in the Tadjura Ridge.

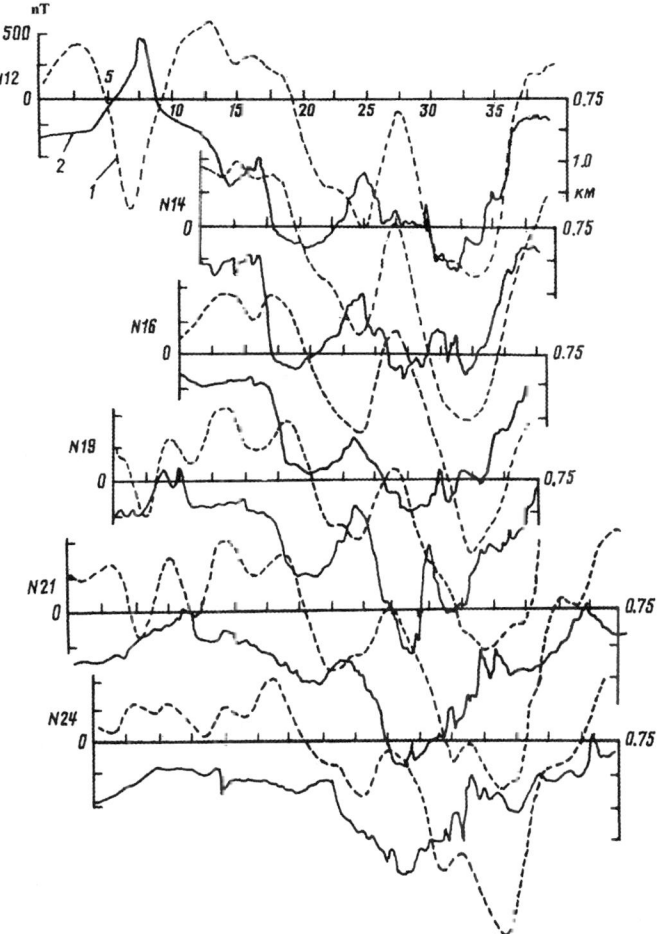

Figure 26 Superimposed profiles of bottom topography and AMF: (1) AMF; (2) bottom topography; Nos. 12 to 24 represent profile numbers.

When performing numerical simulations of the AMF and constructing models of an active magnetic layer, we accounted for the rotation angle of tectonic steps. From geomorphologic estimates, it is 8 to 10°. The technology of this estimate is given in

Valyashko et al. (1985). The best coincidence of the observed and simulated fields was achieved at the following effective magnetization inclination: 18° in the southern, 2° in the northern, and 10° in the axial parts of the study area. From the performed calculations, we obtained the instantaneous and finite spreading rates for the last 2.4 million years (Tables 4 and 5).

Table 3 **Measured intensity of normal residual magnetization I_n of basalts sampled from the Tadjura Ridge axial zone by RAO "Pisces"** [a]

No. of dives and samples	Value I_n, A/m	No. of dives and samples	Value I_n, A/m	No. of dives and samples	Value I_n, A/m
691/1	33.4	813/2	26.7	762/1	43.1
691/3	35.5	821/6	42.2	762/5	71.3
713/4	85.2	850/4	18.1	783/1	10.2
727/5	75.4	851/3	18.4	813/1	71.4
735/2	30.6	691/2	63.1	813/5	32.1
752/2	70.7	712/1	8.1	838/5	42.2
754/2	70.2	727/1	45.7	851/1	28.0
762/3	52.5	732/3	1.0	851/5	28.0
768/3	63.3	742/27	3.1		
793/1	2.4	753/5	43.0		

[a] $I_{n_{av}} = 46.4$ A/m.

Table 4 **Finite spreading rate in the Tadjura Ridge from simulation results**

Time interval, millions of years	Finite rate, cm per year
0.7	1.95
0.8	2.36
1.62	2.28
2.41	2.50

Table 5 **Instantaneous spreading rate in the Tadjura Ridge from simulation results**

Time interval, millions of years	Finite rate, cm per year
0.7-0	1.95
1.0-0.7	1.27
1.83-1.0	2.38
2.41-1.83	1.37

A map of isochrons is presented in Figure 27. The correlation between the trend of the principal fracture zones and of isochrons is a key moment in the understanding of the paleohistory of the region under study. Transform faults may be subdivided conventionally into two groups, namely, the ancient ones that ceased their existence 1.5 to 1 million years ago, and the young active faults that shifted magnetic anomalies to the Bruness epoch. Let us consider the changes that occurred in time. The southern part of the area is bounded by an isochron of 2 million years, the northern by one of 2.4 million years. Their trend is close to the latitudinal one, namely, 272 to 279°. In the central part of the area, younger isochrons 1.8 to 1.5 million years old and younger, have a different trend: 280 to 285°. However, we should note that the trend of old and young (in our terms) transform faults is the same and is 40° in azimuth. So, starting from the period of 1.8 to 1.5 million years ago, the spreading direction began to change, approaching the present-day azimuth of 298 to 310°. Therefore, we may conclude that structural change in the spreading trend and its adjustment (at least 2.4 million years ago) to the plate motion direction with a constant azimuth of 40° is the principal feature of the Tadjura Ridge paleoevolution.

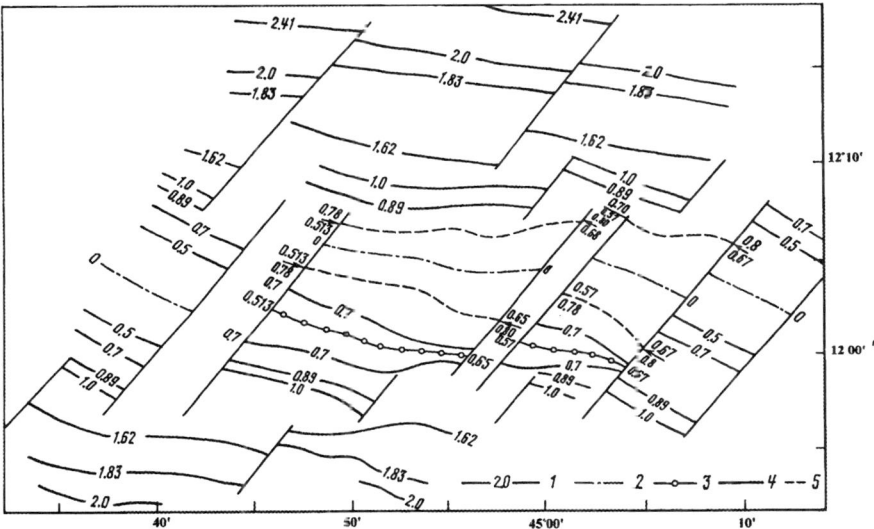

Figure 27 Isochron map of the Tadjura study area: (1) isochrons in millions of years; (2) present-day spreading axis; (3) paleoaxis; (4) transform faults; (5) outer boundaries of active rift branch.

Let us consider the experimental data by Oldenburg and Brune (1975), who watched the film cracking on the surface of the melted wax. By these experiments they showed that when the tension strain increases, there originate centers of opening from which rift evolution starts. A further process of crust splitting under conditions of oblique spreading may result in the formation of a series of echelon steps. Transform faults originate in the process of the further evolution of the spreading axis. They are first set in the areas of partial or echelon junction of rift structures. That is the situation we observe in the case of the Tadjura Ridge. The coexistence of the two subparallel rift zones, overlapping at a distance greater than 20 km, the occurrence of specific transform faults, namely, Shtokman and Keldysh ones, evidence most distinctly the change of the spreading trend in the Tadjura Ridge. Within the segment of a "classic" spreading in the west and east of the area, the opening axis is practically orthogonal to transform faults. This suggests that changes occur, not within the entire rift but in its individual parts only. Therefore, we may conclude that the structural change of the Tadjura Ridge is not completed yet and continues at present.

Let us generalize the factual data available and consider two alternative hypotheses of the melted material supply and formation of a new crust in the Tadjura Ridge. The first supposes that two subparallel spreadings, overlapping and active at present, originated due to oblique spreading and its further change. In fact, comparison of their bottom morphology and magnetic fields with the published data on the East Pacific Rise (Macdonald et al., 1984) supports this hypothesis. If so, such phenomena were first detected at low-spreading ridges with small productivity of basalt volcanism. The second hypothesis supposes that the subparallel spreading axes were formed one after another when the axis jumped for a new position and the old one became extinct during the structural change from the oblique spreading to an orthogonal one and due to westward motion of the rift. We believe the second hypothesis is more reliable due to some indirect reasons. The amplitude of the southern branch of the axial anomaly is two to three times less than that of the northern one. The data of numerical simulations show that the summary width of the anomaly at the time of the Bruness, in the area of subparallel rift zones, corresponds exactly to the width of the anomaly (in the Bruness) in the segments of a "classic" rift. In the framework of this hypothesis, it is possible to estimate the rate of the Tadjura Ridge westward propagation. According to our estimates, it is quite high, that is, 5 to 7 cm per year, on average. Comparison of the obtained data with the results of the research in the Galapagos Rift zone (Hey et al., 1986; Miller and Hey, 1986) allows us to reveal similar elements in their geometry. We may speak about the similar V shape of 0.7-million-year isochrons and of the outer boundaries of the northern spreading axis, which expose more ancient oceanic crust. In both models, the common thing is a supposition that the entire area may be associated with the zone of a complex transform dislocation; in our case, it is the area between the Shtokman and Keldysh faults, including the faults themselves. A higher rate of the westward propagation of the Tadjura Ridge, as compared to that of the Galapagos Rift zone, supports this thesis. However, more thorough comparison of the two structures does not seem possible due to the less sophisticated geophysical survey of the Tadjura Ridge when compared to that of the Galapagos Rift.

Let us consider a 2-million-year paleohistory of the Tadjura Ridge in the context of the second model. At that time, the spreading axis coincided with the southern branch of the rift zones. The rate of crust accretion was high, 2.5 cm per year. By 0.89 million years ago, the rate dropped to 2.36 cm per year, and in the Bruness time it was 1.95 cm per year. Analysis of instantaneous spreading rates showed that changes in the rate occurred about 1.5 million years ago due to reorientation of the spreading trend. As follows from Figure 27, at a time 0.67 to 0.57 million years ago, the existing spreading axis shifted sharply northward for 4 to 10 km, exposing a more ancient crust, and propagated farther to the west at a high rate of 5 to 7 cm per year, whereas the southern axis ceased its activity. For about 0.5 million years the generation of the active spreading axis corresponds to the northern branch of the mapped rift zones.

In conclusion, we would note that on the whole, the Tadjura Ridge is similar to other rift zones of the World Ocean, according to its morphology and magnetic fields. However, detailed geomagnetic studies have revealed a set of principal episodes in its paleohistory. Some of them, we believe, are poorly investigated in the context of modern knowledge of the rifting within low-spreading ridges.

Let us compare the obtained results with the Red Sea spreading characteristics (Zonenshain et al., 1981; Belyaev et al., 1981; Valyashko et al., 1987b, 1988). Hydromagnetic surveys, performed on the third cruise of the R/V *Professor Shtokman*, covered the area of 2500 km^2 (five study areas), thus controlling the Red Sea rift in the vicinity of 18° N.

Figure 28 shows an integrated map of the AMF curves. The rift axis coincides with the zone of the gradient maximum for the central anomaly. Correlation is violated at five segments, thus manifesting the transformed faults. All faults are expressed distinctly in the bottom topography. At four segments, dextral shift of the rift axis occurred. The azimuth of the fault trends is 57°, and all faults are orthogonal to the rift zone axis. Displacements are not great: about 1.5 km. The fault in the vicinity of 18° N is an exception; its displacement is 7 km.

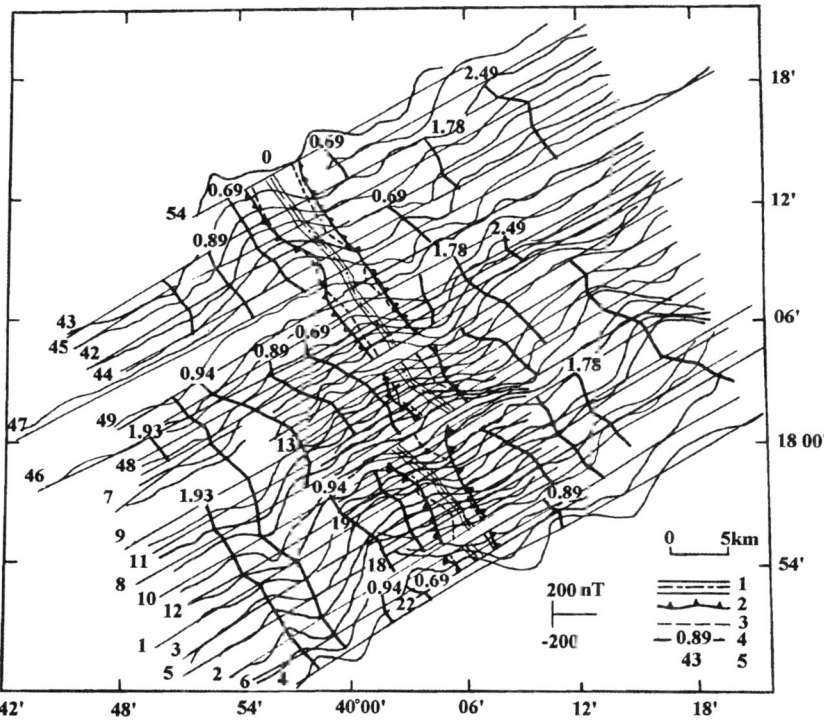

Figure 28 Integrated map of AMF curves of the 18° N study area: (1) present-day spreading axis; (2) boundaries of the central uplift from geomorphological data; (3) boundaries of the central uplift from geomagnetic data; (4) axes of correlated anomalies; (5) numbers of profiles.

Table 6 **Finite spreading rate in the Red Sea from simulation results**

Time interval, millions of years	Finite spreading rate, cm per year		
	African plate	Arabian plate	Finite rate
0.69	0.62	0.52	1.15
0.94	0.82	0.76	1.6
1.378	0.76	0.63	1.4
2.48	0.82	0.72	1.5

Table 7 **Finite spreading rate in the Red Sea from the correlation of magnetic lineations**

Time interval, millions of years	Finite rate, cm per year
0.5	1.15
1.0	1.73
2.0	1.9
3.0	2.0

Results of estimates of finite spreading rates (Tables 6 and 7) showed that formation of a new crust in the Red Sea is an irregular and asymmetric process. The rate of crust accretion is approximately 0.1 cm per year higher from the African plate side, than from the Arabian one. The modern spreading rate is 1.15 cm per year.

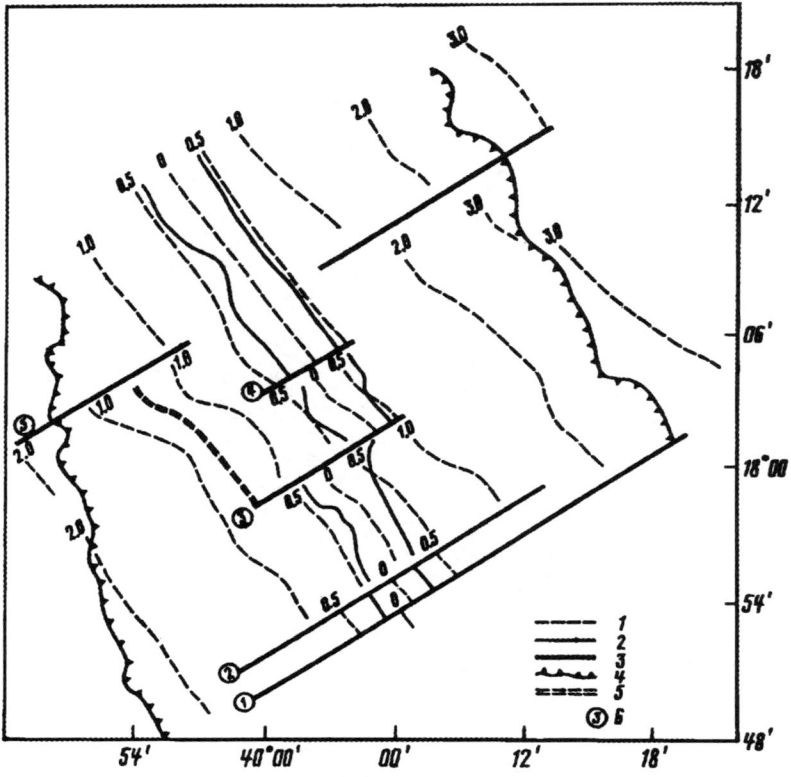

Figure 29 Scheme of isochrons of the Red Sea study areas (18° N): (1) isochrons in millions of years; (2) boundaries of the central uplift; (3) transform faults; (4) salt slopes; (5) spreading paleoaxis; (6) numbers of transform faults.

Figure 29 shows the isochron map. Isochrons up to 3 million years old are mapped for that region. So, judging from the configuration of salt slopes, about 2 million years ago, there appeared windows in the evaporite thickness similar to the depressions that exist in the Red Sea now, for instance, the Atlantis II. The finite rupture of the evaporite thickness occurred 1 million years ago. At that time, the fault of 18° N did not exist, and the rift zone axis extended from the south to the north with no eastward shift. An isochron of 1 million years, located west of the rift paleoaxis, provides evidences of that.

In the time interval of 1 to 0.5 million years ago, a 7-km eastward jump of a spreading axis occurred east of fault 18°. That explains the presence of the second isochron of 1 million years at the African plate and its absence at the Arabian one. So, two principal elements in the oceanic crust formation dynamics have been revealed in the Red Sea, namely, its irregularity and jump-like migration of the rift axis.

Comparing them with the result obtained for the Tadjura Ridge rift, we may conclude that these elements seem to be common in the process of formation of a new oceanic crust in the regions described. Besides, the reduction of the spreading finite rate in the time interval from 3 million years ago until today can be explained by a collision between the Arabian peninsula and Eurasia. We shall note that in the Red Sea, plates move apart perpendicular to the spreading trend, and that a 10% asymmetry in the rates of new oceanic crust formation registered there support the detected axis jump.

In conclusion, we should note that the research in the northwestern Indian Ocean at the Carlsberg Ridge described above (Merkuriev, 1988) showed that the spreading features (irregular character, subparallel echelon structure of rift zones, axis jumps, and spreading trend changes in time and in the process of rift migration) revealed in the Red Sea and Tadjura Rifts are typical of that oceanic region.

VI. THE JUAN DE FUCA RIDGE

The Juan de Fuca Ridge is one of the most thoroughly studied in the Pacific (Vogt and Byerly, 1976; Elvers et al., 1973; Hey and Wilson, 1982). It is very young: only 20 million years old. The evolution of the ridge was accompanied by its propagation in the north-south direction. For the first 10 million years, the ridge had a distinct meridional strike, then and by now, it has changed to NNE-SW (azimuth is 24°). A resultant spreading rate decreased with time, from 7.5 to 5.9 cm per year. Now, the spreading direction is WNW-ESE (azimuth is 294-114°).

In 1986, during the first cruise of the R/V *Akademik Mstislav Keldysh*, a complex geophysical survey was performed in the Juan de Fuca Rift zone near 46° N alongside seismic acoustic studies, direct geological observations, and sampling by man-operated submersibles. The study area was located in the vicinity of the Axis Seamount, in the caldera of which the American researchers found a hydrothermal activity (Canadian-American Seamount Exp., 1985). Thirty three profiles were made in that area spaced 600 to 800 m apart. The Magnavox satellite system and the Loran-C radionavigational system were used for adjustment of profiles. The survey was performed by computer technique; the vessel worked in the autonomic regime. The navigational error for site determination did not exceed 200 m.

From the computer echo sounding, a bathymetric map of the area has been compiled, with isodynams spaced 10 m apart (Figure 30). The Axis Seamount has an asymmetric caldera developing from northwest to southeast. Its strike does not coincide with the general strike of the axial zone of the Juan de Fuca Ridge and approaches it at an angle of 45°. The caldera is about 7 km long, and its eastern wall is destroyed starting from the central part. In the northern part of the caldera, the altitude of its walls does not exceed 50 m; their height increases southward to 150 m. In the southern part of the study area, the depth to the bottom reaches 1800 m. The total altitude of the Axis Seamount over the bottom is 500 m. Geological survey of the caldera by man-operated submersibles "Pisces" has revealed some spreading elements such as gja zones. As the result of that survey, it was determined that the last magma outflow occurred about 5000 years ago. That and some other findings allowed Prof. L. P. Zonenshain to suppose that the Axis Seamount coincides with the axial zone of the Juan de Fuca Rift.

Magnetic lineations are known to be the principal indicator of the spreading process. Therefore, we believe that it is possible to find the exact position of the Juan de Fuca axial zone only in the course of a thorough interpretation of magnetic field anomalies.

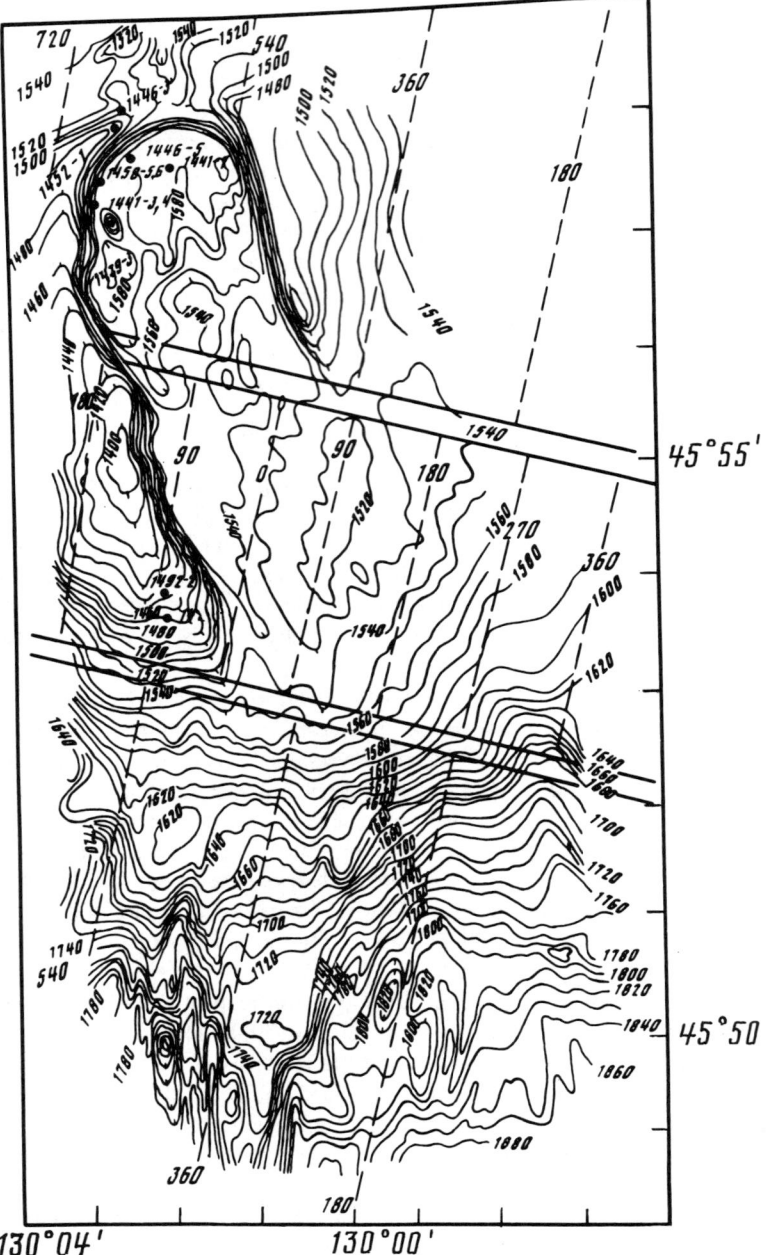

Figure 30 Bathymetric map of the Juan de Fuca rift zone. Contour lines are given in 10 m.

Two profiles (entrance and exit ones), bounding the study area from the north and south, were analyzed for a magnetic chronological adjustment of the area. Figure 31 shows the position of these profiles. The performed simulations showed that the area is located entirely within the oceanic crust corresponding to the Recent (Bruness) time. The obtained spreading half-rates are given in Table 8.

Table 8 **Spreading half-rates of the Juan de Fuca Ridge**

Time interval, millions of years	Spreading half-rate, cm per year
0.00-0.72 (Bruness)	2.86
0.91-0.97 (Charamilio)	3.0
1.66-1.87(2)	2.91
2.91-3.40(2A)	3.09
3.87-3.97(3)	3.6

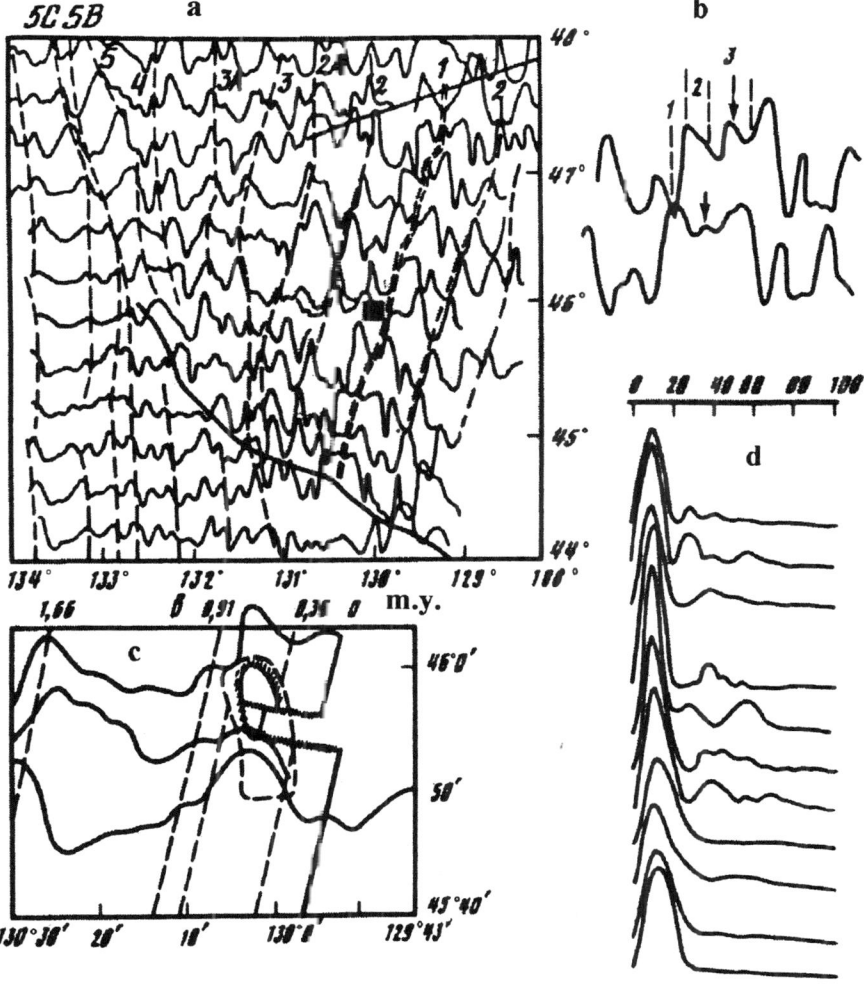

Figure 31 Map of the AMF curves (a) of the study area (Elvers et al., 1973): the position of the area is given by the black square; (b) tracing of magnetic lineations: the figures stand for numbers of segments of axial anomaly covered by detailed survey; (c) position of regional profiles: the position of the rift zone and of the study area, with dashed lines representing isochrones and figures indicating age in millions of years; (d) anomaly spectra registered within the study area.

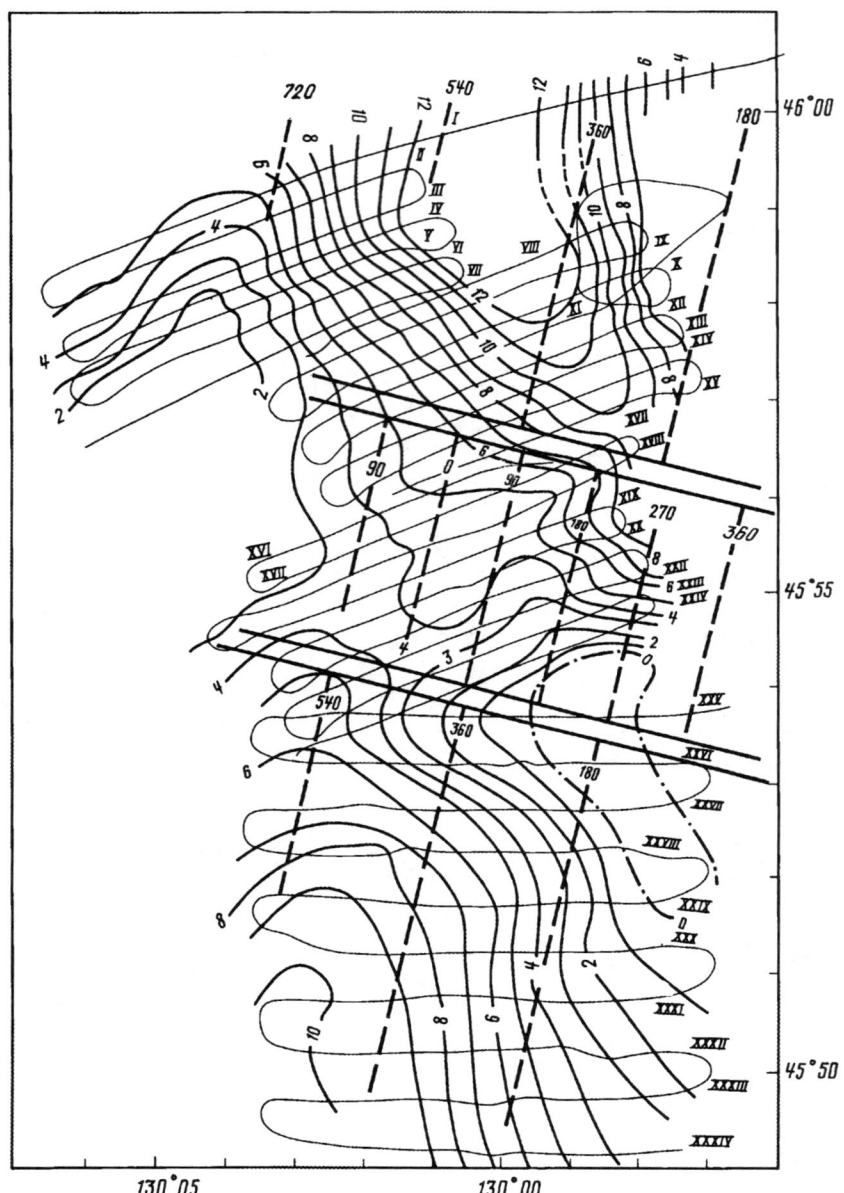

Figure 32 Map of AMF. Isodynams are given in hundreds of nT. Roman numbers correspond to profile numbers. Doubled lines show transform faults, dashed lines show isochrones, figures show the age in thousands of years.

These data evidence once again that the spreading half-rate decreased for the last 4 million years from 3.6 to 2.86 cm per year.

The analysis of the AMF map presented in Figure 32, has revealed a complex and ambiguous pattern of anomaly distribution. The entire study area is located within the domain of intensive positive lineations with the amplitude reaching 1200 nT in the north. In the central area, it decreases to 200 nT, and it increases again in the south, to 1000 nT. Isodynamic lines within the area, in the north and south, have a NNE-SSW trend. Only in the central part of the area, their trend is different.

A set of overlapping profiles (Figure 33) gives more visual correlation of the magnetic field and bottom anomalies within the study area. According to the shape of registered anomalies, all profiles can be subdivided into three groups: from 1 to 16, from 17 to 24, and from 25 to 34. Remarkable change of the anomaly shape occurs between profiles 16 and 17 and profiles 24 and 25. Besides, a number of local changes are also recognized there. One part of them is associated with the topography features of the magnetic basement, for instance, with the caldera walls, and the other part seems to be of a different origin. Here, we should note that the entire study area is located within the anomaly identified with the Bruness time. The shape of that anomaly is complex due to the occurrence there of a large block with rugged bottom topography. Therefore, to find the real effect of local anomalies, we should exclude the effect of the regional background and solve the problem of field discrimination. Thus, a special procedure is necessary to analyze the AMF.

Figure 31d presents spectrum examples of observed anomalies. They are complex ones, with some isolated local maxima. Therefore, adaptive algorithms of filtering and calculation of high field derivatives were used to process these signals (Strakhov and Valyashko, 1984; Valyashko and Strakhov, 1984). The problem of field discrimination was solved as follows. To determine more distinctly the local anomalous effects, the "regional" background was excluded, and secondary derivatives of the AMF were calculated from the remnant field. The noise magnitude present in measurements was determined by the correlation discriminator and then was used when calculating derivatives.

Results of calculations are given in Figure 33 as overlapping curves of the AMF and its secondary derivatives. It is clear from the figure that the caldera is pronounced against the field and is expressed as two local maxima corresponding to its western and eastern sides. Most distinctly, they are correlated in the interval between profiles 7 and 11. Farther southward, the correlation is violated, which may be explained by destruction of the eastern caldera wall. Extremely distinct correlation of the local maximum in the field of derivatives is noted between profiles 18 and 24. This maximum does not depend upon any topographic elements of the magnetic basement. No correlation of local anomalies is registered in the southern group of profiles, between profiles 25 and 34.

In plan, in the transformed field, the angle between the strike of local anomalies associated with the caldera of the Axis Seamount and the strike of the local anomaly correlated along profiles 18 to 24 is 45°. The strike azimuth of the recognized local anomaly is 24° and coincides with the strike of the axial zone of the Juan de Fuca Rift. And at last, Figure 31b shows the curves of the observed AMF over the axial zone of the Juan de Fuca Rift, taken from the work of Elvers et al. (1973). Their comparison with the anomalies observed within the study area (Figure 33) indicates that profiles from 1 to 16 corresponds to segment 1 at the axial anomaly, whereas profiles from 17 to 24 are associated with segment 3, that is, exactly with the axial zone of the rift. Profiles from 25 to 34 correspond to segment 2.

So, displacement of the oceanic crust blocks within the study area occurs along two transform faults. A 10-km eastward jump of the rift axis happened along these faults. The position of the transform faults and the jump of the rift zone axis of the Juan de Fuca Ridge are presented in Figure 31c. The recognition of the axial zone within the area allowed us to compile the map of isochrons, given in Figure 30.

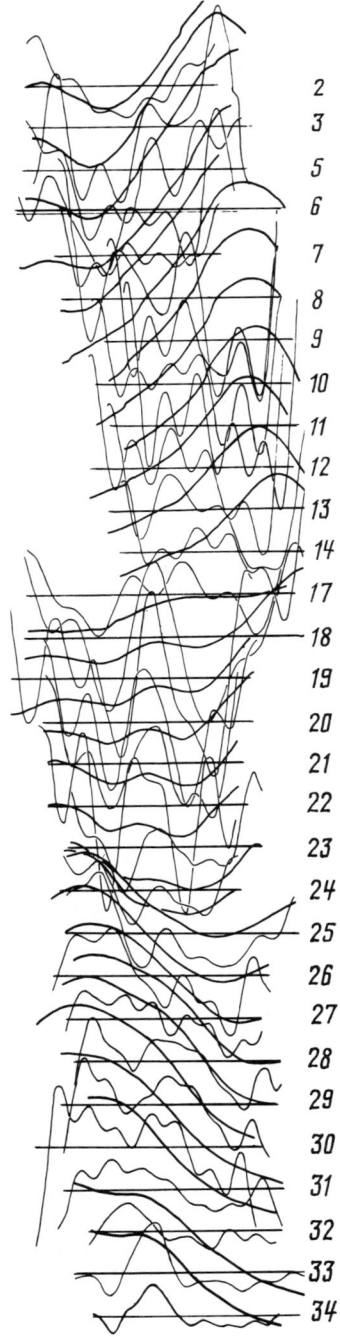

Figure 33 Overlapped curves of AMF and its secondary derivatives.

The analysis of the isochron map shows that the northern transform fault destroys the eastern wall of the caldera, thus shaping it as a broken horseshoe. The second transform fault bounds the caldera from the south and continues farther westward. The age of the oceanic bed surrounding the caldera is 700,000 years to the west and 400,000 years to the east. The rift axis is located in the central part of the study area and corresponds to zero isochron; it extends for 6 km. So, the Axis Seamount is a recent volcano located near the rift axis and separated from it by a transform fault.

The analysis of magnetic properties of the samples uplifted during the dives may tell much about the origin of the Axis Seamount. During the dives of ROV "Pisces", a set of samples was brought up for the geological study of the volcano caldera. The petrographical analysis of those samples was made by M. I. Kuzmin, L. P. Zonenshain, and Yu. A. Bogdanov. Twelve samples of the uplifted basalts presenting various age groups formed the data basis for the magnetic study. The samples present three types of basalt: (1) volcanic tubes 50,000 years old (Nos. 5, 8); (2) nappe basalt from the wall top (Nos. 1, 6, 7); and (3) basalt nappe at the caldera bottom 4500 years old (Nos. 2 to 4, 9 to 12).

Samples for measurement were taken from a distance no less than 1 cm off the glassy crust. The following parameters were measured: natural remnant magnetization I_n, magnetic resonance κ, magnetization saturation I_s, coercive force H_c, remnant destructive field H_{rc}, and Curie points T_{c1}, T_{c2}. Results of the research are given in Table 9. The thermomagnetic analysis showed that magnetic mineral in all samples is presented by titanomagnetite.

A statistical analysis has revealed stable correlation between κ and I_s ($\kappa = 0.75$), κ and T_{c1} ($\kappa = 0.76$), and also between H_{rc} and T_{c1} ($\kappa = 0.79$). This may evidence for the oxidation of titanomagnetite in some samples.

The analysis of the measured magnetic properties allowed us to distinguish between the samples according to the size of titanomagnetite grains, their composition, degree of oxidation and magnetic resistance. The histograma presented in Figure 34 distinctly show the discrepancy in magnetic properties of samples. We divided the set of samples into four conventional groups.

The first group comprises samples 1 and 6 (Table 9). It differs from the others mainly by the following parameters: $T_c = 160$ to $170°$; $I_n = 8.4$ A/m, that is, three to five times less than in the others; $\kappa_{av} = 1.4 \times 10^{-3}$ emu cm^{-3} Oe^{-1}, that is, twice as many as in the others. Values I_n, κ, Q, T_c, and H_{rc} are recognized in the histograms as individual modes. Values I_{st} I_{so}, H_{rc} /H_c the type and shape of the curves $I_s(T°)$ indicate a small degree of titanomagnetite oxidation.

The given values of κ, I_{rs} /I_{so}, H_{rc}, H_{rc} /H_c indicate that the largest grains of titanomagnetite are present in those samples (Table 9). Samples 2, 3, 9, 10, 11, and 12 comprise the second group; it has the greatest discrepancy of values, though the samples are compatible. The average values of I_n, κ, T_s, H_c, and H_{rc} differ from other samples. Average Curie points are $T_{c1av} = 215°$, $T_{c2av} = 260°$. The curve $I_s(T°)$ character reveals the presence of titanomagnetite there. That evidences oxidation of titanomagnetite in some samples of the second group. In some samples, the difference in the Curie point as compared to the first one reaches $100°$ C. This indicates different composition of titanomagnetite that is determined by the thermodynamic regime in magma, and it indicates the depth of a magma chamber. Titanomagnetite grains are small, and judging from I_n, κ, Q, I_{rs} /I_s, they are close to one-domain grains.

The sampling sites of the first two groups extend for hundreds of meters. The difference in magnetic properties seems to result from the difference in time and environment of their formation.

Table 9 **Magnetic properties of basalt samples from the Axis Seamount**

No.	Number of station, short petrographic characteristics	Z, cm	I_n, A/m	κ, 10^{-3} emu cm³Oe⁻¹	Q	I_s A/m	I_{rs} A/m	I_{rs}/I_s	H_c, Oe	H_{rc}, Oe	H_{rc}/H_c	$T°_{c1}$	$T°_{c2}$	I_{st}/I_{so}	I_n/I_i
1	1441-4 Nappe basalt from the caldera	1	9.3	1.4	13							160	185	1.3	
		2	8.2	1.3	12	2670	276	0.10	38	68	1.8	160	175	1.1	
		3	7.4	1.3	12										0.8
2	1446-3 Nappe basalt with a glassy crust; Gja region on the northern seamount slope	1	8.3	0.2	111										
		2	20.8	0.5	87	1960	418	0.21	192	260	1.3	255	300	1.2	
		3	22.7	0.5	91										1.4
		4	23.5	0.6	72										
3	1446-5 Nappe basalt with glass; younger northern part of caldera	1	8.0	0.1	160	872	120	0.14	620	900	1.4	260	265	1.1	
		2	23.5	0.4	117	1690	380	0.22	240	320	1.3	255	275	1.0	
		3	22.8	0.6	72	2050	430	0.21	130	175	1.3	220	270	1.2	1.7
		4	22.0	0.8	54										
4	1447-4 Basalt with glass from rope lava at the northern part of caldera; hydrothermal field	1	52.8	0.6	179										
		2	62.7	0.6	196	2315	595	0.26	135	190	1.4	15	180	1.1	2.8
5	1452-1 Fragment of volcanic tube; basalt with glass; mount waste at the wall; more ancient	1	6.8	1.5	170										
		2	55.9	1.5	193	2136	585	0.27	175	225	173	180	205	1.1	
		3	46.4	1.5	1143										3.6
				1.4											
6	1479-3 Basalt with glass; mount waste near the wall; more ancient; resembles dolerite	1	8.7	0.3	12							170	219	1.2	
		2	7.9	0.4	11	2581	276	0.10	46	77	177	175	195	1.3	
		3	8.4	0.4	11										0.9
		4	9.2		13										

No.	Sample / Description		Z	I_n	I_s	I_{st}	Q	H_c	H_{rc}		T_{c1}	T_{c2}		I_i
7	1492-2	1	26.7	0.6	191									4.9
	Nappe basalt from the mount; more ancient; greatly metamorphosed	1.5	47.4	0.4	243						210	235	1.2	
		2	56.1	0.3	255	2047	0.26	254	310	1.2	200	210	1.1	
				0.7										
8	1492-4	1	24.6	0.2	84.8	2490	0.23	200	256	1.3	200	245		
	Basalt from the tube walls; ancient glass from two sides	2	68.2	0.5	359									
		3	23.1	0.5	140									
		4	60.0	0.7	171									
9	1441-7	1	17.5	0.2	184									
	Nappe basalt with glass	2	24.7	0.5	95	1780	0.22	149	198	1.3	233	258	1.1	
		3	26.6	0.5	106	1780	0.22	124	158	1.3	215	225	1.1	1.8
		4	22.6	0.7	65									
10	1458-1	1	11.1	0.2	101									
	Basalt with glass; nappe lava; region of hydrotherms	2	35.1	0.5	135	1960	0.26	230	288	1.3	245	270	1.2	2.3
		3	32.8	0.6	104									
		4	53.1	0.6	177									
11	1458-6	1	16.4	0.8	40									
	Nappe basalt from the caldera; region of the northern hydrothermal field	2	19.5	0.8	46	1523					210			
		3	18.9	0.8	48									
12	1441-3	1	8.7	0.1	158									
	Nappe basalt with glass	2	34.6	0.7	102	2136	0.24	200	280	1.4	258	303	1.1	
		3	32.3	0.7	90	507					240	280	1.2	2.7
		4	35.2	0.8	888									
		5	46.7	0.6	143									

Note: Z is the distance from the glassy surface; I_n is a natural remnant magnetization; T_{c1} is the Curie point, determined from the function $I_s(T°)$ in the course of first heating; T_{c2} is the Curie point after heating to 600° C; k is magnetic susceptibility; Q is the Koenigsbergen factor; I_{st} is magnetization of saturation after heating to 600° C; I_s is magnetization of saturation (bulk); I_{rs} is remnant magnetization of saturation; H_c is a coercive force; H_{rc} is a remnant destructive field; I_i is an ideal magnetization at a constant magnetic field 1 Oe.

The third group is composed of samples 4 and 5 taken in the region of the hydrothermal field. However, they underwent little oxidation, $T_{c1} = 175°$, $T_{c2} = 180°$. According to the Curie point and, thus, from the titanomagnetite composition, these samples are closest to those of the first group. According to κ, Q, I_{rs}, I_s, H_c, and H_{rc}, showing the value and concentration of magnetic mineral in the rock, they differ by about one order; this may be due to the different environment of titanomagnetite crystallization.

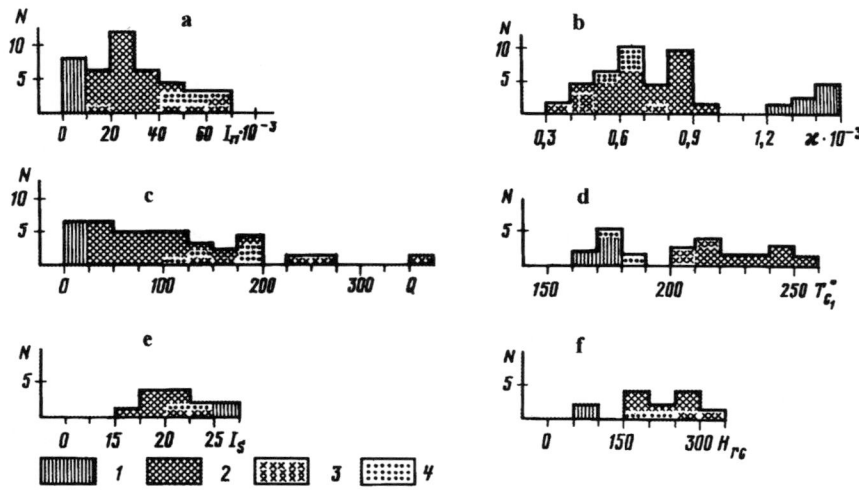

Figure 34 Histograms of magnetic properties distribution for various sample groups: (1) samples Nos. 1, 6; (2) samples Nos. 9 to 12; (3) samples Nos. 4, 5; (4) samples Nos. 7, 8; (a to c) number of samples $N = 40$; (d) $N = 16$; (e to f) $N = 14$.

Samples 7 and 8 of the fourth group seem to have the highest magnetization $I_{n_{av}} = 60$ A/m; the smallest $\kappa = 0.5 \times 10^{-3}$ emu cm^{-3} Oe^{-1}; the highest factor $Q > 200$ (the Koenigsbergen factor). They seem to be the oldest of the entire set. According to their position, basalts of this group are more extended than the others.

The above-mentioned differences show evidence for various composition and environment of crystallization of the effusive magma. This may say something about the cyclic character of magnetism and about the variety of tectonic processes at the time when this caldera was formed, for instance, transform fault setting in the caldera of an active volcano.

The obtained data agree well with the data of Valyashko et al. (1987), according to which the cyclic character of volcanic activity is typical of the high-spreading ridges. As a rule, a collapse caldera is formed after the intensive stage (Ekinyan, 1984). So, the Axis Seamount is a volcano of the central type formed in the rift zone of the Juan de Fuca Ridge due to magma flow from the upper mantle through the principal channels. The age of its formation is approximately 700,000 years. Approximately for that time, it shifted at 10 km off the axis due to the spreading process. However, it turned to be near the rift zone again (and thus near the magma chamber) because of the spreading axis jumps. That seems to be manifested at the youngest stages of volcanic activity of the Axis Seamount.

VII. THE NORTHWESTERN PACIFIC BASIN AND THE SHATSKIY RISE

The Northwestern Pacific Basin, which is a region of special geological-geophysical concern, has been studied extremely irregularly by now. The most thoroughly studied regions are those of island arcs and adjacent deep trenches, as well as also the Northwestern Basin itself, whereas such principal morphotectonic structures as the Shatskiy and Hess Rises are studied poorly.

Systems of the Mesozoic magnetic lineations were revealed in the Northwestern Pacific Basin in the early 1970s during the joint Japanese-American geomagnetic research there. The age of the oceanic crust varies there from 108 to 156 million years as follows from the origin of magnetic lineations recognized from the deep drilling data. That geomagnetic survey made it possible to extend the magnetic geochronological scale for the late Mesozoic to 165 million years and to form the basis for the identification of the observed anomalies (Heyes and Pitman, 1970; Larson and Chase, 1972; Larson and Pitman, 1972; Hilde et al., 1978; Sharman and Risch, 1988; Handschumacher et al., 1989; Sager et al., 1988; Nakanishi et al., 1989).

This research also allowed a supposition that in the early Cenozoic and Mesozoic Pacific, besides the Pacific plate, there existed the Kula, Farallon, and Phoenix plates (Atwater and Grow, 1970; Larson and Chase, 1972). In some works, the time of the Kula plate existence is limited by 85 to 50 million years. It is supposed that prior to that time (157 to 85 million years), there was an ancient plate named Izinagi (Engebretson et al., 1984). Hilde and co-authors believe that the Shatskiy Rise was formed 142 to 116 million years ago along the transform fault between the Pacific and Farallon plates, the latter being subducted under the Pacific plate. The reconstructions of the Pacific plate motion based on the geomagnetic survey of the Japanese, Hawaiian, and Phoenix systems allowed a supposition that their formation occurred within the paleolatitudes 6 to 15° N, while the Shatskiy Rise, at the time when anomalies were set, was located within the Pacific-Kula-Farallon plate triple junction and was of the ridge-ridge-transform type (Hilde et al., 1976; Kogan et al., 1983).

From the strike of magnetic lineations, we may conclude that the main reorganization of the spreading at the Kula/Pacific Ridge occurred between anomalies M16 and M15 (132 to 130 million years). From 157 to 111 million years, the Kula plate experienced the northwestward displacement, relative to the Pacific plate, at an average rate of about 9 cm per year. During the time of 111 to 71 million years, which corresponds approximately to the period of a Quiet magnetic field, the Kula plate moved northwestward relative to the Pacific plate at an average rate of about 5 cm per year (Kononov, 1984).

A 40° northward clockwise rotation was verified by paleomagnetic measurements of the cores from the DSDP boreholes in the central and northwestern parts of the ocean (boreholes 45, 52, 52, 195, 303, 304, 35, 307, 310, 430, 432, 433, 464, 465, and 466). So, in boreholes 303, 304, and 307, the remnant magnetic declination of basalt corresponds to paleolatitudes of 6 to 15° north or south of the equator.

These reconstructions can adequately explain the geometry of the Mesozoic magnetic lineation pattern within the Northwestern Pacific and, therefore, are assumed by many researchers. However, the geological origin of the Shatskiy Rise is still speculative. Some believe that it was formed as the result of tectonic syntaxis (Puscharovsky and Melankholina, 1981; Melankholina, 1988).

A great scope of complex geological-geophysical works, geomagnetic survey included, was performed by the 12th, 21st, 23rd, 29th, and 42nd R/V *Dmitry Mendeleev* cruises as well as the 53rd and 56th cruises of the R/V *Vityaz* in the Northwestern Pacific and in the vicinity of the Shatskiy Rise to study the deep structure of the crust and lithosphere in those most ancient parts of the World Ocean. The program of research comprised seismic observations, gravity survey, and geological and geomorphologic studies (Gorodnitsky et al., 1984) (see Figures 35 and 36).

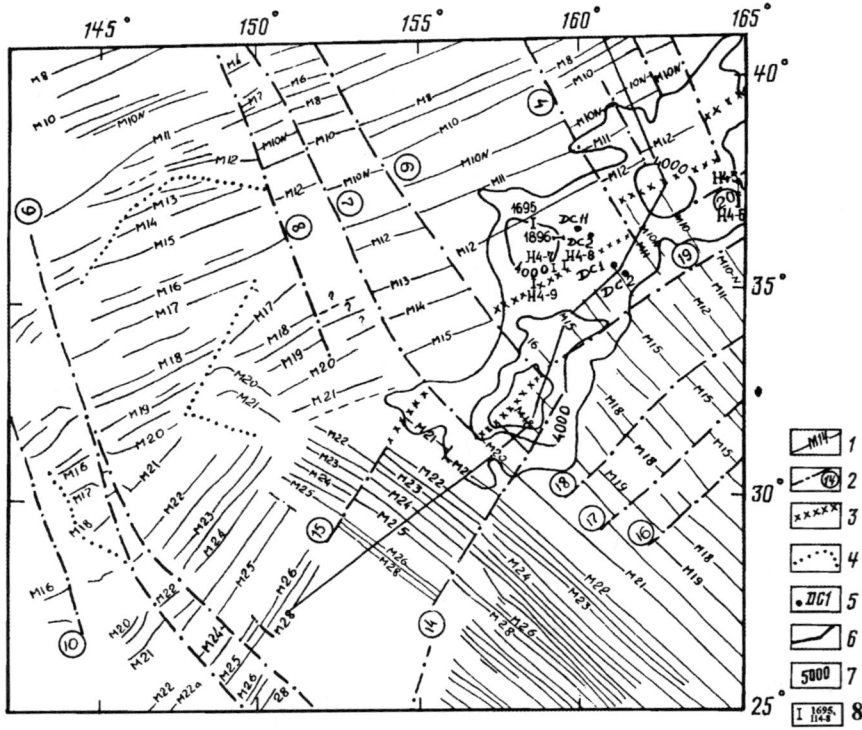

Figure 35 A scheme of the geophysical state of art in the Northwestern Pacific Basin: (1) magnetic lineations and their numbers; (2) transform faults and their numbers; (3) suture zone between plates within the Shatskiy Rise; (4) pseudo-faults (5) bottom stations of DSS; (6) seismic and geomagnetic survey geotraverse of the 29th cruise of the R/V *Dmitry Mendeleev*; (7) isobaths; (8) station of dredged rocks.

The Shatskiy Rise is a large morphostructure elongated from the northeast to the southwest. Its dimensions are 1200 by 300 km within a 5000-m isobath. The genesis and geological nature of the structure is still speculative. Morphologically, it comprises three domes, namely the Northern, Central, and Southern ones separated by tectonic discontinuities. The Southern dome is the most tectonically isolated. An anomalous high heat flow has been registered near its southern boundary. Further in the text, we shall assign the Central and Northern domes to the northern block (Gorodnitsky, 1985).

Within the Shatskiy Rise, the sedimentary cover structure is studied in detail from the cores of boreholes and CSP data (Structure ..., 1984; Twing et al., 1968; Init. Rep. DSDP, 6, 1971; 20, 1975; 32, 1980; 66, 1982, Ludvig and Houtz, 1979). The comparison of the CSP and DSD data has revealed that the Mesozoic-Cenozoic sedimentary

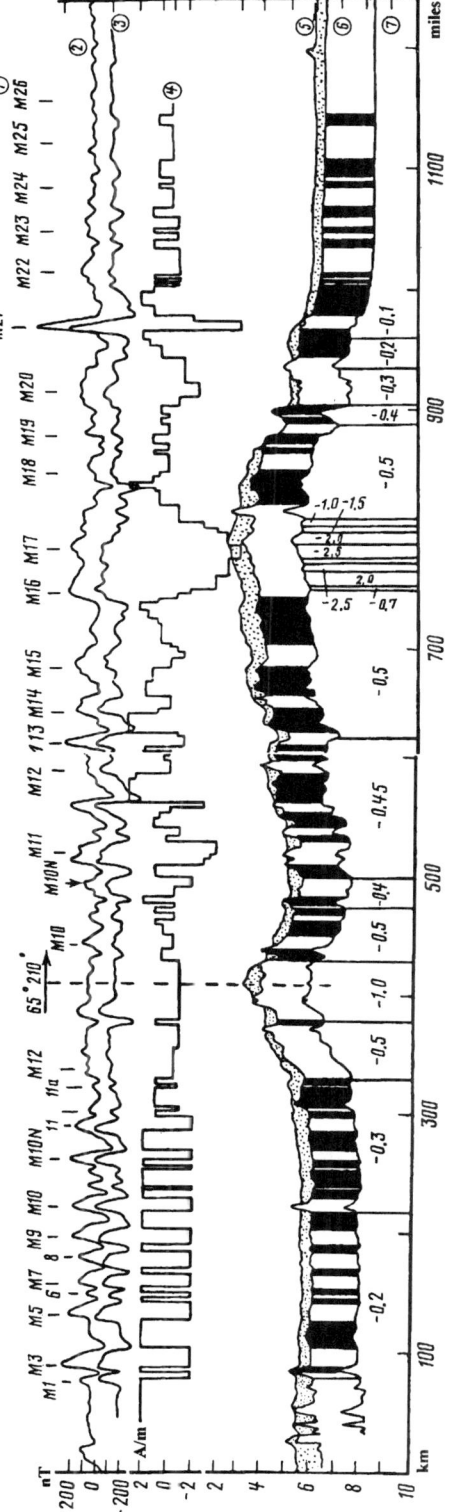

Figure 36 Geomagnetic profile across the Shatskiy Rise: (1) numbers of magnetic lineations; (2), (3) observed and simulated magnetic fields; (4) distribution of calculated effective magnetization; (5) sedimentary cover; (6), (7) upper and lower active magnetic layers, respectively.

cover in the southern part of the Rise is up to 1 km thick. The most complete column was obtained from boreholes 305 and 306. Borehole 50 reached the basalt basement. The age of the sedimentary cover bottom there was dated back to the Valanginian-Hauterivian (138 to 125 million years) (Init. Rep. DSDP, 1971). Four stratified layers are recognized on the CSP recordings.

The bottom depth in the Cretaceous time was several hundred meters (Douglas and Mouhade, 1972). It agrees with the data on the shallow or even subaerial outflow of basalts prior to the time when the Shatskiy Rise was subsiding gradually at 2 to 3 km as it moved northward. The analysis of the lithological and micropaleonthological data also evidences a higher hypsometric level over the abyssal basins (at 1 to 1.5 km) and supposes a tropical climate at the Shatskiy Rise during the Maestrichtian time.

A ENE depression with a sedimentary cover up to 800 m and the near fault ridge of the same strike in the southeastern termination of the area are recognized in the bottom topography between the Central and Southern domes of the Shatskiy Rise (Figure 37; Table 10; Figure 38d). This depression is believed to be a boundary between two plates, judging from the data comparison of CSP, hydromagnetic studies, and DSS.

The first DSS works were performed in the Northwestern Pacific Basin in 1957 to 1959 in the Kuril-Kamchatka trench and adjacent basin. They were continued in 1961 during the 28th cruise of the R/V *Vityaz* (Mikhno, 1964). From the results of the Japanese and American works (Den et al., 1969, 1973), the Earth's crust section was compiled in 1967 to 1969 for the southern part of the Shatskiy Rise and the basin adjacent to it from the west. DSS works were carried out during the *Vema* and *Conrad* sea expeditions in 1965 to 1976, both in the Northwestern and Northeastern Basins with crusts of different ages (Houtz et al., 1976, 1980).

The works at the Shatskiy Rise and in the Northwestern Basin were continued on the 21st, 23rd, and 29th cruises of the R/V *Dmitry Mendeleev* (Structure ..., 1984). The Earth's crust sections are given in Figure 39. In the upper part of the section, according to Neprochnov data, lies a sedimentary layer 0.8 km thick and P-wave velocity of 1.8 km/s; below it lies oceanic Layer 2, with P-wave velocity of 5.4 km/s and 1.3 km thick; and below it lies Layer 3, 5.6 km thick and with P-wave velocity of 6.8 km/s. The M boundary lies at a depth of about 13 km and has P-wave velocity of 8.2 km/s. The thickness of the crust there is about 8 to 9 km.

Within the central part of the Shatskiy Rise, between the Central and Northern domes, DSS works were performed at stations DC-9, 11, 2 (Figure 35), and the following seismic column was obtained. Layer 1 is 0.5 km thick with a P-wave velocity of 2.0 km/s; Layer 2 is subdivided into Layers 2a and 2b, respectively, which are 1.7 and 4.7 km thick, and have P-wave velocities of 4.5 and 5.4 km/s. Layer 3 is subdivided into Layers 3a and 3b characterized, respectively, by thicknesses of 3.8 and 4.7 km and P-wave velocities of 6.8 and 7.4 km/s (Structure ..., 1984). For DC-1 we have Layer 1, which is 0.8 km thick and has a P-wave velocity of 2 km/s; Layers 2a and 2b, which are 1.2 and 2.3 km thick and have P-wave velocities of 4.5 and 5.4 km/s; and Layer 3, which is about 5 to 6 km thick and has a P-wave velocity of 6.7 km/s.

So, within the Shatskiy Rise, the M-boundary depth in the Japanese sequence is 20 to 22 km and that in the Hawaiian sequence is 15 km, which is considerably more than in the neighboring basins.

The Earth's crust section for the southern block of the Shatskiy Rise (Den et al., 1969) gives a general idea of its structure. Figure 35 presents this section after its complex interpretation with the analysis of geomagnetic and CSP data. Transform fault 11, which is distinctly manifested in the magnetic field by the displacement of

Table 10 **Coordinates of profiles of Figures 37, 38**

Profile No.	φ	λ	Profile No.	φ	λ	Profile No.	φ	λ
1	42 08.8	147 36.4		36 02.5	156 22.5		37 41.0	165 39.4
	42 11.4	148 34.0		36 01.7	157 18.7		37 40.6	165 52.1
	42 11.4	150 08.9		36 03.1	151 54.7		37 41.0	166 41.8
	42 14.2	151 06.2		36 26.2	158 57.8		37 48.3	167 41.5
	42 04.4	152 44.6		36 51.5	159 18.7		37 49.9	169 20.8
	41 18.6	153 38.1	7	33 56.7	146 30.5		37 56.7	170 14.2
	40 53.6	154 00.3		34 04.6	147 47.1		37 48.2	171 17.2
	39 58.9	154 41.8		34 15.4	149 19.3		37 32.2	172 12.1
	38 16.3	155 19.6		34 26.3	151 09.0		36 56.7	173 12.5
	37 55.9	156 11.3		34 35.0	152 06.0	11	36 17.7	158 24.5
	37 12.7	156 53.2		34 41.8	152 38.7		36 45.1	159 02.9
	36 33.3	157 40.3		34 47.4	153 10.9		36 59.0	159 37.4
	36 12.6	158 18.5		35 05.8	155 53.3		37 07.9	160 16.3
2	40 36.7	154 37.4		35 23.4	156 33.9		37 23.6	161 14.7
	40 10.8	155 07.9		35 52.7	157 36.9		37 49.3	162 17.0
	39 29.5	155 58.7		36 11.4	158 09.7		38 06.4	163 12.0
3	39 40.4	156 33.2		36 27.2	159 01.8		38 14.9	163 39.7
	39 16.3	157 13.2		36 28.8	159 30.0		38 23.0	164 00.3
	38 29.0	158 28.8	8	30 31.4	155 51.8		39 16.3	164 48.2
	37 47.3	159 27.5		30 40.8	156 47.3		39 33.3	165 02.1
	37 26.7	159 52.5		30 45.2	157 10.2		40 40.0	165 48.6
	36 42.5	159 30.6		30 51.6	157 47.1		40 59.6	165 54.5
4	36 57.8	159 07.1		31 00.9	158 78.0		42 17.0	166 45.0
	37 17.5	158 29.6		31 36.3	158 06.4		42 52.5	167 06.5
	37 35.3	158 16.3	9	43 33.0	157 53.0		43 33.3	168 16.9
	37 38.6	157 47.2		43 41.3	158 25.6		43 43.0	169 02.8
	36 58.4	158 12.5		43 53.4	159 30.3		43 51.0	169 57.5
	37 21.4	157 37.0		43 41.7	160 53.8		43 45.5	170 35.2
5	28 18.7	168 28.3		43 31.2	161 55.3		44 09.7	171 17.9
	28 23.7	167 24.9		43 19.4	162 55.3	12	35 03.3	159 03.8
	28 28.7	165 23.6		43 06.5	164 04.0		35 27.3	159 06.7
	28 29.7	164 29.9		42 52.3	165 33.1		35 17.8	159 10.0
	28 33.5	163 05.3		42 47.3	166 02.7	13	35 56.7	159 20.5
	28 35.1	162 23.5		42 41.0	166 39.1		35 41.7	159 09.5
	28 30.8	161 34.0		42 32.9	167 23.4		35 28.8	159 12.0
	28 42.0	160 31.1		42 23.9	168 17.0		35 17.9	159 12.7
	29 19.2	158 54.4		42 05.2	169 56.9	14	35 42.9	158 45.5
	29 43.4	157 38.7		41 54.0	170 54.3		35 28.3	158 46.9
	29 51.8	157 12.2		41 39.2	172 11.6		35 17.7	158 47.0
	30 06.8	156 25.0		41 24.7	173 19.8		35 06.6	158 46.3
	30 25.3	155 37.8	10	36 34.1	159 08.0		35 00.0	158 45.3
	30 38.1	154 56.0		37 17.6	158 40.1	15	35 09.4	158 59.7
	31 01.2	153 33.5		37 27.0	159 11.7		35 07.2	159 07.7
6	35 07.9	144 27.2		37 23.2	159 46.5		35 00.9	159 28.9
	35 21.6	146 31.3		37 33.2	160 43.7	16	35 41.9	159 03.9
	35 32.8	148 12.9		37 33.8	160 55.2		35 29.7	159 00.9
	35 36.4	150 16.5		37 35.6	161 17.2		35 14.4	158 59.7
	35 38.7	151 41.7		37 36.3	162 24.2	17	35 37.9	159 04.8
	35 40.1	152 29.8		37 38.3	163 01.2		35 28.5	159 03.9
	35 44.3	153 21.4		37 40.3	163 36.8		35 15.6	159 00.2
	35 55.3	154 43.3		37 40.2	164 27.3		35 02.5	158 57.5
	35 58.5	155 27.6		37 39.5	165 05.8			

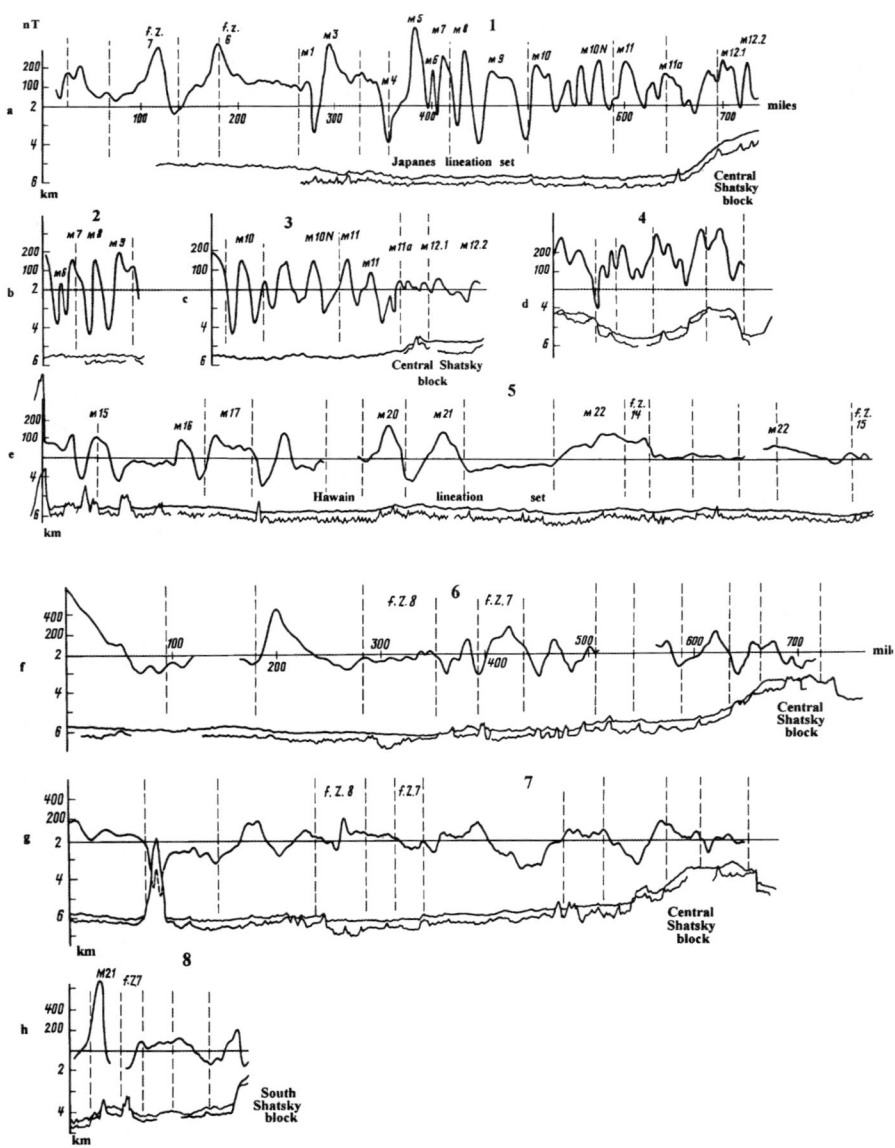

Figure 37 Bottom topography and AMF at the Shatskiy Rise profiles; observation points are from Table 10.

magnetic lineations of the Japanese sequence, runs between Stations 1 and 2 on the profile (see Figures 35 and 39). From the CSP data, it corresponds in the bottom topography to the graben with an amplitude of 500 m (Figure 39). From the data of Den et al., a small change in velocity of P waves for Layers 2 (from 5.2 to 5.25 km/s) and 3 (from 6.89 to 6.95 km/s) is observed there. This velocity change seems to reflect the increasing difference between two blocks contacting along fault 11.

Further along the profile, between Stations 4 and 5, the velocity pattern changes considerably. Two structural sublayers can be distinguished there for Layers 2 and 3

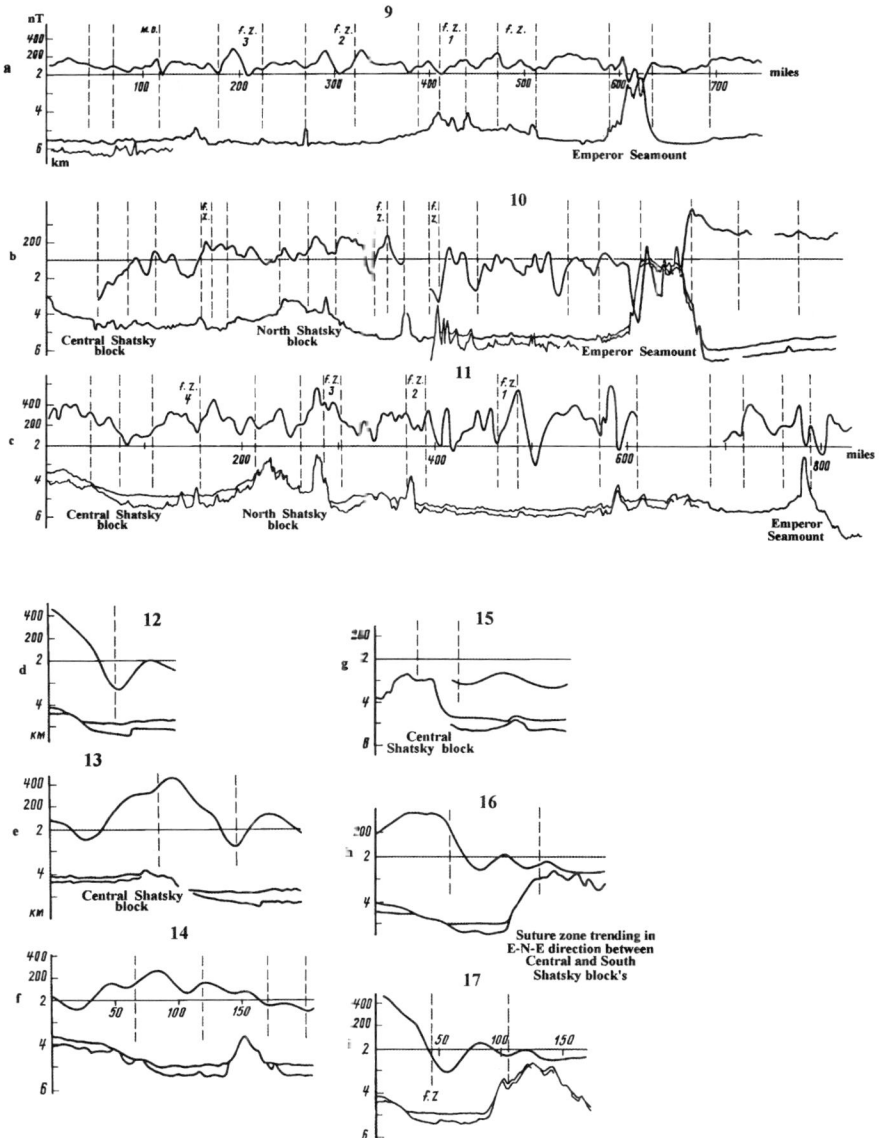

Figure 38 Bottom topography and AMF in the Northwestern Pacific Basin and Shatskiy Rise.

(Figure 39), namely, 2a and 2b, and 3a and 3b. They are characterized, respectively, by the following values: their thicknesses are to 2 km, 2.2 km, 7.5 km, and their velocities are 5.1, 5.70, 7.0 and 7.52 km/s.

When compared to the map of magnetic lineations (Figure 35), the given part of the crust between Stations 5 and 6 can clearly be assigned to the area controlled by the system of the Hawaiian anomalies, whereas that between Stations 1 and 4 belongs to the Japanese system of magnetic lineations. Transform fault 10 is traced along the boundary between Stations 6 and 7. It has a SE-NW strike and is traced from the abyssal basin toward the southern block of the Shatskiy Rise (Figure 35), where it is expressed by an escarpment in the acoustic basement topography (Figure 39).

Den et al. distinguish, for Stations 7 to 8, a solid Layer 2 with P-wave velocity of 5.05 to 5.31 km/s and 3 km thick; Layer 3a with P-wave velocity of 6.95 to 6.73 km/s and 6.2 km thick; and Layer 3b with velocity of 7.79 km/s. The observed changes in P-wave velocity and depth to the top of Layer 3 may be explained by differences in the age of the blocks of the crust that are verified by magnetic data in accordance with which the blocks are assigned to different plates with different ages of the crust. The profile strike changes for a submeridional between Stations 9 and 12. According to the map of magnetic lineations (Figure 35), this part of the profile falls at the boundary between two systems of magnetic lineations. The Southern Shatskiy transform zone of the NNE strike is registered at the boundary between Stations 12 and 11. Displacement of anomalies from M28 to M16 occurs along this zone. So, the areas of Stations 12 and 11 are assigned to the crust of different ages, and that is reflected in the velocity pattern (Figure 39). For Layer 2a, the velocity changes from 4.74 to 4.49 km/s; for Layer 2b, it varies from 5.56 to 5.43 km/s; and for Layer 3, it changes from 7.30 to 6.91 km/s.

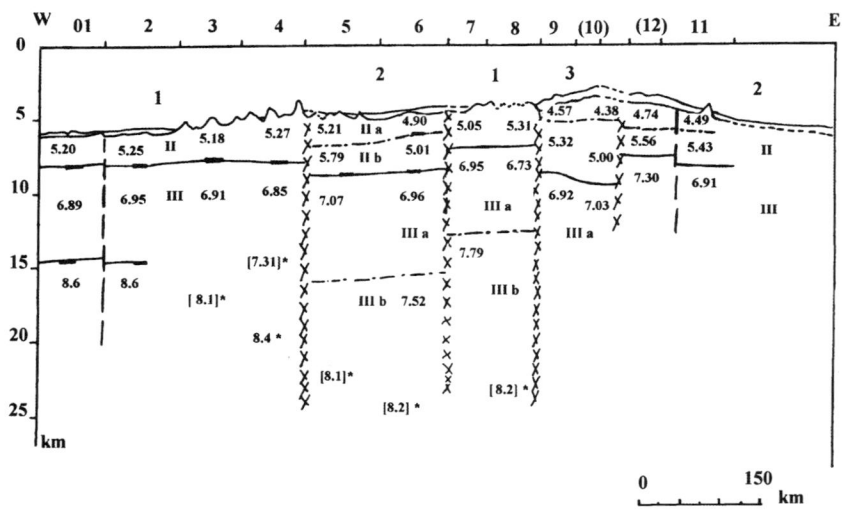

Figure 39 Cross section of the Earth's crust in the southern block of the Shatskiy Rise from data of Den et al. (1969) with account of geomagnetic data interpretation: (1) area of the Japanese magnetic lineations; (2) area of the Hawaiian magnetic lineations; (3) suture block.

A comparison of seismic data and results of the interpretation of magnetic lineations allows us to draw a boundary between the Kula and Farallon plates within the Shatskiy Rise. Besides, it seems expedient to pay attention to the increase of the crust thickness under the domes of the Rise.

Petromagnetic analysis has been made for the dredged rocks from the Central and Northern domes of the Shatskiy Rise (Figure 35). Blocks and fragments of tholeiitic basalt and hyaloclastite covered by a thick ferromanganese crust were uplifted from the northern escarpment of the Central dome (St. 1695). Basalt is characterized by abundant pores and amygdaloids (up to 40% of their body volume). Basalt has a porphyritic texture with inclusions of plagioclase and olivine substituted by smectite. On the whole, the rock is greatly metamorphosed (Rudnik et al., 1981). The remnant magnetization varies within 3.6 to 4.1 A/m, with factor Q being about 18 to 30. Remnant magnetization in the basalt dredged from the eastern frame of the Central dome (St. 1896) is considerably less and varies between 0.7 to 1.7 A/m, with factor Q being 1.2 to 9.4 and 3.1, on average (Lin'kova and Raikevitch, 1989).

Rocks, dredged from the NEE ridge sited in the depression between the Southern and Central domes (Figures 35 and 38), are basalts metamorphosed to various degree, reflected in their magnetic properties. So, rocks that underwent barite-apatite mineralization have magnetic susceptibility κ and remnant magnetization $I_n < 0.08$ A/m (St. H 4 to 7, R/V *Akademik Nesmeyanov*). Basalts from St. H4-8, H4-9 are less changed; their magnetic susceptibility is $\kappa < 10^{-5}$ emu cm^{-3} Oe^{-1} and $I_n < 1.9$ A/m with factor Q being 24. Rocks related to alkaline basalt and overlying hypsometrically the tholeiite one were uplifted from the Southern and Northern domes of the Rise (Kashintsev and Suzyumov, 1981). Basalts from the lower part of the slope (St. H4-5) are characterized by a wide discrepancy in magnetic parameters. So, magnetic susceptibility varies from (92 to 440) $\times$ 10^{-5} emu cm^{-3} Oe^{-1}), average being 240 $\times$ 10^{-5} emu cm^{-3} Oe^{-1}; remnant magnetization varies from 6.3 to 15.1 A/m, $I_{n_{av}}$ with being 6.8 A/m; factor Q varies from 4.8 to 207, with Q_{av} being 57.5. That discrepancy is less for the basalts uplifted from the middle part of the slope (St. H4-6). Respectively, values are as follows: (200 to 290) $\times$ 10^{-5} emu cm^{-3} Oe^{-1}, average being 220 $\times$ 10^{-5} emu cm^{-3} Oe^{-1}, for magnetic susceptibility; from 2.2 to 7.8 A/m, at $I_{n_{av}}$ being 4.3 A/m, for remnant magnetization; and 3.9 for factor Q.

As follows from the diagram by Zeidervilde, two anti-parallel directions of magnetization coexist in the rocks sampled from the NEE ridge located between the Central and Southern domes of the Shatskiy Rise (St. H4-9). One of these directions is destroyed at 20 to 200°C (opposite I_n). The second magnetization direction which determines I_n is destroyed within the temperature interval from 200 to 500°C.

A comparison of these data with CSP, DSS, and the AMF T_a allows a supposition that the rocks dredged at Stations H4-7, 8, and 9 were changed greatly during the interaction of the two plates, probably as the result of secondary heating of the rocks due to the sliding of one plate relative to the other with a partial subduction, and in the result of magnetization loss due to barite-apatite mineralization. Alkaline basalts located in the area of the junction of the two plates (St. H4-6) have experienced small secondary temperature changes. Comparison of the calculations based on the results of the magnetization data and petromagnetic analysis of the rocks uplifted from the Central dome of the Shatskiy Rise allows an assumption on the occurrence of serpentinite on the northern escarpment (St. 1695), where natural remnant magnetization is about 4 A/m, and in the area of Station H4-5, where natural remnant magnetization reaches 15 A/m.

As mentioned earlier, the Shatskiy Rise is located in the junction area of two systems of magnetic lineations, namely the Hawaiian and Japanese ones, the oceanic crust in the adjacent basins being 118 to 165 million years.

According to the latest works (Sager et al., 1988; Nakanishi et al., 1989), two principal trends of magnetic lineations of the Japanese sequence can be recognized north of 25° N and west of 150° E. The first has an azimuth of N42E south of the point (31° N and 147° S), and the second has an azimuth of N66E north of that point. At the anomaly M20n-1 boundary, we can observe a clockwise reorientation of the anomaly by 24°

Let us consider a sequence of anomalies with an azimuth of N66E. Such a trend of anomalies is typical of the areas north and northwest of the Shatskiy Rise. Anomalies

M1 to M20 and transform faults 4 to 13 are recognized in that area (Figure 35) (Hilde et al., 1976; Sager et al., 1988; Nakanishi et al., 1989). On the published maps of the lineation curves, the most reliable are anomalies M1 to M10. Identification of more ancient anomalies is not so exact due to insufficient density of profiles and scanty knowledge on their strikes, which in some cases is close to the strike of the anomalies proper. So, in the mentioned works, anomalies are identified to M20, in the oceanic crust block between faults 9 and 10, and to M18, between faults 7 and 9.

Analysis of the data obtained in the course of our survey required a reidentification of some anomalies according to the scale by Harlend and Cox (1985). So, the most ancient anomaly in the considered crustal block between faults 7 and 9 is M12.2, according to our identification. Similarly, anomalies in the block between faults 9 and 10 were also reidentified. Lineation M15 is the most ancient one that can be recognized with confidence. The recognition of more ancient anomalies seems to be problematic. Identification of anomalies between transform faults 10 and 12 is given in Figure 35, according to Sager et al. (1988). The spacing of lineations between faults 4 and 7, given in Figure 35, also agrees with the data by Sager and others, but for the position of the most ancient anomaly M12.1, which we recognize in the Northern dome of the Shatskiy Rise. And, at last, north of the Shatskiy Rise, a magnetic bending M1-M12 formed as a product of the Japanese/Hawaiian lineation junction has been identified (Sherman et al., 1988; Sager et al., 1988; Nakanishi et al., 1989).

The published maps on that region do not seem to identify with confidence the mentioned anomalies as a magnetic bending due to an insufficient number of profiles and their trends. We believe there are no such triple junctions but a simple northward displacement of anomalies along the series of transform faults. The presented data (Hilde et al., 1976; Sherman et al., 1988) cast some doubt as to the correct identification of magnetic anomalies of the Japanese sequence in the block between transform faults 3 and 4. When comparing profile X (Hilde et al., 1976) of the block by which identification was made with profiles O, P, Q, and R of the neighboring block, we can see that the authors had to reduce the spreading rate down to 2.2 to 2.9 cm per year to identify anomalies M5 to M10 at profile X, whereas the rate in the neighboring block is 6 cm per year.

When comparing the CSP data with the bathymetric data, it is clear that the region is related to the zone of an acoustic basement splitting. The dislocations recognized there are spread over the sedimentary cover as well (Figure 38). A comprehensive interpretation of these data allows a supposition that the anomalies identified there (Sherman et al., 1988) are of tectonic origin.

East and southeast of the Shatskiy Rise, the Hawaiian sequence of magnetic lineations of N50W strike are recognized there (Hilde et al., 1976; Sager et al., 1988; Nakanishi et al., 1989; Sherman et al., 1988; Valyashko et al., 1990b). In all cited works, there is a certain disagreement in the identification of magnetic lineations and of transform faults along which their displacement occurs. Therefore, we had to check up the identification of magnetic lineations M1 to M29 of the Hawaiian sequence east of the Shatskiy Rise. The procedure was based on the published maps of magnetic lineation curves (Nakanishi et al., 1989), bathymetric data (Sherman et al., 1988), and our own profiles.

The amplitude of these lineations is about 600 nT in the zone of young anomalies and decays down to 80 nT in the ancient part of the region. An average amplitude of the Hawaiian anomaly sequence is one half that in the Japanese one. Identification of magnetic lineations and of the locus of transform faults 11 to 27, along which their segments are shifted, was made from the model simulations.

South of the Shatskiy Rise, there is a bending of magnetic anomalies M29 to M21. A distinct junction of magnetic lineations can be traced to anomaly M24b; it is not definite in the case of more ancient lineations. The strike of magnetic bending is N15W. The magnetic lineation pattern on that part of the map, given in Figure 35, coincides with the data by Handshumacher et al. (1988). However, from our data, transform fault 22, which has acquired the name of the Southern Shatskiy transform zone, continues beyond the southern block of the Rise toward the latitude of 32.5° N, and displacement of magnetic lineations of the Hawaiian sequence occurs along it from M28 to M17.

In the vicinity of the Southern dome of the Shatskiy Rise, anomaly M21 is pronounced against the background of small-amplitude anomalies M17 to M28. Anomaly M21 has an amplitude greater than 600 nT (Figures 35, 36, and 37a) and is correlated with the tectonic escarpment 1 km high located at the southern slope of the dome. High heat flow is registered there. Such local increase of anomaly M21 amplitude may be explained by a process of serpentinization of the lowermost oceanic crust and its enrichment by magnetite, which is verified by the model simulations which have recognized a block with an estimated magnetization of about 6 A/m.

Let us consider the interpretation results of geomagnetic survey along the regional profile spaced between transform faults 4 and 7, and crossing the Japanese sequence of lineations from CL to M12.1, and extending along the Shatskiy Rise trend. The profile crosses the Southern dome and decays in the Northwestern Pacific Basin, where it crosses the Hawaiian sequence of lineations to M28 (Figures 35 and 36).

Anomalies M10 to M17, with an amplitude of 100 to 200 nT, half of the Japanese anomaly sequence, were identified at the profile between the northern and southern blocks. Magnetic simulations along the profile were based on the principle of a direct problem solution in a two-dimensional context by trial and error of equivalent models as long as they agree optimally with the estimated and observed anomalies. The trial of an active magnetic layer considered as well the data of DSS (Neprochnov et al., 1984). Magnetization for the model was tried from the range of values obtained during the analysis of magnetic properties of the rocks sampled by dredging (Lin'kova and Raikevitch, 1989).

Two ways of simulation were taken up. The first considered a one-layer model of an active magnetic medium presented by basalt, whereas the second considered a "two-layer" model with respect to the magnetic effect of the underlying crustal strata. As a result of the first simulation, segments of normal and high effective magnetization were distinguished. As follows from Figure 36, the average effective magnetization of the basalt layer in the northern part of the profile, at the Japanese lineation sequence within the Northwestern Pacific Basin, is 2 A/m. Anomaly effective magnetization up to 6 A/m was registered at some segments of the northern and southern blocks of the Rise. High values, up to 4 A/m, were registered in the zones of tectonic displacements between domes, to which anomalous high heat flow (Gorodnitsky, 1985) and intensive sedimentary cover dislocations (from CSP data) are attributed. High magnetization is not typical of the basalt of the given age, which allows a supposition on a greater thickness of an active magnetic layer than the standard one of Layer 2, in some parts of the Shatskiy Rise.

To check that supposition, we calculated an active magnetic layer in the context of a two-layer model; the lower boundary of the former was assumed from the DSS data. An effective magnetization of 1.5 to 2.0 A/m is sufficient in that case to get the best coincidence of the estimated and observed anomalies, which is close, on the whole, to the average magnetization of basalts (from experimental data). That points out a considerable effect of the lower oceanic layers upon the AMF. From the latest research, Layer 3b, composed by serpentinitized peridotite of the upper mantle, seems to be such a second active magnetic layer in the lowermost crust.

The AMF in the Shatskiy Rise area is strongly affected by tectonic dislocations, part of which are either a continuation of the transform faults recognized in the adjacent basins or coincide with the plate boundaries (Figure 38d). That can partly explain a magnetic inhomogeneity of basalt uplifted from that zone: from $I_n < 0.08$ A/m (St. H4-7) to the normal one for basalt, that is, 1.9 A/m (St. H4-8, H4-9) (Lin'kova and Raikevitch, 1989). We determined the locus of the tectonic suture associated with the plate boundary (Figure 35) from a comprehensive interpretation of the available geological-geophysical data. A displacement of the suture zone between the plates occurs along the extension of transform faults 1 to 10 within the Shatskiy Rise. The suture zone trend within the Southern dome is N32E, and within the Central and Northern domes, it coincides in principle with the strike of the lineations of the Japanese sequence, placed north and northwest of the Rise, and is N65E.

It should be noted that high heat flow against thick lithosphere (Gorodnitsky, 1985), anomalous upgrading of the crust within the domes of the Rise due to Layer 3, sharp decay of the Bouguer anomaly, and intensive dislocations of the sedimentary cover may provide evidence in the region of the Shatskiy Rise for shear and compressing stresses, which lead to the formation of the nappe-thrust structures. In the latter, the overlying crustal blocks slide over a plastic serpentinite layer in accordance with the mechanism of a two-layer tectonics (Lobkovsky, 1988).

From the experimental data of a component survey (Gorodnitsky et al., 1980), a series of model curves have been calculated at a subsequent change of a magnetic inclination angle with an interval of 5° as long as they showed the best coincidence with the observed curves. From the estimates, the best coincidence of the simulated and experimental curves is reached at a paleomagnetic inclination I_n equal to $-15°$.

The obtained data indicate that the Shatskiy Rise was set near the equator and then shifted northward at an angular distance of about 35°. Such migration correlates with the dredging data performed on the 21st and 23rd cruises of the R/V *Dmitry Mendeleev*, which evidence a subsurface or subaerial character of basalt outflow on the Rise (Melankholina et al., 1981).

Quantitative estimates show that vertical subsidence of the Shatskiy Rise from the sea level to its present depth agrees well with its horizontal displacement in the northwestward direction. The performed simulation allows a supposition that within the Shatskiy Rise, an active magnetic layer has a two-stage structure, thus reflecting a stage-by-stage tectonic evolution of the oceanic lithosphere in that region.

That is verified by the paleomagnetic research. It was determined that basalts dredged from the Central dome of the Rise were formed during the epoch of reverse polarity, whereas a partial thermal remnant magnetization was formed as a result of a secondary heating of rocks during the period of direct polarity when plates moved relative to each other.

Simulation results showed clearly that the rocks composing the lowermost oceanic crust played a considerable role in the AMF formation, in particular in the zones of tectonic displacements, when an intensive process of serpentinization started within their boundaries.

The performed identification of magnetic lineations has cast some doubt on the occurrence of a triple junction of M12-M1 northeast of the Shatskiy Rise, and it allows a supposition that the ridge-ridge-transform system existed until the end of the Cretaceous Quiet Zone.

Chapter 4

Geomagnetic Study of Transform Faults

S. V. Lukyanov, G. M. Valyashko, and A. M. Gorodnitsky

CONTENTS

   I. Estimation Technique 145
  II. The Mendocino Fracture Zone 146
 III. The Murray Fracture Zone 152
  IV. The Clarion Fracture Zone 155
   V. The Heezen Fracture Zone 158
  VI. The Kurchatov Fracture Zone 162
 VII. The Emperor Fracture Zone 165

Up to now, transform faults have been scarcely known as compared to the axial zones of mid-oceanic ridges. However, they are of special geological concern since deep parts of the oceanic crust within these structures are often accessible to researchers and their examination provides a key to understanding the processes at the boundaries of the conjugated lithosphere blocks.

A program comprising a series of cruises was implemented in the P. P. Shirshov Institute of Oceanology, Russian Academy of Sciences, to study the Pacific transform faults, namely, Heezen, Kurchatov, Galapagos, Emperor, Mendocino, Murray, Clarion, and some others. In the fracture zones, a large scope of geological sampling was made besides the complex geophysical survey, and by now, petromagnetic analysis of a set of the sampled bedrocks has been performed (Gorodnitsky et al., 1984; Popov et al., 1989).

The Pacific faults are characterized morphologically and geologically by a considerable shear constituent, difference in the bottom depth, and lithosphere thickness of the young and ancient sides of the fault (Gorodnitsky, 1985). The younger side of the fault is uplifted as compared to the ancient one, and a fault-line ridge is located on the young side. The difference in the depth to the bottom of the sides is proportional to the difference in their age (Sorokhtin, 1973). On the basis of modern techniques of simulating the anomalous magnetic field (AMF) and from the rich *a priori* geological-geophysical information, obtained in the expeditions, we have attempted to investige the relations between the AMF and the crust morphology and structure in the transform zones and to estimate the probable effect of the lower oceanic crust upon the AMF.

I. ESTIMATION TECHNIQUE

A two-dimensional magnetic simulation of the layers, with real topography and thickness known from the deep seismic study, was made. We also used a technique of generalized linear inversion together with the adaptive reparameterization of the model that allowed us to get stable results in agreement with the *a priori* information and data of hydromagnetic survey and petromagnetic analysis of samples. The technique used is described in Chapter 2. We shall emphasize here that the use of an adaptive approach allows us to estimate the representativeness of the primary data, that is, to get a geologically valid model.

It is common knowledge that solution of an inverse magnetometric problem that is the basis of the simulation is ambiguous, and formal use of any techniques of simulation cannot give a definite answer to the problem in question (Strakhov, 1981). This made it necessary to attract additional geological-geophysical information. Its usage narrows the ambiguous solution and allows us to compare the models of the medium structure obtained for various regions of the ocean.

According to the available data, mean magnetization of an active magnetic layer is 2 to 4 A/m within oceanic basins. Numerous data of the AMF simulation performed to identify magnetic lineations are based on the assumption that the basaltic layer of the oceanic crust is the only source of magnetization, its thickness being 0.5 to 1.0 km. From that assumption, magnetization of the layer should be 10 to 15 A/m, which exceeds the mean value obtained by petromagnetic analysis.

There are numerous publications that generalize the satellite, hydromagnetic, and petromagnetic data that have revealed the effect of the lower crustal layers upon the formation of magnetic anomalies, in particular, of serpentinized peridotite (Won, 1985; Gorodnitsky et al., 1987, 1990; Pecherskiy et al., 1986).

Therefore, when calculating, the roof of the active magnetic layer was associated with the crystalline basement top, and the layer thickness was assumed equal to that of the crust obtained from DSS data.

In that case, if the real thickness of the active magnetic layer is considerably less than that of the oceanic crust, the estimated effective magnetization should be considerably lower. This is true for the opposite situation. To estimate mean effective magnetization in the conjugated blocks of the crust, in each region, calculations were performed along the profiles parallel to the fault strike. Then, they were compared with the magnetization value obtained when calculating along profiles across the fault strike. Estimated magnetization was controlled by petromagnetic data.

Below, we present the results of geomagnetic study for individual transform faults.

II. THE MENDOCINO FRACTURE ZONE

The Mendocino fracture zone is the largest in the Pacific and extends for 6800 km from the Mendocino Cape to 170° E. It is expressed in the bottom topography as a series of ridges, depressions, and scarps. In some places, the zone width reaches 80 miles. The zone has been examined the most thoroughly in its eastern part; west of 135, it has been examined poorly (Carres et al., 1988; Rea et al., 1983; Raff et al., 1961; Klitgord and Mamerickx, 1982).

During the 42nd cruise of the R/V *Dmitry Mendeleev*, two study areas were surveyed within that zone, namely, M-4 and M-5. Study area M-4, with the center in the point of 137° N and 165° E, is located west of lineation 32a, where the age of the crust is 80 million years old. We use here only the petromagnetic data obtained within that study area.

Eleven profiles of complex geophysical survey were made within study area M-5, with the center in the point of 39°40 N and 145° W. Magnetic lineations 15 and 13 were mapped by two latitudinal profiles on the northern side of the fracture. So, it became possible to determine the age of the oceanic crust on the northern side of the fault as 39 to 37 million years old. The age of the southern side is dated back to 61 to 55 million years from anomaly 27 located near the western border of the study area and from anomaly 26 located near the eastern border of the area.

In the bottom topography (Figure 1A), the fault is expressed as a huge scarp of latitudinal elongation with a general depth change of 2 km. The altitude of the fault-line ridge, located at the younger northern side, is about 1 km. A trough about 5.5 to 6.0 km deep runs parallel to it. A seamount with a relative elevation of 2.2 km is recognized on its southern side. According to paleontological analysis made by Saidova, the age of the sediments uplifted from the seamount summit is 55 million years old, which agrees well with the age determined from magnetic anomalies.

In the magnetic field, the transform fault is associated with a sign-reversed anomaly (Figure 1B). Its extremum is confined to the ridge crest or is shifted toward the younger side. In some places, its amplitude exceeds 600 nT. The extremum sign corresponds to the polarity of the magnetic lineation approaching the fault from the younger northern side. The difference in the strike of magnetic lineations on the southern and northern sides of the fault does not exceed 2°, and they are practically orthogonal to the fault strike.

Figure 2 presents the estimates from latitudinal profiles 5-N and 5-S. On the northern side of the fault, the effective magnetization I_c reaches 2 and 3.5 A/m in the vicinity of anomalies 15 and 13. It is possible to recognize there distinct local changes in magnetization corresponding to a fine spatial structure of the field, which is verified by gradient measurements. Amplitude of these changes is about 1 A/m. On the southern side of the fault, a seamount with effective magnetization up to 5 A/m is pronounced in the western part of profile 5-S, whereas in its eastern part, anomaly 26 is located (event 26n). Mean magnetization along the profile is about 1.5 A/m, and mean effective magnetization for both sides of the fault does not exceed 2 A/m.

At the meridional profiles 5-1 and 5-3, estimated I_n does not exceed 2.5 A/m and agrees well with its mean values on the latitudinal profiles. Most intensive anomalies over the fault-line ridge exceeding 600 nT originate at the contact of different-sign anomalies along the fault. So, profile 5-1 is located in the reversed event 12r on the northern side of the fault and in the normal event 26n on its southern side; profile 3 is located in the normal event 13n on the northern side and in the reversed event 26r on the southern side of the fault.

Figure 2 illustrates the simulation estimates along profile 5-3 showing the extremum northward displacement over the fault-line ridge toward the younger side of the fault, which may be explained by faulting due to uplift of the fault-line ridge.

Analyzing the simulation results, we can conclude that estimated effective magnetization on the fault sides (no more than 3 A/m) corresponds to the values typical of the conjugated basins.

Within the study area, the northern side of the fault and the seamount on its southern side were dredged. Pillow lava fragments of different ages comprising mainly and exclusively the second layer of the oceanic crust were uplifted there.

The principal petrographic characteristics are given in Tables 1, 2. Petromagnetic analysis detected that titanomagnetite in the rock underwent one-phase low-temperature oxidation. Curves of the thermomagnetic analysis show the titanomagnetite decay with a magnetite generation. A sharp increase of magnetization saturation is observed at the secondary heating. Basalts from both sides are characterized by the same Curie temperature (T = 300 to 310° C), but by different ferromagnetic concentration. So, in the rocks from the southern side, it is 1.5 times less than in the northern one, which is indicated by I_s values; dimensions of the ferromagnetic grains are less there, which follows from the I_{rs}/I_s, H_c, H_{rc} ratio (Table 1).

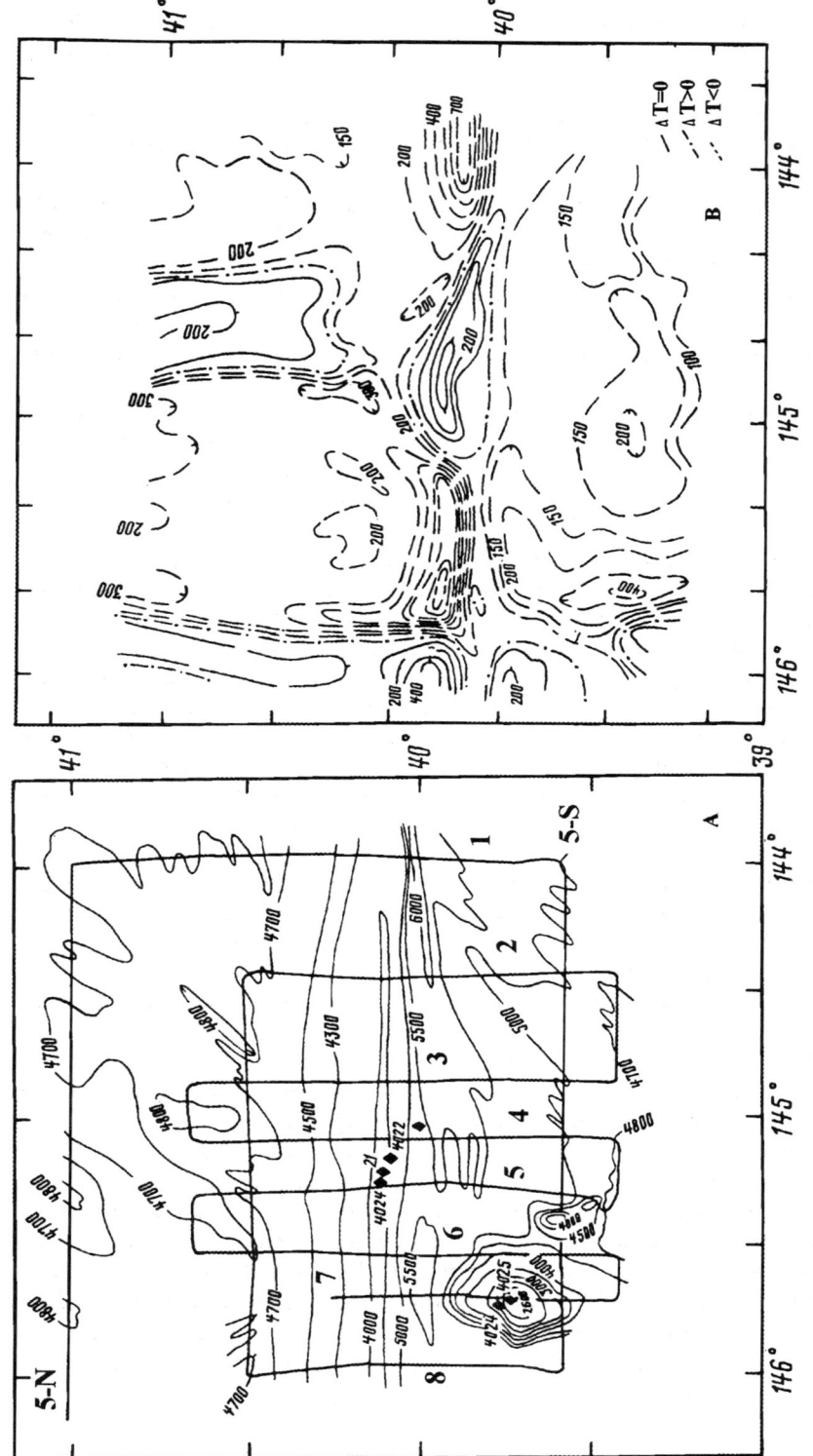

Figure 1 Mendocino fracture zone area M-5: (A) scheme of profiles and bottom topography map of AMF, (B) isodynams are drawn in each 100 nT.

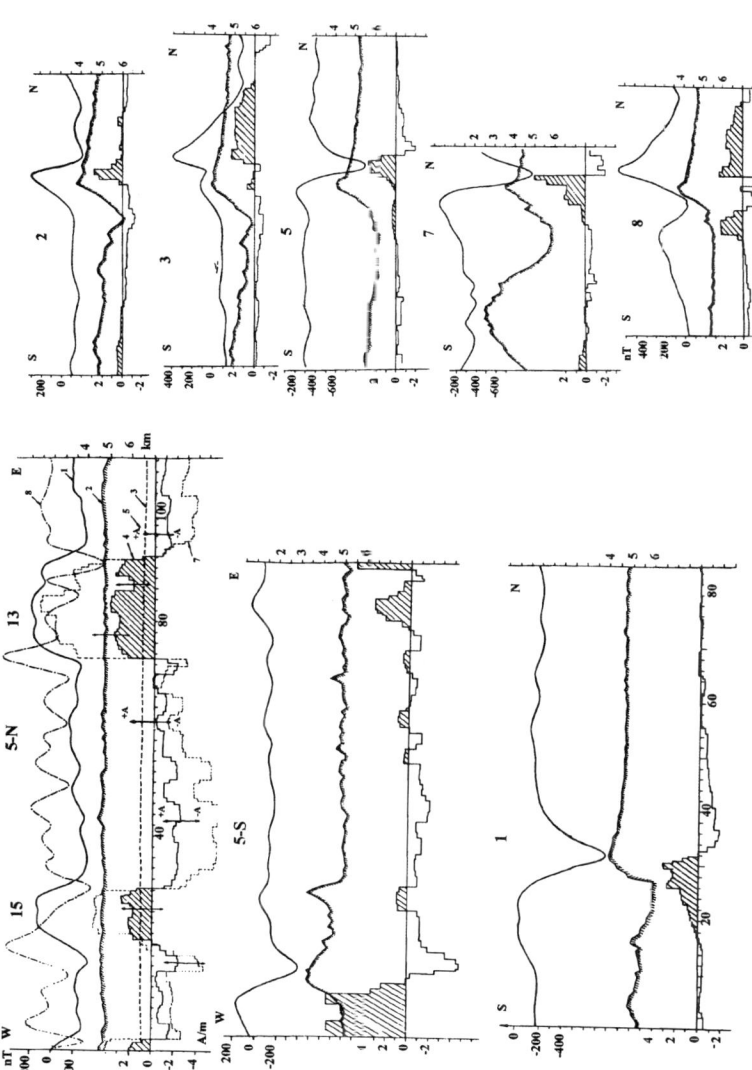

Figure 2 Estimated effective magnetization I_c: Numbers above graphs are profile numbers. Numbered lines within graphs: (1) AMF; (2) bottom topography; (3) annihilator; (4) effective magnetization I_c in A/m; (5) confidential interval; (6) effective magnetization I_c at active magnetic layer thickness H = 2 km; (7) measured horizontal gradient (base = 150 m).

Table 1 Mean petromagnetic characteristics of rocks sampled in the Mendocino fracture zone (M-4 and M-5), the Murray fracture zone (M-6), and the northern side of Clarion fracture zone[a]

Number of area; age crust, millions of years	N of samples	I_n, A/m	$\kappa \times 10^{-3}$ emu cm³ Oe⁻¹	Q	$I_s \times 10^{-3}$ A/m	I_{rs}, A/m	I_{rs}/I_s	H_c, Oe	H_{rc}, Oe	H_{rc}/H_c	$T°_{c1}$	I_{st}/I_s
M-4; 80	15	3.00	0.67	14.2	0.40	0.08	0.23	138	217	1.65	380	2.07
		1.95	0.44	14.1	0.13	0.03	0.08	94	139	0.17	36	0.84
	11	1.49	1.88	1.9	0.97	0.08	0.08	84	150	1.8	513	0.86
		1.56	0.89	2.7	0.27	0.02	0.04	44	87	0.22	49	0.25
	8	2.03	1.59	4.6	0.80	0.09	0.14	110	200	1.86	596	0.79
		1.98	1.11	6.6	0.51	0.03	0.06	45	70	0.16	12	0.19
	3	0.11	0.04	3.1	0.08	0.005	0.07	150	300	2.0	575	1.15
M-5	5	1.12	0.14	16.0	0.24	0.05	0.22	238	434	1.86	304	1.37
NB; 40		0.62	0.03	7.6	0.1	0.03	0.12	133	253	0.27	64	0.12
SB; 60	11	0.44	0.07	15.4	0.16	0.055	0.36	362	546	1.52	310	3.79
		0.26	0.05	6.6	0.05	0.01	0.06	171	270	0.13	43	1.06
M-6	24	1.12	0.16	16.5	0.2	0.07	0.37	307	434	1.43	340	5.44
SB; 30		844	0.11	13.8	0.06	0.03	0.06	39	57	0.16	21	1.36
NB; 46	12	1.15	0.11	24.4	0.14	0.05	0.32	292	427	1.47	386	3.15
		0.77	0.03	12.6	0.04	0.02	0.07	34	43	0.16	31	0.99
M-7, 2490		0.34	0.01	7.1	0.06	0.011	0.165		235		475	2.19
M-7, 2485		1.68	0.51	6.9	0.30	0.014	0.487		256		367	5.05

[a] I_n is a natural remnant magnetization; T_{c1} is the Curie point, determined from the function $I_s(T°)$ is the course of first heating; T_{c2} is the Curie point after the heating to 600° C; κ is magnetic susceptibility; Q is Koenigsberger factor; I_{st} is magnetization of saturation after heating to 600° C; I_s is magnetization of saturation (bulk); I_{rs} is remnant magnetization of saturation; H_c is a coercive force; H_{rc} is a remnant destructive field; I_i is an ideal magnetization at a constant magnetic field, 1 Oe. NB northern side of the fault; SB, southern side of the fault.

Table 2 The northern side of the Clarion fracture zone

No. of area	I_{rp} A/m	10^{-3}emu cm^3 Oe^{-1}	Q	I_s A/m	I_{rs} A/m	I_{rs}/I_s	H_{rc} Oe	$T°_{c1}$	I_{st}/I_s
2490-4-A	0.28	0.08	7.1	0.03	0.003	0.100	195	500	1.33
2490-4-B	0.39	0.11	7.0	0.08	0.019	0.230	275	450	3.05
2495-1	1.86	0.64	5.8	0.51	0.166	0.330	140	370	4.30
2495-12-A	1.96	0.66	6.0	0.28	0.141	0.500	175	400	2.70
2495-12-B	1.37	0.42	6.5	0.22	0.134	0.610	240	375	8.00
2495-12-V	0.95	0.39	4.9	0.29	0.124	0.430	285	375	6.20
2495-12-C	2.43	0.67	7.3	0.37	0.140	0.380	165	275	4.30
2495-12	2.07	0.32	13.1	0.17	0.096	0.560	465	375	5.20
2495-27	1.10	0.45	4.9	0.26	0.157	0.600	325	400	4.64
ev. 2495	0.34	0.09	7.1	0.06	0.011	0.165	235	475	2.19
ev. 2495	1.68	0.51	6.9	0.30	0.137	0.487	256	367	5.05

III. THE MURRAY FRACTURE ZONE

The Murray fracture zone extends from the American continental slope rise to the Hawaiian Ridge. Compared to the Mendocino fracture zone, it is straighter, its strike is 70°, and it has complex morphology. East of the study area, regional geophysical research was carried out by American scientists (Malachoff et al., 1976).

Bottom topography within area M-6, with coordinates of 33°36 to 32°14 N and 135°40 to 134°20 E, is more complex than that of the Mendocino zone. Several echelon-like scarps are distinguished there with a depth of 3500 to 4500 m and oblique inclination to the fault strike (Figure 3A). The altitude of the fault-line ridge crest along the southern side of the fault varies from 750 to 200 m from the east to the west. The altitude of the southern side varies from 300 to 1700 m with respect to the position of the profile on the fault-line ridge (whether it crosses the ridge on its slope or on its summit). A system of scarps with various altitudes and tilted southward is registered along the profile (due to echelon-like strike), the altitude of their northern slope varies from 500 to 1700 m (Figure 4, prof. 4). All the above comprise a rugged character of the fracture zone topography. The width of the fracture zone is about 25 miles, and the zone is characterized by a sign-reversed magnetic field (Figure 3B). The northern side of the fault is gentle, with slightly stratified bedding and depths varying from 4800 to 4900 m. Against it, a seamount range, with depths over their summits of about 3800 m, is registered in the eastern part of the area (Figure 3A).

From DSS data, the thickness of the crust has been determined to be about 7 km on the northern side of the fault and 4 km on its southern side. No rocks of deep oceanic crust layers have been dredged within study areas M-4, M-5, and M-6 on the 42nd cruise of the R/V *Dmitry Mendeleev*. All samples uplifted are related to the 2-d layer of the oceanic crust. Mean petromagnetic characteristics, given in Table 1, indicate that the basalt and dolerite analyzed had experienced practically all types of secondary changes, namely, from a low-temperature one-phase (extremely oxidized in many cases) to a high-temperature (no less than 250°) heterophase changes with subsequent oxidation of the decay products of the primary titanomagnetite. Analyzing these data, it is possible to say that almost all rocks sampled from the Mendocino and Murray fracture zones, except for the basalt of group 4 and dolerite of groups 2 and 3, within area M-4, have considerable remnant magnetization. Positive correlation ($K = 0.84$) is observed between anomaly intensity in the sites of sampling and measured magnetization.

According to the performed geomagnetic survey (Figure 3B), anomaly 20r, about 46 million years old, has been mapped on the northern side of the fault and anomaly 10r, about 30 million years old, on its southern side. The southern side of the fault is characterized by an echelon-like system of sign-reversed anomalies with amplitudes of 200-280 nT to 300-550 nT, when crossing the southern side in the north-south direction. At the given map of AMF, the level of survey has been lowered by 250 nT. Figure 4 shows the estimated I_n at the corresponding thickness of the Earth's crust at both sides of the fault. The dashed line shows the annihilator for profile 4, the level of which is 1 A/m. The annihilator is 0.8 A/m for the southern scarp and 0.5 A/m for the northern one. Mounts at the northern side are marked by a lowering of the annihilator to 0.7 A/m, whereas in the anomalous magnetic field they have amplitudes up to 150 nT. A simultaneous analysis of the magnetic field, CSP data, and estimated I_n supposes that these seamounts are set at the tectonic suture parallel to the trough between two crustal blocks with different magnetization.

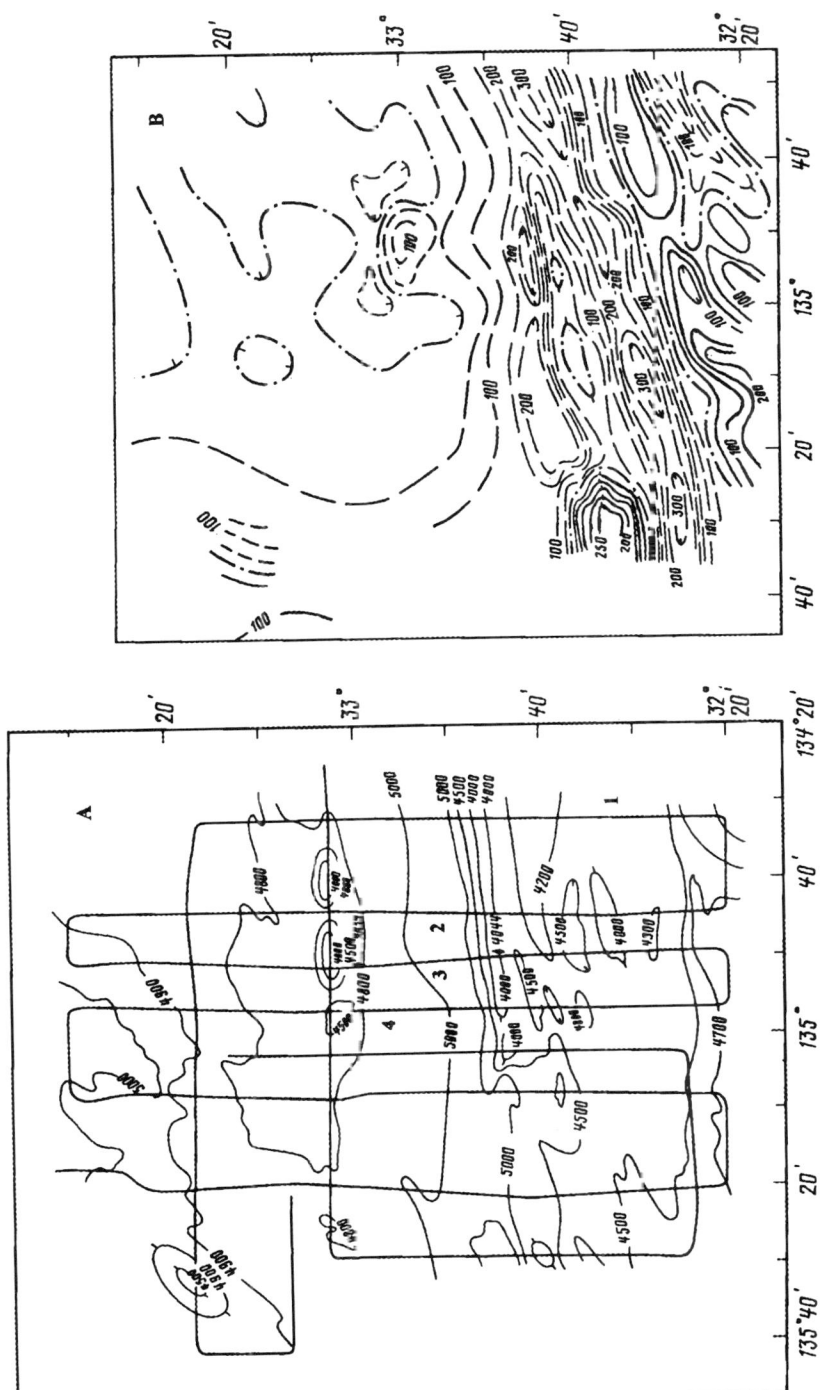

Figure 3 Study area M-6, Murray fracture zone: (A) profile pattern and bathymetry; (B) AMF map.

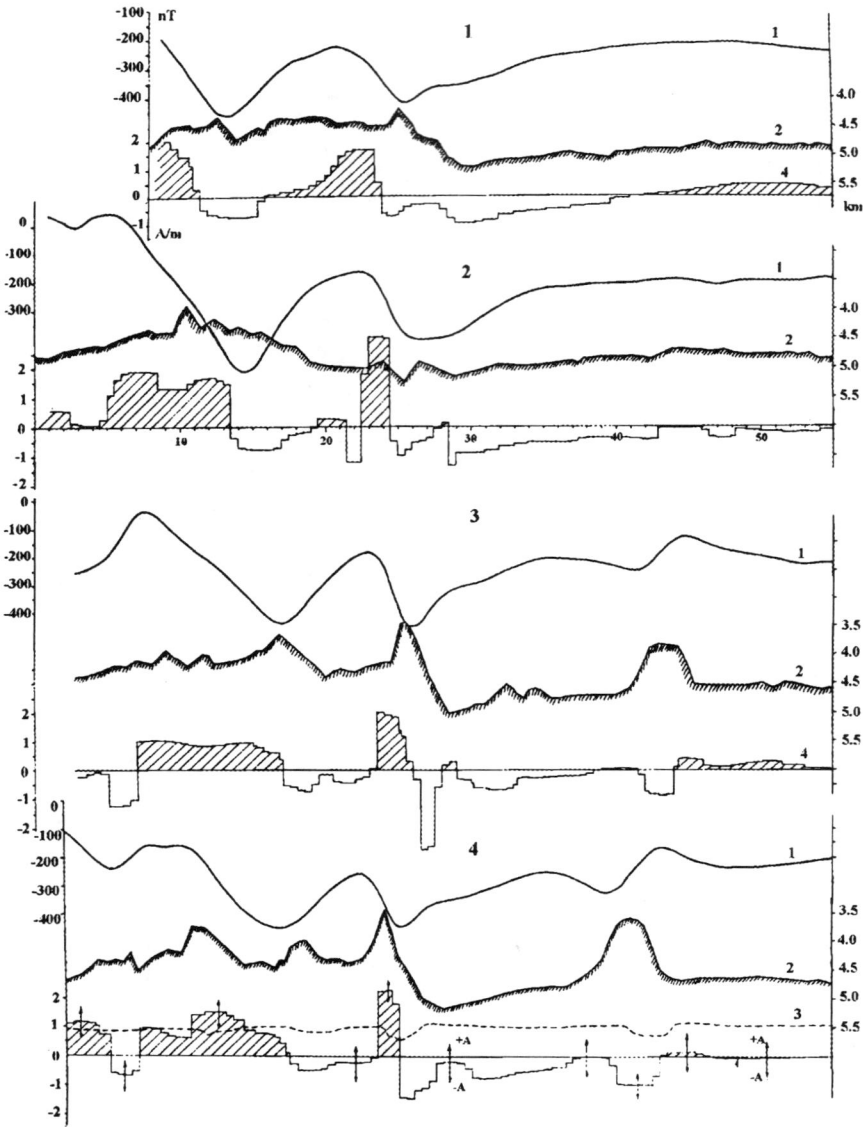

Figure 4 Estimated effective magnetization I_c at the Murray fracture zone, area M-6. For notations see Figure 2.

The northern block is characterized by slightly positive (to 0.5 A/m) or even zero values, whereas the southern block has negative values (I_n = 0.5 to 0.8 A/m). The seamount range itself has values of −1 A/m. A scarp 200 m high in the basement topography corresponds to that boundary (Figure 4, prof. 1 to 3). The above said allows a supposition on the heterogeneous structure of the northern side. However, most pronounced tectonic heterogeneity is expressed on the southern side of the fault, where curves of I_n have some peaks of 2 to 3 A/m confined to tectonic discontinuities, and also blocks with I_n = 1 to 2 A/m, which are attributed to echelons in the bottom topography.

Comparing petromagnetic data with the solution of the inverse problem on estimation of magnetization, we obtain good agreement of the estimated and measured values of I_n for the sites of dredging.

IV. THE CLARION FRACTURE ZONE

The Clarion fracture zone is located at the boundary of Cenozoic magnetic lineations and quiet magnetic field. The survey was performed on the 28th cruise of the R/V *Dmitry Mendeleev* along 12 profiles, with a total extension of 330 miles; area M-7 has coordinates 14°03.5-14°50.5 N and 146°33-147°20 W.

The fault has a complex morphology. The valley has a W shape, with a maximum depth of 6.5 km. A fault-line ridge, elevated by 700 to 1000 m over the abyssal basin, is traced along the southern side of the fault. The southern slope is deformed by echelon-like northward displacements for 15 to 17 km. The southern side of the fault is complicated by a series of steps in the western and central parts of the area. Their altitude reaches 700 m for the lower scarp and 1000 m for the upper one (Figures 5A and 6). In the central part of the area, following the displacement observed between profiles 1 and 9, the lower scarp disappears in the bottom topography, and we can trace only the upper one, the altitude of which reaches 850 m. The gap between the Clarion fault walls decreases from 23 to 13 miles. In the eastern part of the area (Figures 5A and 6), the transform fault valley has a V shape.

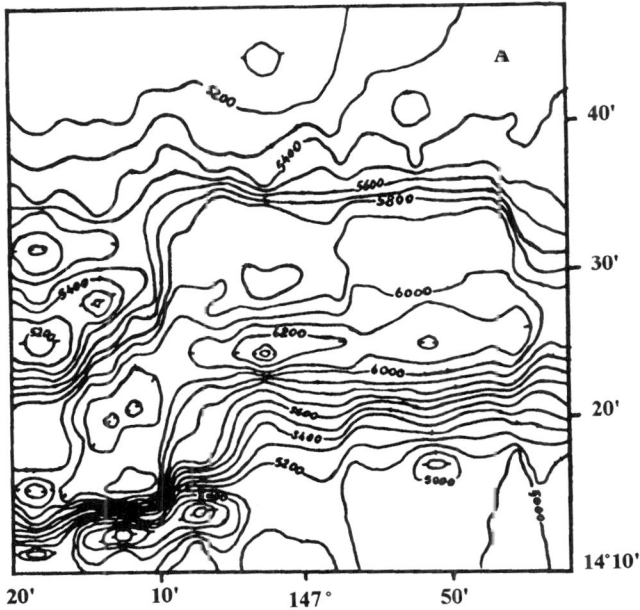

Figure 5 Area M-7, Clarion fracture zone: (A) bathymetrical map of the area; (B) AMF map and profile pattern.

From the analysis of geomagnetic data obtained in the study area (Figure 5B), we can estimate the age of the northern side of the fault as 72 to 78 million years old. The southern side is somewhat younger, but it is not possible to identify the anomaly exactly.

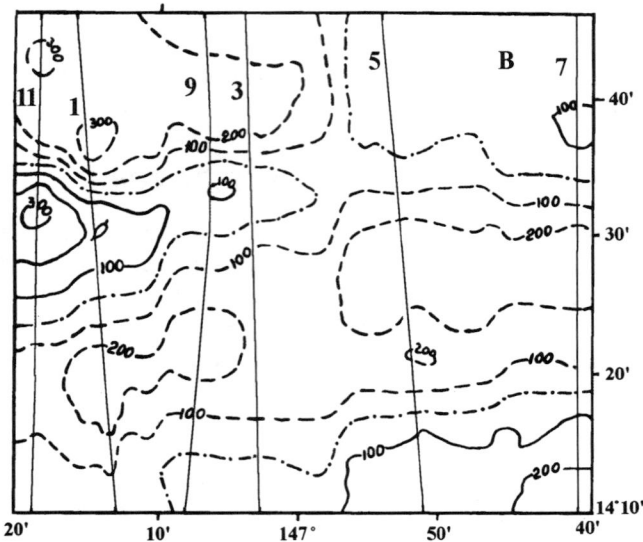

Figure 5 (continued). Area M-7, Clarion fracture zone: (A) bathymetrical map of the area; (B) AMF map and profile pattern.

The AMF of the area has a small amplitude in general. On the southern side, in the vicinity of the fault-line ridge, a sign-reversed anomaly up to 150 nT is recognized. The magnetic field distinctly reflects the echelon-like displacement. In the trough vicinity, a negative anomaly area is detected, which coincides with its strike. A positive anomaly of about 400 nT is traced along the northern side of the Clarion fault in the western and central parts of the study area. It disappears in the eastern part of the area.

The DSS data show the thickness of the oceanic crust in that region to be 5 km (Kosminskaya et al., 1987). From the estimates given in Figure 6, it follows that mean magnetization for the northern side of the Clarion fault is about 1 A/m, and for the southern side, it is 0.7 A/m. The estimated magnetization of 0.7 to 1.5 A/m characterizes the fault-line ridge at the meridional profiles (9, 3, 5, 7), whereas a negative anomaly confined to the fault trough is characterized by reversed magnetization of the active magnetic layer with I_n to 1.5 to 2.5 A/m. According to our estimates, the most intensive positive anomaly to 400 nT is caused by a narrow block with effective magnetization of 3.5 to 6 A/m.

Almost a complete sequence of the crust has been exposed by dredging within study area M-7 (Melanhkolina et al., 1983). Basalts have experienced all types of secondary changes, and their remnant magnetization is not great, 0.7 to 1 A/m and sometimes 2 A/m. Petromagnetic analysis of the dredged rocks is given in the work by Popov et al. (1989) Table 2. When comparing the estimated magnetization with the data of petromagnetic analysis of the dredged rocks, we may say that they agree well with magnetization of serpentinite (3 to 4 A/m), found within the area. We shall note, however, that serpentinite has a mean magnetization of 3 to 4 A/m, has a Koenigsbergen factor of 1, and may be assumed to be the source of magnetization.

Generalizing the results of estimations and petromagnetic analysis of the dredged rocks, we may conclude the following. In the faults, where shear with compression predominates, the registered magnetic anomalies are the superposition of the marginal effects of magnetic lineations that contact along it. The sign of the anomaly over the fault-line ridge is determined by the sign of magnetic lineation approaching from the younger side. Most intensive anomalies are observed at the contact of anomalies of opposite polarity. This is most pronounced in the Mendocino fracture zone within area M-5.

157

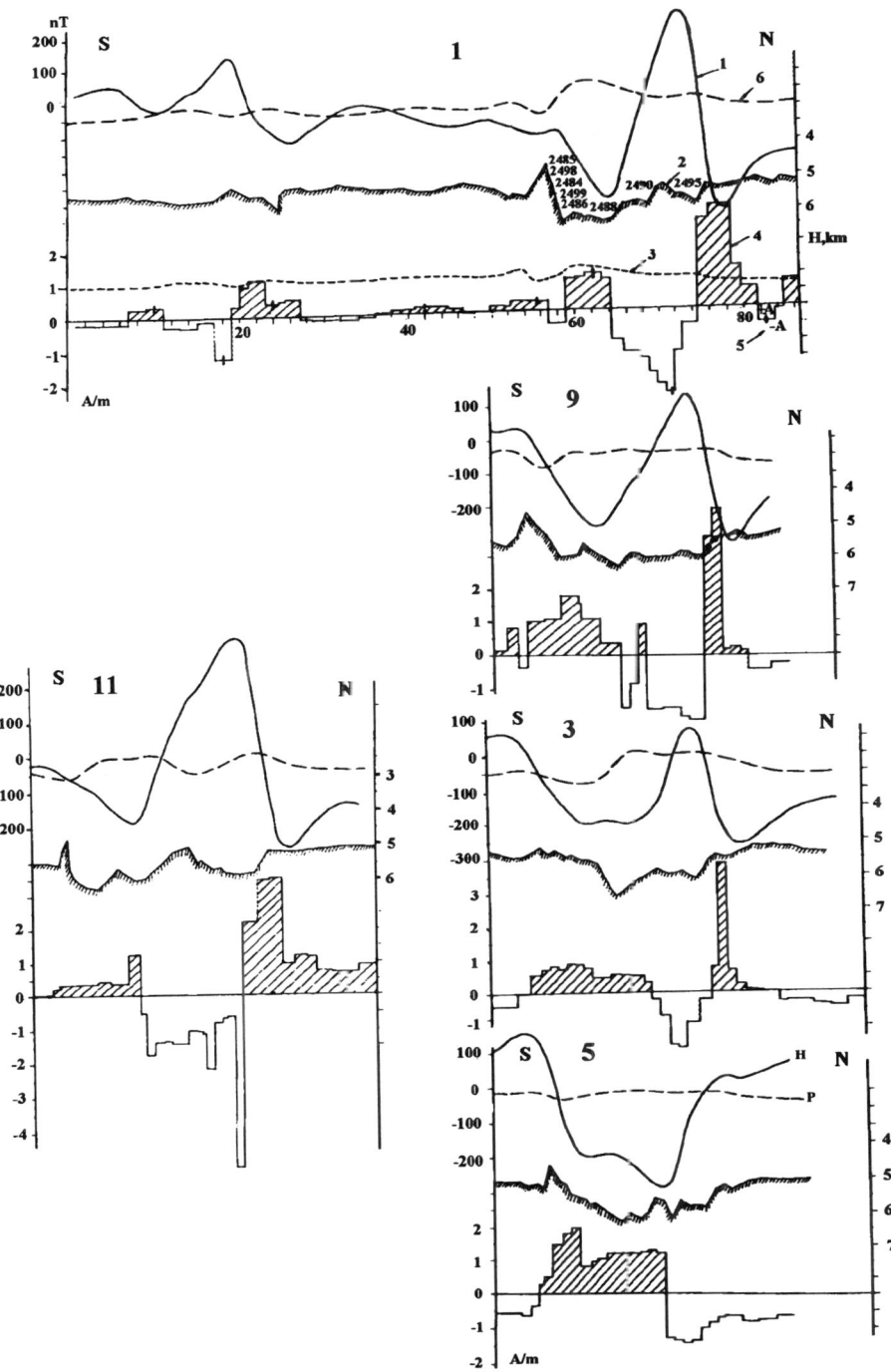

Figure 6 Estimated effective magnetization I_c and numbers of stations of dredging at profile 1. For notations see Figure 2.

In faults that have an oblique dip toward the spreading direction, where an echelon-like system of fissures facilitating the water penetration is observed, an active magnetic layer may be built from below due to serpentinization, which may affect the magnetic field itself. Study area M-6 within the Murray fracture zone, may serve as an example of this.

In faults that have predominating extension strain and a considerable gap between their walls, the abovementioned regularities are preserved, but anomalies over their trough are the function of their local sources, in particular, serpentinized peridotite of the lower oceanic crust. This is illustrated well by simulations within area M-7 in the Clarion fracture zone.

V. THE HEEZEN FRACTURE ZONE

On the 24th cruise of the R/V *Akademik Kurchatov*, complex geological-geophysical study was performed in the active part of the fault within study area M-8, with coordinates of 54°30 to 55°45 S and 124°30 to 127° N (Gorodnitsky et al., 1980). At present, the main results of that expedition have been published. More recently, however, American scientists performed a detailed bathymetric survey of that fault using a precessional multi-beam echo-sounder (Lansdale, 1986). A simultaneous analysis of these data, the results of the geomagnetic study of the 24th cruise of the R/V *Akademik Kurchatov,* and the AMF simulation results allowed us to approach the estimation of the possible effect of the lower oceanic crust upon the formation of anomalies.

In reviewing the published results of the geomagnetic studies, we note that magnetic anomalies are recognized on the northern and southern sides of the fault. The southern side is the younger of the two. Anomalies 2-2a of 1.67 to 3 million years old are identified there, whereas anomalies 3a-5 of 5.4 to 8.9 million years old are recognized on the northern side. In the topography, the fault is expressed as a huge scarp with a trough as deep as 6.1 km. The summit of the fault-line ridge in some places is located at a depth of 600 m (Figure 7).

The anomaly amplitude at the northern side is 200 to 300 nT, whereas on the southern side, it is 300 to 400 nT. Sign-reversed anomalies are confined to the fault-line ridge, and their amplitude reaches 1000 nT. In plan, they coincide with the ridge strike. The western margin of the study area extends only 25 miles from the rift zone located between the Heezen and Tharp transform faults. Considering Lansdale's data on the Eltanin fracture zone, it is interesting to analyze the integrated latitudinal profile drawn in Figure 8. It starts from the rift zone axis located 25 miles to the west of the study area. As a basis we took Lansdale's (1986) bathymetric data obtained by a multi-beam echo-sounder and supplemented them with data on the bottom topography and geomagnetic measurements. Considering the function reflecting the lithosphere plate subsidence with age, it is possible to conclude that the subsidence depth along the southern side of the transform fault is greater than along its northern side since the depth of the spreading ridge crest of the Heezen fault is 250 m deeper than that of the EPR. As a result, the area of equal depth on both sides of the fault is shifted by 50 km northwestward of the middle of the active part of the transform fault. The general regional subsidence of the oceanic crust is about 1800 m along the active part of the fault. The fault-line ridge is elevated at 2.8 km over the abyssal plain. From the multi-beam echo-sounding (Lansdale, 1986), the fault-line ridge is traced on both sides of the spreading ridge between the Heezen and Tharp transform faults, and it is not traced along the EPR and the Pacific-Antarctic Ridge.

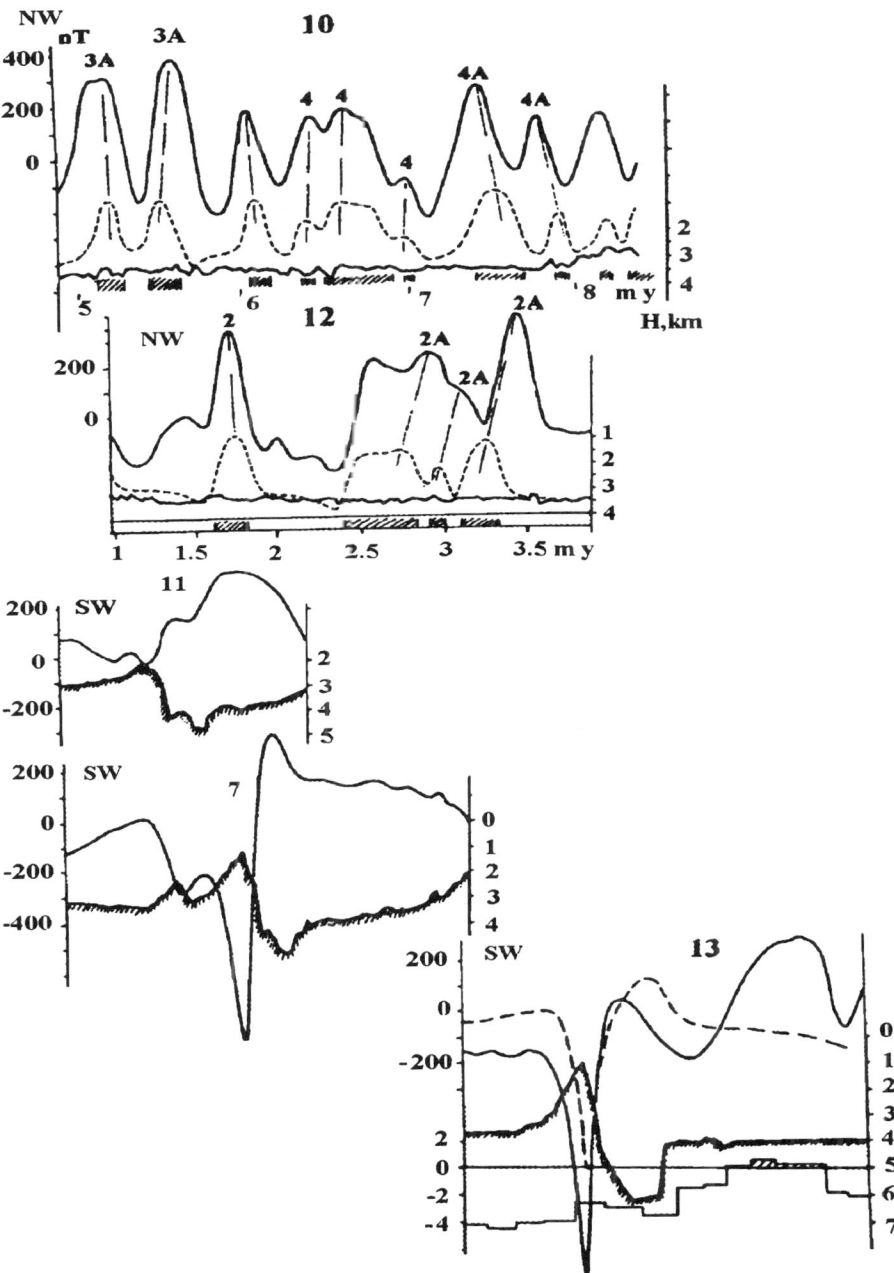

Figure 7 Area M-8, Heezen fracture zone, and estimated effective magnetization I_c. For notations see Figure 2.

It is common knowledge that in the Eltanin fracture zone, starting from anomaly 7 (26 million years old) to anomaly 3 (4 million years old), the spreading reorganization happened ten times and that 4 million years ago the Heezen fault opened.

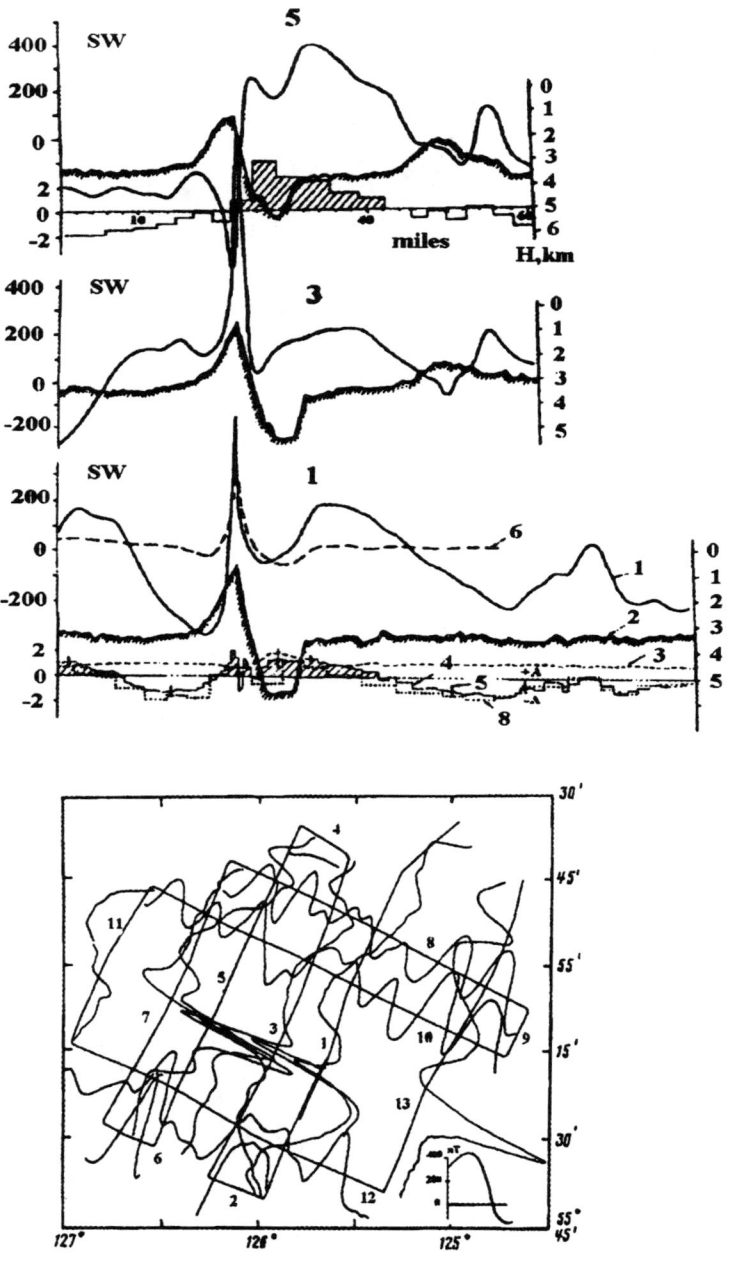

Figure 7 (continued).

It was the moment, as follows from the figure, that the fault-line ridge was set. Knowing the spreading rate, it is possible to calculate the vertical formation rate of the fault-line ridge and trough. As follows from the figure, formation of the fault-line ridge was completed by 0.5 million years ago. The bulk of the ridge was formed already 1 million years ago, and the ridge reached its maximum height by 2.5 million years (anomaly 2a). Therefore, the formation rate of the fault-line ridge may be estimated as 2 cm per year.

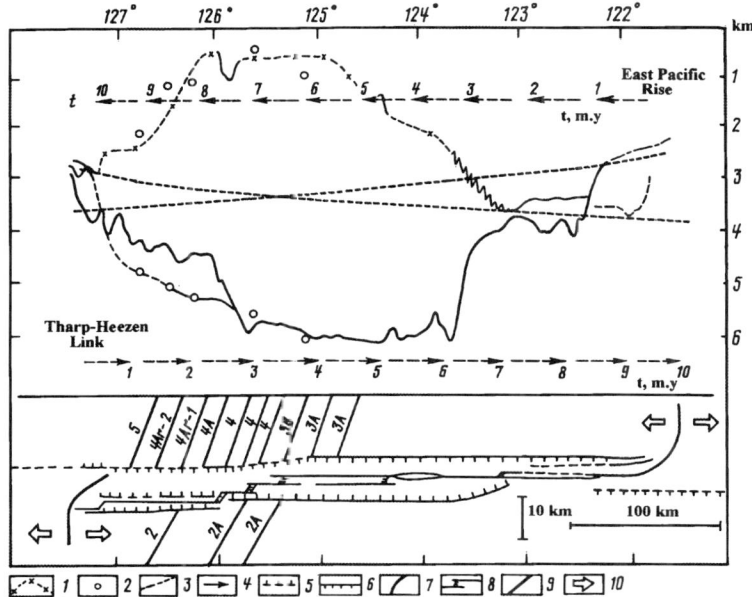

Figure 8 Integrated latitudinal profile by Lansdale (1986), supplemented by new data on the bottom topography and magnetic measurements: (1) profile of fault-line ridge topography; (2) measured depths on southern and northern sides; (3) estimated curves of lithosphere block subsidence; (4) time scale; (5), (6) fault walls; (7) rift zones; (8) fault thalweg; (9) magnetic lineations; (10) spreading trend.

According to the data by Bonatti (1976) and in the assumption that the fault-line ridge was formed due to serpentinization exclusively, which is 50% of the rock body, the uplift rate of the ridge is 1 cm per year.

Considering this, we note that the fault-line ridge formation might occur due to the summary activity of two processes, namely, serpentinization of peridotite from marine water penetration and shear strain originating from the spreading process reorganization. Dredging results and data on petromagnetic properties of rocks uplifted from the fault-line ridge and trough speak in favor of the latter (Kashintsev and Rudnik 1984; Gorodnitsky et al., 1984). According to the data by Lansdale, the serpentinite outcrops are associated with small uplifts about 300 m high extending along the fault trough.

A complete sequence of the crust from the southern wall of the Heezen fault in its active part was analyzed for petromagnetic characteristics. In the uppermost part of the sequence, along profile 1, St. 2175, 2179 (662 to 1000 m), porphyritic and olivine pyroclastic basalts were uplifted. Their natural remnant magnetization is 2.8 A/m (Gorodnitsky et al., 1984). Lower in the section, at St. 2174, dolerite with natural remnant magnetization lower than that of the basalt (I_n = 1.3 A/m) at Q = 5, was recognized. At St. 2172 (2300 to 3000 m), gabbroids, which had experienced greenstone metamorphism, were uplifted. According to their mineral composition, apoharzburgite serpentinite predominate there; serpentinization process has captured 80% of rock there. The rock had experienced secondary hydrothermal changes, with the fissures filled by carbonate. Magnetic examination detected the occurrence of two magnetic minerals in the rock, which is reflected in the occurrence of two phases there, with T_c = 300°C (maggemite) and T_c = 580°C (magnetite). The Koenigsbergen factor for serpentinite is 1, and remnant magnetization is 1.5 A/m. That supposes considerable heterophase changes of the rock at several phases of their formation.

Amphibolite schists uplifted from a depth of 5200 to 5400 m (St. 2173) had formed in the environment of the transition from epidote-amphibolite to amphibolite phase at the basalt metamorphism. Schists are distinctly laminated. Intensive schistosity of rocks indicate the occurrence of compression processes there and local increase in pressure in the zone of their formation. High magnetic susceptibility and minimum Q factor (0.19) are typical of the schists, which reveals predominance of the induced magnetization in them. Magnetic characteristics of amphibolite schists evidence the coexistence of several magnetization directions that are associated with various minerals, and their summary effect produces small I_n values. When considering the rocks down the section, we can note the lowering of the Q factor from 28 to 0.19, increase in magnetic susceptibility and its anisotropy (Gorodnitsky et al., 1984), which may evidence great pressure during the period of formation or metamorphism of magnetic minerals in the rock.

Results of geomagnetic study and simulation (Figure 7, prof. 1) allow a conclusion that high-amplitude sublatitudinal anomalies over the fault-line ridge were formed due to superposition of magnetic lineations on both sides of the fault. The topography effect upon the anomaly formation is great; from our estimates, it is over 600 nT in these high-amplitude anomalies. Annihilator over the summit of the fault-line ridge reaches 0.2 A/m. It increases to 1.8 A/m in the trough vicinity, which evidences a great topography effect of the active magnetic layer. On the fault sides, annihilator value is about 1 A/m. Magnetization over the fault-line ridge summit shows some features due to instability of the solution of the inverse problem for that point. Magnetization in the fault trough and on the fault-line ridge does not exceed 2 A/m, and approximately the same values are registered at its sides. Similar results were obtained for profile 3. However, at profiles 5 and 13, the obtained effective magnetization of the rocks in the trough reaches 3 to 4 A/m. Analyzing the obtained results, we associate these values (3 to 4 A/m), confined to the fault trough, with the possible effect of the rocks from the lower oceanic crust, in particular serpentinite, upon the AMF formation. In the Heezen fracture zone trough, their effect upon the AMF does not exceed 20%.

So, in the active part of the Heezen fault, the principal contribution to the formation of its AMF is ascribed to the effect of a superposition of magnetic lineations, contacting along the fault, and its topography, the formation of which is associated with the change in the spreading direction, and also with serpentinite of the lowermost oceanic crust.

VI. THE KURCHATOV FRACTURE ZONE

On the 24th cruise of the R/V *Akademik Kurchatov*, a grid survey was performed in the region of the earlier unknown fault, which later acquired the name Kurchatov. The described study area had coordinates of 123°50 to 121°58.5 W and 37°49 to 36°38 S (area M-9, Figure 9). The general strike of the fault coincides with that of the transform faults crossing the EPR at its western side to magnetic lineation 6, where the Agasis fault with a N75E strike approaches it.

A fault-line ridge with the depth to the summit of 700 to 2800 m follows the northern side of the Kurchatov fault. The fault trough has a V-shaped geometry, and its depth reaches 6600 m. The southern side is characterized by a depth of 4000 to 4200 m. In the eastern part of the area, the fault has experienced an echelon northward displacement for about 20 miles (Figure 9B).

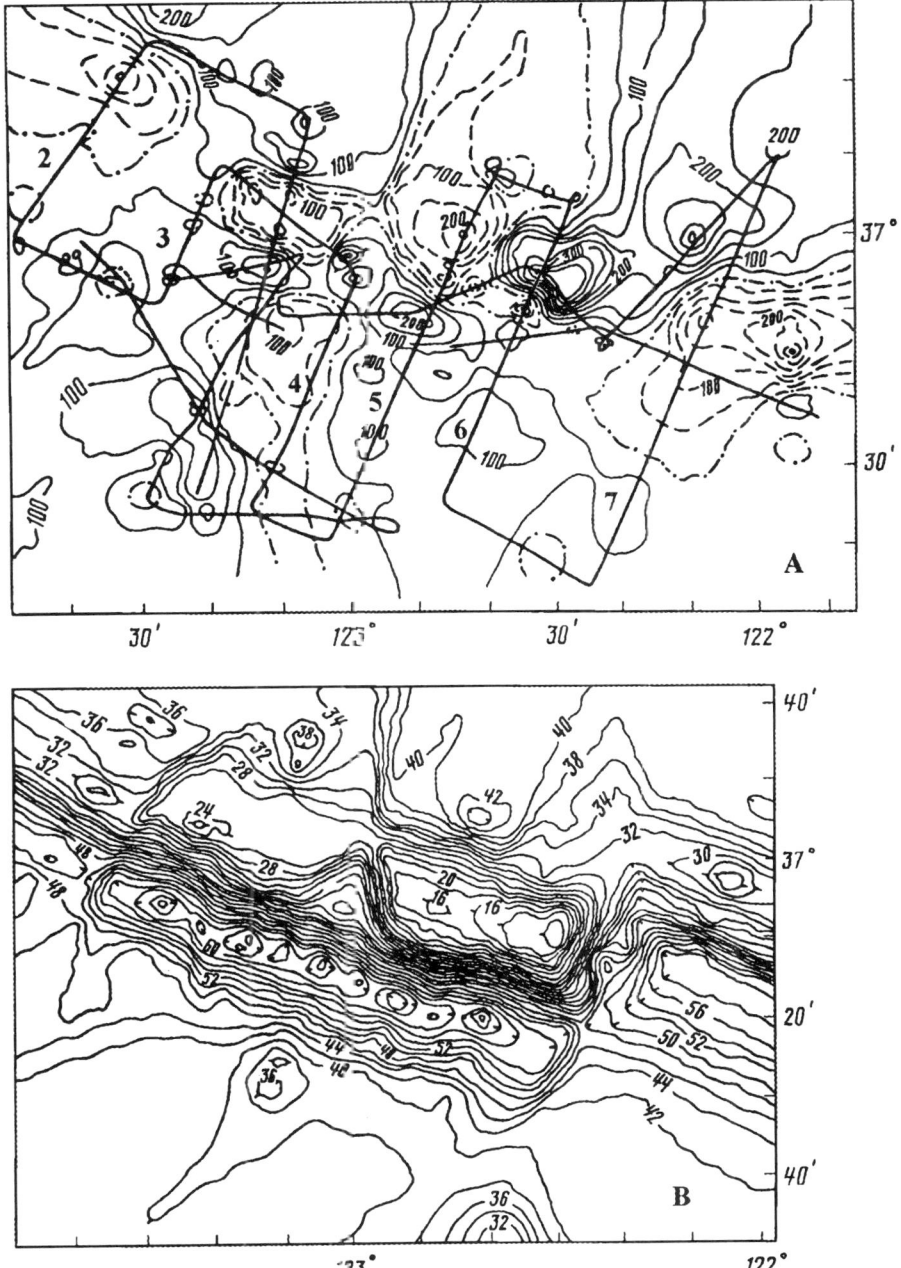

Figure 9 Area M-9, Kurchatov fault: (A) AMF map; (B) bathymetry of the area. Isobaths are given in hundred meters.

In the Agasis transform fault vicinity, SW-NE magnetic lineations from 29 to 7 have been mapped. Their strike coincides with that of N26W of the lineations on the southern side of the fault.

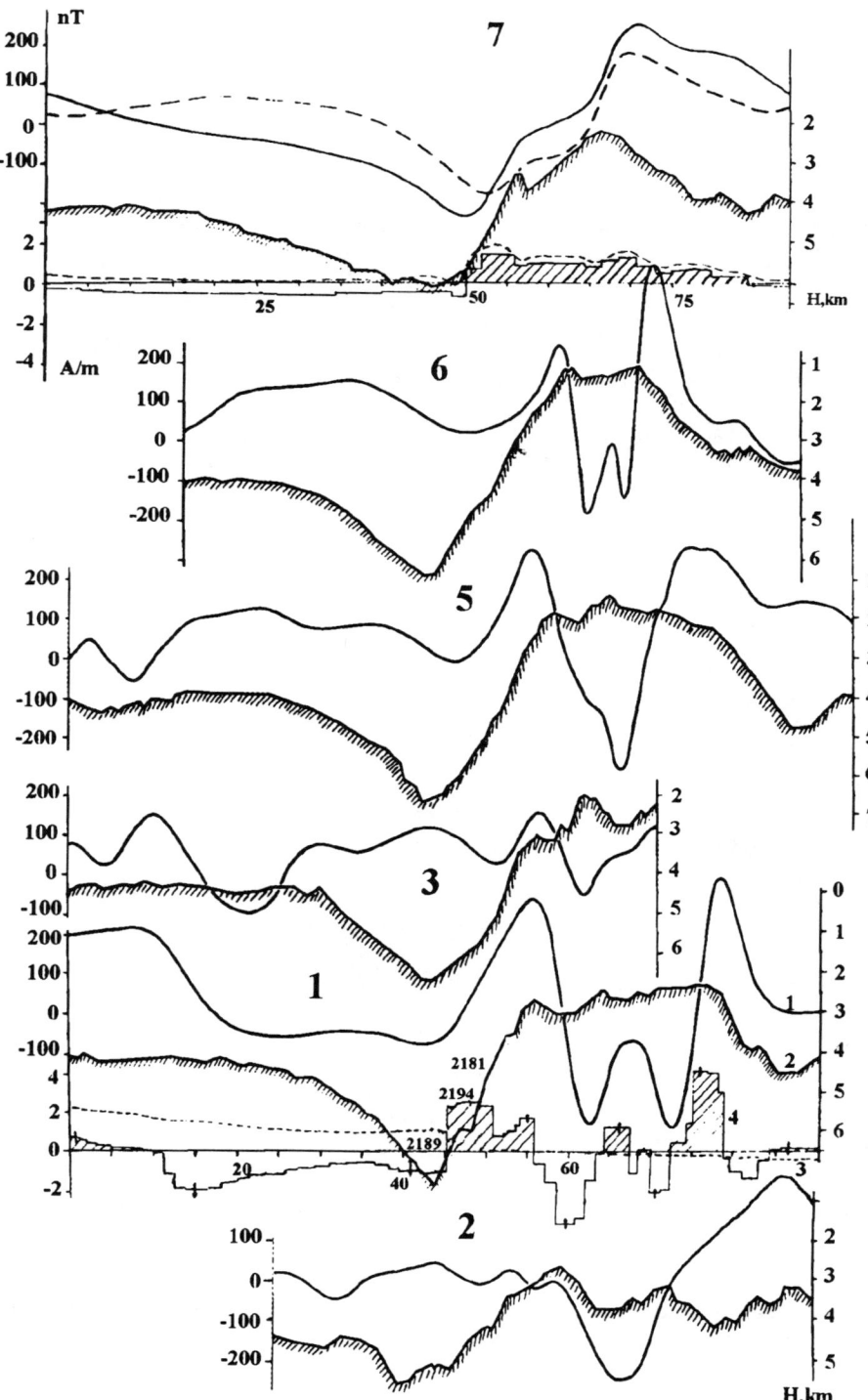

Figure 10 Estimated effective magnetization I_c and station numbers. For notations see Figure 2.

On the northern side of the fault, magnetic lineations of N12E strike have been mapped that correspond to the strike of the Cenozoic anomalies 2-5c on the western EPR flank. It is difficult to identify magnetic anomalies within the area due to short profiles performed. Most probably, the age of the southern side of the fault does not exceed 27 to 26 million years.

In the AMF (Figure 9B), the southern side of the fault is expressed by anomalies with intensities up to 200 nT. The northern side of the fault is expressed by more intensive anomalies. Sign-reversed anomalies reaching 400 nT and more are traced over the crest of the fault-line ridge. These anomalies have specific features that indicate the volcanic origin of the objects generating them (Figure 10). So, anomalies that are typical of volcanic constructions and that usually map a volcano's caldera are registered at profiles 1 and 6. Similar anomalies are also registered at profiles 2 and 5. One may suppose that the fault-line ridge is a chain of volcanic constructions.

To examine the deep structure of the Kurchatov fault, three profiles of deep seismic sounding were performed there. Besides two latitudinal profiles, there is a meridional one, which crosses the opposite sides of the fault (Figure 10, prof. 6) (Neprochnov et al., 1984). On the southern side of the fault, the oceanic crust thickness is somewhat greater (to 8 km), due to lower parts of the crust with velocity over 6.5 km/s. In the trough, the crust thickness is considerably less (4.5 km).

To examine the matter composition of the rocks comprising the Kurchatov fault, dredging was performed that revealed manifestations of intensive volcanic activity on its northern side.

Simulation estimates are given in Figure 10. The figure presents profile 7 spaced in the eastern part of the area, in the vicinity of an echelon fault northward shift. From reconstructed magnetization, the northern side is characterized by positive magnetization of 1 A/m, whereas the southern side has negative values of 0.5 to 0.8 A/m. The simulated field is shown in the figure by a dashed line. It is estimated for the northern and southern blocks of uniform magnetization but different polarity. A good qualitative coincidence shows that the observed field is mainly determined by a topography of an active magnetic layer. On profile 1, crossing the volcanic construction in the central part of the area, the southern side of the fault is characterized by effective magnetization of 1.5 A/m. The northern side of the fault is composed of more magnetic rocks of about 4 A/m. Caldera walls have magnetization of 4 to 2 A/m. Eruption of the rocks comprising the central stock seemed to occur in the epoch of reversed polarity; their magnetization is about 1.5 A/m.

Thus, analyzing the obtained magnetization, it is clear that in the southern part of the area, the values are close to the mean values for the ocean as a whole, but in the regions of superimposed volcanism, they are considerably higher. Considering high magnetization values in the fault trough and thinning of the crust there (from DSS data), we cannot ignore the possible effect of the deeper crustal layers upon the formation of magnetic anomalies. We shall note that volcanic activity generating a thick layer of alkaline effusive rocks is the principal feature of the Kurchatov fault.

VII. THE EMPEROR FRACTURE ZONE

The Emperor fault is the least known structure of the Pacific It extends southeastward from the Aleutian arc for over 2000 km. On the 23rd and 29th cruises of the R/V *Dmitry Mendeleev*, complex geological-geophysical study was performed in the southern part of the study area with coordinates of 178°30 to 176°40 W and 37°54 to

39°40 N, and 177°20 to 175°20 W and 37°00 to 38°20 N, respectively (area M-10). Figure 11 shows the position of profiles of the complex geophysical survey.

As follows from the study, the fault is a deep trough accompanied by a series of echelon depressions over 9500 m deep and two to five miles wide along its strike. Relative depth of the trough is 2 to 5 km, with the walls dipping at 30. The walls are often complicated by scarps 2 to 3 km high, which are of the thrust type; that is reflected in the sedimentary cover. The trough topography varies considerably along its strike from a V-shaped geometry to a W-shaped one (Figure 11). A crest in the basement topography is traced along both sides of the fault. For the northeastern side, it is higher by 0.15 to 1 km than for the southwestern one. The general elevation of the fault-line ridge over the abyssal plain reaches 2 km. The ridge descends toward the latter by steps 0.4 to 0.8 km high. The basement topography of the northeastern side is more rugged than that of the southwestern one; sedimentary thickness there reaches 0.3 km.

To examine the structure of the Earth's crust, DSS was performed within the areas. The following section of the crust has been detected there: sediments take up 0.3 km of the column, and are characterized by P-wave velocity of 1.8 km/s; layer 2 is 2.5 km thick, P-wave velocity being 5.5 km/s; layer 3 is 3 km thick, P-wave velocity being 6.8 km/s. Geological data show that the upper part of layer 2 is presented by oceanic tholeiite. Dolerite from the lowermost part of layer 2 is a fragment of a dyke complex, and its composition is similar to that of the basalt. The change from dolerite to gabbroid corresponds to the transition to layer 3. We should note the hypsometric difference in the bedding of gabbroids on the opposite sides of the fault. Total thickness of the crust there is 5 km (Neprochnov et al., 1984).

According to petromagnetic analysis, basalt is characterized by magnetization susceptibility of 100×10^{-3} emu cm^{-3} Oe^{-1}, I_n of 0.3 to 0.7 A/m, and Q factor of 7.5 to 13.9, whereas dolerite has greater values of I_n (from 1.2 to 5.9 A/m) at $Q = 5.3$.

In geomagnetic plan, most of the Emperor fault is spaced in the quiet magnetic field zone, the age of the crust there being 110 to 80 million years old. So, more detailed dating of that structure by geomagnetic data does not seem possible. According to the character of the AMF, the Emperor fault differs considerably from the Pacific transform faults described above.

To determine the paleolatitude of the fault formation, a simulation of the magnetic field anomalies mapped within these areas was performed with due account of the linear character of the structures accompanying it. The interval of 5 to 8° N seems to be the most probable for the Emperor fault formation. So, it turned out that the huge structure of the fault has shifted northward for more than 40°.

Figure 11 (profiles 8, 9) presents the magnetic effect as a function of fault topography. We shall note that it is not high. The performed simulation shows that anomaly magnetic field depends mainly upon the magnetization distribution in the active magnetic layer (Figure 11).

It follows, from the analysis of the geomagnetic survey profiles drawn in Figure 11, that the southwestern side of the fault is characterized in principle by negative values, whereas its northeastern side has positive anomalies. It should be outlined that the most intensive anomalies with amplitudes to 350 nT sometimes are shifted toward its limbs and do not coincide morphologically with its trough. At some profiles, the trough is not practically expressed, though on profiles 2, 4, and 7, the trough topography is considerably complex.

The most intensive anomalies are registered at profiles 8, 9, and 11. Morphologically, these anomalies coincide with the Emperor trough only on profiles 9 and 11.

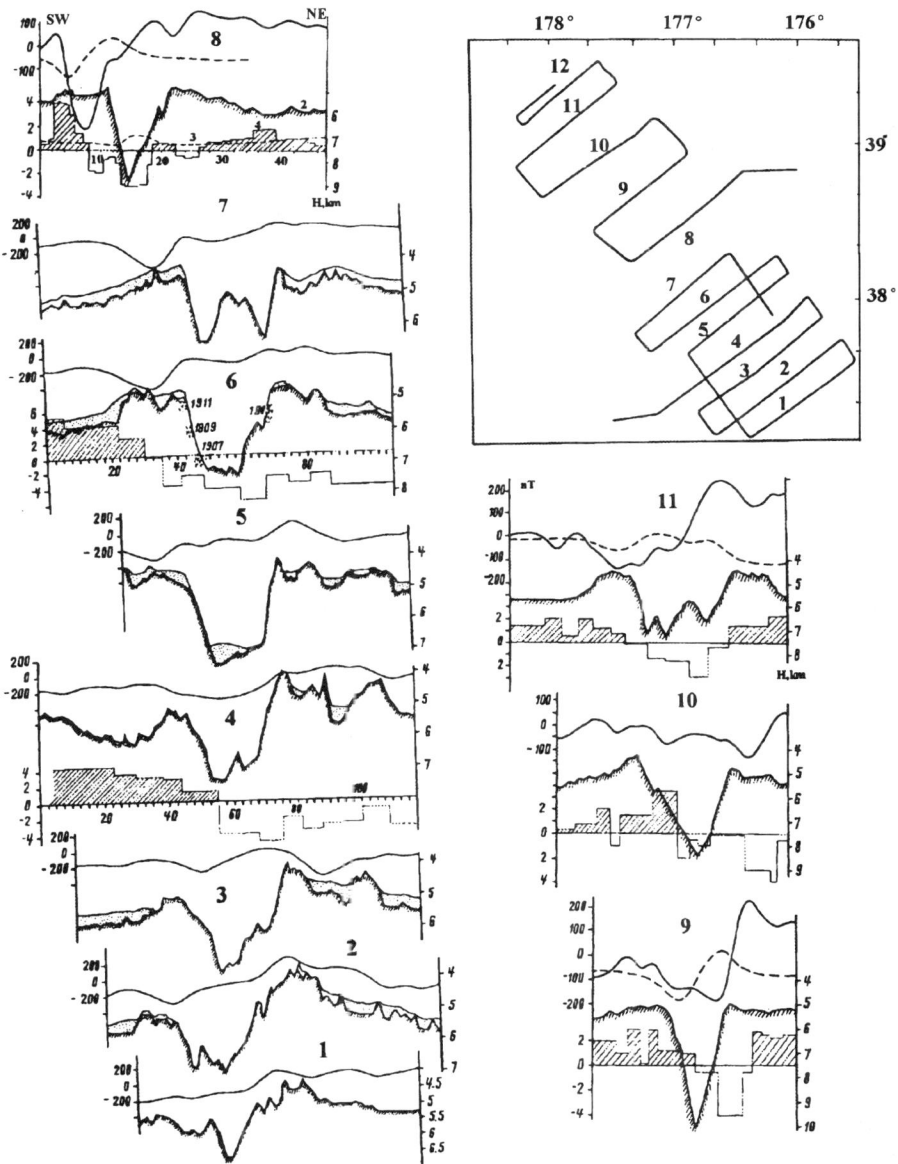

Figure 11 Scheme of profile pattern within area M-10 of the Emperor fault and estimated effective magnetization I_c. For notations see Figure 2.

Analyzing the effective magnetization estimated for that region, it is clear that its high values (up to 4 A/m) are typical of the southern part of the area, whereas values of only 2 A/m characterize its northern part. On the whole, the trough of the fault is characterized by negative values of the estimated magnetization.

As was noted earlier, a number of distinct anomaly extrema are shifted toward its limbs for 10 to 15 km. Lithological analysis of bottom sediments, sampled by ground tubes, was performed by Levitan for the area of the fault. It was detected that concentration of volcanic glass increases sharply 15 km from the fault wall. That confinement of the magnetic field anomalies to the peripheral parts of the fault can be explained by manifestations of the intraplate volcanism that did occur in the epoch of normal polarity, which is reflected in the magnetization distribution there.

According to the character of magnetic anomalies, the Emperor fault differs greatly from the previously considered great Pacific faults, thus indicating its alternative origin. A typical echelon topography of the trough evidences the absence of great horizontal displacements in the region.

The problem of the fault origin is still debatable. Some authors suppose it to be a transform fault, and others consider it an intraplate trough. So, according to the paleo-reconstructions by Zonenshain et al., (1987b), the Chanuk zone is assumed to be an ancient spreading center, and the Emperor fault is believed to be a transform fault with dextral displacement. The poles of the absolute motion of the Pacific plate given in the same paper made it possible to calculate the displacement of the plate blocks at which study areas are located for the time interval of 80 to 110 million years ago. The latitude of 0 to 2° N agrees well with the estimate of the possible paleolatitude of the formation of the magnetic field anomalies obtained from geomagnetic data. The most interesting hypothesis, we believe, was postulated by Sorokhtin (1993). Using the concept of a membrane tectonics by Turcotte and Oxburgh (Turcotte, 1978; Turcotte and Oxburgh, 1978), he ascribes the generation of the fault to northward migration of the Pacific plate. According to this concept, lithosphere plates should adapt themselves to the varying curvature of the rotation ellipsoid of the Earth while they are sliding over the hot mantle surface, generating excessive extension or shear strains in the bodies of large plates. So, extension strains increase gradually in the body of the Pacific plate while it moves from the low latitudes toward the high ones. As a result, they lead to the rupture of the lithosphere shell. Sorokhtin believes that the Emperor trough, which has an echelon structure, was formed due to the initial extension of the plate which, however, was not realized completely.

The given geomagnetic data on the transform faults and, primarily, data of magnetic simulation allow a preliminary conclusion on a probable geological nature of the magnetic field anomalies and on deep structure of the oceanic crust in these specific areas.

It turned out that the basalt layer thickness at its real magnetization value of 2 to 5 A/m is not sufficient to provide for an agreement between the estimated and observed anomalies over entire conjugated blocks of the oceanic lithosphere of different ages in the areas contacting along the fault, and primarily, between magnetic lineations. We must assume that either mean magnetization of the basalt layer reaches 8 to 10 A/m, which contradicts numerous experimental data on magnetization of oceanic basalt, or (as was realized in the model) the active magnetic layer thickness is adequate to the mean thickness of the oceanic crust.

Comparison of the simulated results with the materials of petromagnetic analysis of the lowermost oceanic crust evidences the real effect of serpentinized hyperbasite upon the anomalous magnetic field (Gorodnitsky, 1985).

That agrees well with theoretical estimates of the thickness yield of accretion for the serpentinite layer of the oceanic crust (Lobkovsky, 1988). From these estimates, the thickness of the serpentinite layer under the undeformed blocks of the oceanic crust is limited by plastic properties of newly formed loop serpentinite sealing the fissures and does not exceed 3 km. Therefore, the total thickness of the active magnetic layer in the oceanic lithosphere, in the absence of ferromagnetic minerals in the upper mantle, might be correlated in the first approximation with the thickness of the oceanic crust. The results of magnetic simulation drawn earlier provide evidence that in the Pacific faults analyzed, in the zones with predominating shear with compression (the Mendocino, Heezen, Murray, and Kurchatov faults), the magnetic field anomalies are superposition of the marginal effects of magnetic lineations at the conjugate sides of the fault. The most intensive anomalies are observed at the contacts of the blocks with opposite polarity (areas: M-5 in the Mendocino fault and M-8 in the Heezen fault).

In the faults, where extension strain is great, there is a direct correlation between the magnetic anomalies and objects located in the lowermost parts of the crust, from which areas serpentinized peridotite were sometimes dredged. Such local anomalies were detected earlier at the Clarion fault, where serpentinite has magnetization of 3 to 4 A/m against the low basalt magnetization (Popov et al., 1989), and also in the Emperor fracture zone, where the regime of compression changes for the extension against the echelon structures. In the active part of the Heezen fault, where serpentinite was dredged from the lower part of the trough, a high growth rate of the fault-line ridge was detected, which seemed to result from the formation of a serpentinite layer of low density under it.

The performed simulations prove to be very useful in combination with seismic and petromagnetic studies, and they may be used to estimate the possible effect of the deep layers of the oceanic crust. One should remember, however, that the applied technique does not give an unambiguous answer to this question, and by now, we have no technique that can solve this problem completely. At the same time, the introduction of *a priori* information and selection of formal solutions, combined with the use of independent geological and petrochemical techniques, narrowed this unambiguity and provided magnetization values that in some places reflect the real effect of the deep parts of the crust. Arising from the obtained estimates, the effect of the serpentinite upon the AMF formation may reach 20%.

Chapter 5

Geomagnetic Study of Seamounts

Yu. V. Brusilovsky, A. M. Gorodnitsky, and A. M. Ivanenko

CONTENTS

I. The Northern Atlantic Seamounts 172
   A. The New England Seamounts 173
   B. The Gorringe Seamount Range 174
   C. The Canary Basin Seamounts 180
II. The Somali Basin Seamounts (The Indian Ocean) 183
III. Seamounts of the Tyrrhenian Deep-Sea Basin 187
   A. The Maghnaghi Seamount 193
   B. The Vavilov Seamount 195
   C. The Marsili Seamount 198
   D. Geological Interpretation of the Magnetic Simulation 201
IV. The Pacific Seamounts 202
   A. The Marcus Wake Seamount Range 202
   B. The Caroline Guyots 210
   C. The Magellan Seamounts 213

Seamounts are believed to be vital subjects for geomagnetic study in the ocean. They are abundant within various ocean floor morphostructures and are generally of volcanic origin. Seamounts and volcanic islands were formed at various geological times. According to geochronological data, their age may be estimated in the interval from the Jurassic through the Quaternary. That period is characterized by global-scale horizontal motions of lithosphere blocks and considerable seafloor spreading. So, to study the plate kinematics, it is essential to carry out an appropriate paleomagnetic interpretation of the obtained geomagnetic and bathymetrical survey over the volcanic seamounts composed of basalt that has a thermal origin. A paleomagnetic analysis of the AMF structure over seamounts implies a simultaneous solution of a direct magnetometry problem on the expected anomaly over the mount, when its topography and magnetization are known, and comparison of the simulated and observed anomalies.

Such analysis allows an estimation of the value and direction of the total magnetization in the effusive rocks comprising the mount.

Examination of magnetic properties of rocks sampled from seamounts when dredging or in the course of deep-sea drilling has revealed that their natural remnant magnetization considerably exceeds the induced one (Gorodnitsky, 1985). The Koenigsbergen factor for these rocks varies from 5 to 57, being 35 on average. That allows a more certain assumption that the calculated factor of summary magnetization corresponds in fact to the remnant magnetization of the basalt comprising seamounts. In its turn, it allows a determination of paleolatitudes and coordinates of the virtual paleomagnetic pole. The modern technique of magnetic simulations of seamounts, applied to examine their deep geological structure, volcanic activity, and paleomagnetic history, is formulated in Chapter 2. Meantime, such simulation requires certain assumptions.

One of them is an assumption about the uniformity of the magnetization intensity of seamounts. It is based on the statement that formation of seamounts and volcanoes on the oceanic lithosphere took an interval of several millions of years, whereas duration of a constant magnetic polarity epoch reaches tens of millions of years. The best correlation between the observed and estimated anomalies was obtained for the Cretaceous seamounts when the number of reversals is considerably less as compared to the Tertiary time (Heirtzler et al., 1968; Larson and Pitman, 1972). The most favorable period for paleomagnetic reconstructions of seamounts is the extended time interval of the early and late Cretaceous characterized by predominance of a direct magnetic polarity at the extremely low frequency of reversals. Therefore, we can admit the assumption on a constant magnetization intensity of the Cretaceous seamounts. As for the Tertiary period, when the duration of a constant polarity varied from 0 to 46 million years, there is a great discrepancy between the observed and estimated anomalies (Francheteau et al., 1970). That supposes a reversible magnetic polarity of the majority of the Tertiary seamounts, which hinders their paleomagnetic analysis (Harrison and Ball, 1975).

A constant value of magnetization in a seamount is also an important assumption, though sometimes it is not observed. In particular, there is a great discrepancy between the estimated effective magnetization in seamounts and that in the rocks sampled from the seamount top when dredging. Harrison and Bonatti (Harrison, 1971) supposed that hyaloclastite with no ferromagnetic material may be generated by fragmentation and hydration near the seamount summit. In that case, the anomaly observed over a seamount may sometimes have a greater period than the estimated one because the depth to the magnetic object would be greater than the depth to the seamount summit. Such a pattern is observed in particular in the Pacific at the Muzykanty Seamounts, Bezlunnye Seamounts, namely Show, Cone, Riofu, Sysoev, and Maiko (Harrison et al.,1975). For estimates in the ancient parts of the ocean it is important to know the real — not observed — topography of the seamount magnetic basement buried under the loose sediments, the thickness of which may sometimes reach several hundred meters. So, data of seismic profiling should be used for correct simulations.

The technique of magnetic simulation of seamounts described in Chapter 2 was used to study their geological structure and to estimate the Mesozoic and Cenozoic kinematics of the plates containing the oceanic lithosphere. The problem was solved in two aspects. The age of the Northern Atlantic Seamounts, comprising the North American plate, was estimated from the paleomagnetic data and comparison of their paleomagnetic poles with the corresponding poles of the North America. To reconstruct the position of the Pacific plate which does not contain any continents and, in particular of the Somali plate of the Indian Ocean, and also to study the nature and character of the volcanic activity there, the paleomagnetic data on seamounts were analyzed in combination with the paleomagnetic interpretation of magnetic lineations and paleoclimatic data. Magnetic simulation of the Tyrrhenian seamounts was performed to study the space-time evolution of basalt volcanism.

I. THE NORTHERN ATLANTIC SEAMOUNTS

To perform magnetic simulation for the Northern Atlantic, we used the materials of geomagnetic and bathymetric survey over the New England Seamounts extending as a NW-SE range for 800 km from the region of the Jorges Bank (Man Bay) to the Bermudian and Angular Rise. We also used the data of the detailed geomagnetic study of the Gorringe Seamount range as a part of the Horseshoe Seamount system in the eastern Azores-Gibraltar fracture zone, and some seamounts of the Canary Basin.

A. THE NEW ENGLAND SEAMOUNTS

To estimate the value and direction of the natural remnant magnetization, a paleomagnetic analysis was made of the geomagnetic and bathymetric materials obtained in the detailed geomagnetic survey (Gorodnitsky et al., 1977). The "Mount-01" program was used for this interpretation. Paleomagnetic latitudes and also coordinates of the virtual paleomagnetic poles were determined for each group of seamounts recognized from the estimated magnetization declination and inclination in 43 simulations of the New England Seamounts, Bermudian and Angular Rises. The geometrical mean values of the paleomagnetic pole coordinates for seamounts of the mentioned provinces are 65° N and 159° W (Figure 1). When we compare these coordinates with the geometric mean of the Cretaceous North America paleomagnetic pole (66° N, 193° E), it became clear that they were close to each other (Gorodnitsky et al., 1978). Such coincidence allows us to suppose that the New England Seamounts were formed in the Cretaceous period. As follows from the geomagnetic survey data, the majority of seamounts have normal magnetic polarity, and thus we may conclude that they were formed mainly during the late Cretaceous, which is characterized by predominantly normal polarity of geomagnetic field.

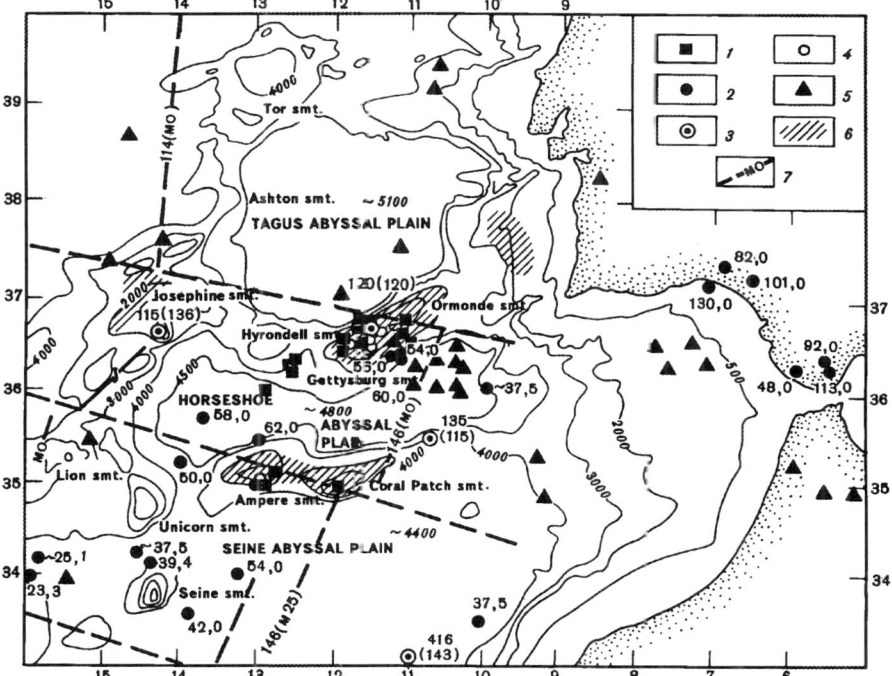

Figure 1 Scheme of geological-geophysical study of the eastern Azores-Gibraltar fracture zone and the Horseshoe Seamount range system: (1) sites of bedrock dredging; (2) sites of heat flow measurement and heat flow density in mW/m^2 (from literature); (3) deep drilling boreholes with a corresponding number (figure stands for the age of exposed rock, in millions of years); (4) sites of heat flow measurement and heat flow density in mW/m^2 (from data of the 12th cruise of the R/V *Vityaz*); (5) earthquake epicenters; (6) anomalous gravity areas; (7) identified magnetic lineations with a corresponding numbers (figure in brackets corresponds to the age in millions of years; isobaths are spaced in 1000 m).

Geomorphologically, the New England Seamounts are presented by two isolated groups northwestward and southeastward of the Rebot Seamount. The estimated position of virtual poles and their detailed comparison with the curve of the North America paleomagnetic pole migration, allows an assumption that the first northwestern group was formed 70 to 120 million years ago. The southeastern seamount group and seamounts of the Angular Rise seemed to be formed a bit later, that is, about 60 to 70 million years ago. As for the Bermudian Seamounts, the presumable age of their formation is about 50 million years ago (Gorodnitsky, 1977). It is worth mentioning that the intensity of magnetic anomaly over the northwestern part of the New England Seamounts is considerably higher (up to 1000 nT) than that over its southeastern part. That may be explained by the fact that seamounts of the northwestern part were formed during the late Cretaceous period when geomagnetic field polarity was constant. A comparatively low intensity of anomalies in the southeastern part of the New England Range may be the consequence of reversals of the geomagnetic field polarity during the period of its formation and, as the result, different magnetic polarity in the individual layers and parts of the volcanic construction. That corresponds to the existing view of magnetic polarity reversal in the Maastrichtian-Paleocene (60 to 70 million years ago).

Geological data on the late Cretaceous continental margins of North America made it possible to suppose that deep faults along which the formation of the New England Seamounts followed did exist already at the early stage of the Northern Atlantic opening started in the Cretaceous. The Northern Atlantic opening and associated late Cretaceous activation of basalt volcanism (80 to 100 million years) has stimulated the subsequent evolution of the New England Seamount Range. The maximum altitude of the New England Seamounts (no more than 3 km) indicates that they were formed at a young, relatively thin oceanic lithosphere and seem to be composed predominantly of tholeiite basalt. The subsequent activation of the faulting occurred there from west to east, and so the New England Seamounts must be older in the west and younger in the east. That agrees well with the data of detailed analysis of the estimated paleomagnetic pole positions and is also verified by the dredging and deep-sea drilling data, and also by a geohistorical analysis of the magnetic field lineations (Donets et al., 1975; Northrop et al., 1962; Uchupi et al., 1970).

Thus, paleomagnetic analysis of geomagnetic and bathymetric data on the seamounts of the New England, Bermudian, and Angular Rises made it possible to estimate their age and to gain insight into the evolution history of the volcanic system which follows the northwest fault system.

B. THE GORRINGE SEAMOUNT RANGE

The Gorringe Seamount Range is a part of the Horseshoe Seamount system located in the eastern Azores-Gibraltar fracture zone at the Eurasian/African plate boundary. The range itself is morphologically related to the northern branch of the Horseshoe Mountain chain and extends from the west to the east for 180 km, its width being about 74 km (Figure 1). In the north and south, the range is bounded by the Tagus and Horseshoe Basins. The range elevation over them is 5100 and 4800 m, respectively. The range has two summits, namely, the Gettysburg, with a minimum of 25 m, and Ormonde, with a minimum of 42 m.

Geological structure and tectonic evolution of the range is complex. From the research of many specialists (Verzhbitsky et al., 1989a; Prichard and Gann, 1982), the western summit — the Gettysburg — is composed of gabbro-dolerite, and of alkaline basalt in its uppermost section. The hyperbasite/gabbro interface is detected at the northeastern slope of the Gettysburg Seamount at a depth of about 2300 m. Sea research with submersibles (Aurence et al., 1977) showed that the Gorringe Range in its western part (Gettysburg) is composed of serpentinitized harzburgite and gabbro-dolerite and in its eastern part (Ormonde) by gabbroid and alkaline basalt. The dating of the gabbro, dredged from the Gettysburg Seamount slopes, was performed by K/Ar method. It was stated that the age of some mineral phases embraces three concordant groups. The oldest of them is 135 ± 3 million years. Besides, the age of 105 ± 3 million years, associated probably with a thermal heating, and the age of 82 ± 3 million years, presumably associated with the phase of cutting and deformation due to transform motion along the plate boundary, was detected from the plagioclase feldspathoid.

From the relic forms and foraminifera, the oldest sediments exposed within the Gorringe Seamount Range by borehole 120 (Figure 2) is dated back to 120 million years, minimum (Barremian). At the same time, the series of alkaline basalt dredged from the Ormonde Seamount slope (sampled by the submersibles and obtained from the drilling by the R/V *Robert Conrad*) were dated back to 50 to 60 million years by the K/Ar method.

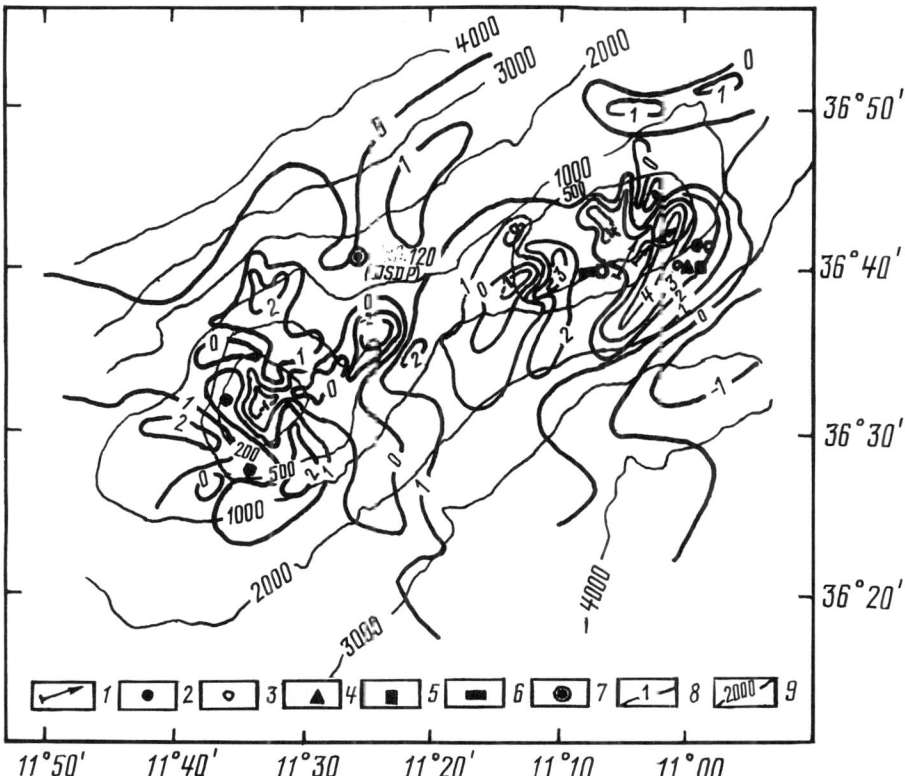

Figure 2 Map of the AMF isodynams for the Gorringe Seamount Range: (1) sites of dredging; (2) serpentinites; (3) alkaline basalt; (4) gabbroids; (5) greenstone metamorphites; (6) limestones; (7) position of borehole 120 (*Glomar Challenger*); (8) isodynams of ΔT_a in A/m; (9) isobaths in m.

According to modern concepts, the Gorringe Seamount Range itself has a tectonic origin. As follows from the deep-to-shallow sedimentary facies transition revealed in the course of deep-sea drilling (borehole 120), the oceanic crust block uplift started not earlier than 20 million years ago and stopped about 5 million years ago.

It should be noted that the age of the oceanic crust under the Gorringe Range is 135 million years; that is the most ancient one for the oceanic crust detected by isotopes. Magnetic lineations in that part of the Atlantic are rare and have no reliable identification (Figure 1). Meantime, they show that the oceanic crust south of the Azores-Gibraltar fracture zone is older than the crust north of it and that the spreading between Europe and North America started later than that between Africa and North America. The opening between the African and North American plates seems to have started about 180 million years ago (Pitman and Talwani, 1972). The age of 135 million years, mentioned above for the Gorringe Range, supposes that the range is a fragment of the oceanic crust adjacent to the African plate. The magnetic anomaly following the western boundary of the Horseshoe Basin (Figure 1) is correlated well with the young part of the anomaly and speaks in favor of the younger age of the oceanic crust adjacent to the African plate (Laughton and Witmarsh, 1974). This makes it possible to conclude that the Gorringe Range was cut from the African plate during the last stage of compression (Purdy, 1975).

We have no data on the latitudinal position of the crustal block of the Gorringe Range during the period of its formation; these are essential for reconstruction of the tectonic evolution of the range itself and the adjacent region (Azores-Gibraltar tectonic zone).

Detailed geomagnetic survey performed on the 12th cruise of R/V *Vityaz*, resulted in compilation of the AMF map (Figure 2), which shows in particular the points of sampling. The profiles are spaced five miles apart, on average. The root-mean-square (RMS) of the survey within the study area is 22 nT. The map illustrates that the AMF over the Gorringe Range has a rugged topography and high intensity of local signvariable anomalies attributed to the Ormonde and Gettysburg summits mainly (Brusilovsky and Gorodnitsky, 1990). Anomalies with amplitudes over 700 nT and periods of 1.0 to 1.5 km were registered over the Gettysburg summit. A negative anomaly with an intensity of 400 nT was registered over the summit itself. The area of intensive short-period anomalies is bounded by a 400- to 500-m isobath. The Ormonde summit is characterized by a series of magnetic field anomalies with amplitudes of about 100 nT and a period over 2 km. The area of these anomalies is also bounded by a 500-m isobath. In plan, two principal systems of lineaments may be distinguished. The first, which embraces the entire area of the study, has a northeastern strike with an azimuth of 20 to 30°. It differs from the general strike of the range isobaths (60 to 70°). The second system of lineaments is orthogonal to the first and has an azimuth of 110 to 130°. When comparing the geomagnetic survey materials with those of the side-scan sonar survey (Verzhbitskiy et al., 1989a) performed within the Gettysburg summit area, the first system of lineaments is correlated well with the ranges of serpentinite detected there. The second system is correlated with the vertical escarpments of the slopes of the corresponding strike. The two mentioned systems of lineaments are typical of both short- and long-period anomalies.

Geomagnetic parameters of rocks sampled by dredging from the Gettysburg and Ormonde summit slopes were analyzed simultaneously with the geomagnetic survey over the study area of the Gorringe Seamount Range. The result of this analysis, namely, of the magnetic susceptibility κ, remnant magnetization I_n, the Koenigsbergen factor Q are given in Table 1 (Gorodnitsky et al., 1988).

Table 1 **Magnetic parameters of rock samples from the Gorringe Seamount Range**

Coordinates	No.	Short petrographical characteristic[a]	I_n, A/m	$\kappa \times 10^{-3}$ SI unit	Q	$C=5\kappa/4\pi$ ×100%
36° 42.6	1	Basalt, hawaiite rareporphyritic	0.0017	0.26	0.16	0.01
11° 11.8	2	Alkaline basalt, porphyritic olivine	14.7	6.0	72.0	0.24
	3	Basalt, hawaiite (composed of abundant leucite plagioclase)	0.0012	0.30	0.12	0.01
36° 41.8	4	Alkaline basalt		15.0	6.6	0.59
36° 42.7	5	Tholeiite basalt		0.4	3.7	0.016
11° 03.6		A	0.06			
11° 02.1		B	0.04			
	6	Same		1.5	0.6	0.06
		A	0.04			
		B	0.02			
36° 42.3	7	Alkaline basalt	1.9	15	3.7	0.59
36° 44.5	8	Same	2.5	20	3.7	0.79
11° 03.5	9	Same	14.2	15	27.9	0.59
11° 02.4	10	Same	1.9	15	3.7	0.59
36° 31.0	11	Serpentinite				
36° 31.3		A	6.45	77.2	2.5	3.06
11° 34.4		B	4.35		1.7	
11° 31.5	12	Same A	1.05	18	1.7	
		B	0.56		0.9	0.71
11° 30.9	13	Same A	3.9	10	11.5	0.40
11° 33.4		B	2.4	15	4.7	0.59
36° 40.9	14	Same A	2.7	23.5	3.4	
36° 04.3		B	4.4		5.5	0.93
	15	Weathered serpentinite	0.1	12	0.25	0.48
	16	Gabbroid	0.158	6.0	0.75	0.24
	17	Same	0.11	3.0	1.08	0.12
11° 02.4	18	Same	0.047	0.1	13.8	0.004
11° 03.5	19	Dolerite				
36° 42.3		A	0.024	0.5	1.4	0.019
36° 44.5		B	0.027		1.6	
	20	Xenolith amphibole	2.5	5	14.2	0.19

[a] A and B are two measured cubes of the same sample.

It is clear from the table that massive serpentinite comprising the Gettysburg summit is of great magnetic susceptibility, 10 to 77.5×10^{-3} SI unit, which corresponds to the magnetite concentration in a rock (0.4 to 3%). Remnant magnetization of serpentinite varies within the range of 0.56 to 6.4 A/m at an average value of 3.2 A/m. The Koenigsbergen factor Q is 0.9 to 11.5. The obtained data provide evidence that serpentinite has the principal effect upon the anomalies of magnetic field over the Gettysburg Mount.

About 100 basalt fragments were uplifted in the course of dredging within a 500-m isobath from the slopes of the Ormonde Seamount summit composed of basalts which are subdivided into two groups (see Table 1). The first group is composed of magnetic differences with a remnant magnetization of I_n = 1.9 to 14.7 A/m. Their mean magnetization is 7.6 A/m (from six samples), the Q factor is 3.7 to 72.0, the magnetic susceptibility $\kappa = 6$ to 20×10^{-3} SI unit. The second group comprises "nonmagnetic" differences ($I_n < 0.06$ A/m and $\kappa < 1 \times 10$).

From the petrographic study, "nonmagnetic" samples are presented as a rule by paleotype tholeiite basalts which underwent lengthy stages of metamorphism (until the greenstone variables) and low-temperature oxidation. The group of "magnetic" basalts is presented by alkaline basalts of younger age (Prichard and Gann, 1982) to which intensive anomalies of the AMF over the Ormonde summit seem to be related (Figure 2).

As follows from the study of magnetic properties of rocks comprising the Gorringe Seamount Range, where the oceanic crust block comes to the bottom surface (Purdy, 1975), the primary magnetization of the second oceanic crust layer, presented by basalt of tholeiite composition, is missed due to the processes of metamorphism and weathering of the Gorringe Ridge summit. At the same time, the chemical magnetization of serpentinite turned to be more resistant to secondary changes. That determined the situation when it was serpentinite that served as the principal disturbing object within the ancient crustal block, that is, of the Gorringe Ridge.

Meantime, the primary magnetization of alkaline basalt, which seems to have been generated by the superimposed volcanism about 60 million years ago (Prichard and Gann, 1982), is preserved sufficiently well.

Examining magnetization and magnetic susceptibility of the gabbroid and dolerite dredged from the Ormonde summit below a 400-m isobath, it turned out that they are practically nonmagnetic ones. Their remnant magnetization does not exceed 0.02 to 0.15 A/m.

Using the map of the AMF, compiled from the detailed geomagnetic survey data, and analyzing the bathymetric data, the Gorringe Seamount Range AMF has been simulated.

An optimal model of a seamount range has been obtained by an exhaustive search of numerous variants of the inverse problem solution. It provides for the best correlation of the observed and estimated anomalies. The particular solutions and parameters of regularization were chosen with an account of *a priori* information on rock magnetization obtained from the study of magnetic parameters of the dredged rocks. Within the range area under study, ten smaller areas were distinguished with normal and reversed polarity for which magnetization vector components and paleomagnetic characteristics were estimated. The RMS was about 10%.

For the areas of normal magnetic polarity, the mean paleolatitude Φ_c was 11° N, and the mean effective magnetization I_c was approximately 0.5 A/m. For the areas of reversed magnetic polarity, the mean I_c was –0.2 A/m and Φ_c was 6° S.

At present, the Gorringe Seamount Range is located at 36° N. So, the northward drift of the range since the moment of its formation is about 25°.

As mentioned earlier, the age of the Gorringe Range is about 135 ± 3 million years, whereas the Ormonde summit, composed of alkaline basalt, is only 60 to 65 million years old. From the global paleodynamic reconstructions (Zonenshain et al., 1984), it follows that the African plate drift from 135 million years ago through the present time is approximately 25°, whereas the reconstruction of the African plate for 60 million years ago practically coincides with its present position.

In this connection, a special simulation has been performed for the Ormonde Seamount. The results are very close to those obtained when performing simulations for the range as a whole: I_c is 1 A/m, Φ_c is 8° S.

For that model, the northward drift was somewhat similar, that is 28°. That is close to the drift of the entire massive of the Gorringe Range but does not coincide with the African plate motion starting from 60 to 65 million years ago (that corresponds to the age of alkaline basalt from the Ormonde summit) and until the present time. A special simulation has been performed for the Ormonde Seamount bounded by a 2-km isobath to determine paleomagnetic characteristics of its summit part composed of alkaline basalt. A RMS of the exhausting search for that model was 9%. The model is also an alteration of areas with normal and reversed magnetic polarity. Mean I_c is 2 A/m and Φ_c is 10° N.

There, we can recognize the area bounded by a 1.5-km isobath, which coincides in plan with the alkaline basalt outcrops of 60 to 65 million years old. Effective magnetization I_c there is 1 A/m, paleolatitude Φ_c is 22° N, the latter being rather close to that for the African plate (25° N) determined from the geodynamic 60- to 65-million-year reconstructions (Zonenshain et al., 1984; Scotese et al., 1987). That allows a supposition that 60- to 65-million-year-old alkaline basalt from the Ormonde summit was formed as a result of a superimposed volcanism at the paleolatitude of about 22° N.

The RMS of the observed and simulated fields at the modeling was 10%.

From the magnetic simulations, a mean rate of an absolute northward drift of the African bearing plate was estimated for the early Cretaceous through the present time. It was 2.0 cm per year that is compatible with the mean rates of the African plate drift in the mentioned geological time calculated from the paleomagnetic data (Zonenshain et al., 1984).

So, magnetic simulations of the Gorringe Range allowed the following principal conclusions:

1. The Gorringe Seamount Range has a complex heterogeneous structure and is composed of positive and negative magnetized rock massives that evidence its many-stage formation corresponding to the epochs of normal and reversed magnetic polarity.
2. The bulk of the Gorringe Range massive composed of rocks with effective magnetization of 0.3 to 1.0 A/m was formed near the equator within the latitudinal range of 8 to 12° N, apparently near the Jurassic/early Cretaceous boundary in the Berriasian and Valanginian ages characterized by reversals of the geomagnetic field polarity that may explain the occurrence of positive and negative magnetized rock massives. Magnetic lineations M12 and M11 correspond to the considered geological epoch (Laughton and Witmarsh, 1974), which is verified by the position of the Gorringe Range between anomalies M25 and J (Figure 1).
3. Alkaline basalt of 60 to 65 million years old and I_c of about 1 A/m from the

Ormonde summit were generated by a superimposed volcanism at the paleolatitude of about 20° N, whereas the bulk of the range massive under the Ormonde summit is uniform with the entire Gorringe Range.

4. Since the time of its formation, the Gorringe Range, which is now located at a latitude of 36.5° N, has drifted northward at an angular distance of about 25°. When comparing the obtained data with paleogeodynamic reconstructions, this seems to correspond in the first approximation to the northward drift of the African plate from 135 million years ago through the present time, the mean rate of the meridional drift being approximately 2.0 cm per year.

C. THE CANARY BASIN SEAMOUNTS

Magnetic simulation, based on detailed geomagnetic survey data (obtained in the northwestern Atlantic by the "Sevmorgeologia" institution) and bathymetric material, has been performed for several seamounts located at the eastern slope of the mid-Atlantic Ridge between the Azore Islands and the northern Canary Basin. Among the seamounts under study there are the Cruiser, Jer, and Erving as well as the unnamed mount north of the Cruiser Seamount with coordinates of its summit of 33°56 N and 28°21 W (Figure 3). According to the existing geomorphological zoning, they comprise the Azores group of seamounts and a volcanic chain having a common basement elongated from the southwest to the northeast with an azimuth of about 40°. The seamount range altitude varies within 3 to 3.5 km. The results of geomagnetic survey and of geological study show that all of them are volcanic in origin and are composed of tholeiite and alkaline basalt. Together with the Great Meteor, Atlantis, and Plato Seamounts, they comprise the system of the Azores seamount group located at the Azores volcanic plateau south of the Azores Islands.

The study of the Canary Basin Seamounts, attributed to the northwestern oceanic margin of the African plate, has the following goals:

1. To study the stage-by-stage evolution of basalt volcanism
2. To estimate the age of the seamount formation
3. To determine paleolatitudes corresponding to the time of seamount formation and to derive the absolute northward drift rate of the bearing lithosphere

Using the exhausting search of numerous variants of the inverse problem solution, geomagnetic models have been simulated for each seamount; these comprise the set of areas with normal and reversed polarity with various effective magnetization I_c and estimated paleolatitudes Φ_c.

A joint analysis of magnetic simulations for four seamounts showed that all of them have very close paleomagnetic characteristics and effective magnetization I_c.

Similarity of the estimated I_c and Φ_c and analysis of geomorphological data allow us to consider the Jer, Erving, and Cruiser Seamounts as a single volcanic massive of the southeast strike, within which a series of transform faults orthogonal to the general strike can be recognized from the geomagnetic survey data and from the bathymetry data (Figure 4).

All the seamounts studied have normal magnetic polarity and mean paleolatitude of the formation as $\Phi_c = 12°$ N. At present, mountains are located at 31 to 33° N.

So, we can assume that the oceanic part of the African plate, at which these seamounts are located, has drifted northward at an angular distance of about 18 to 20° since the moment of their formation.

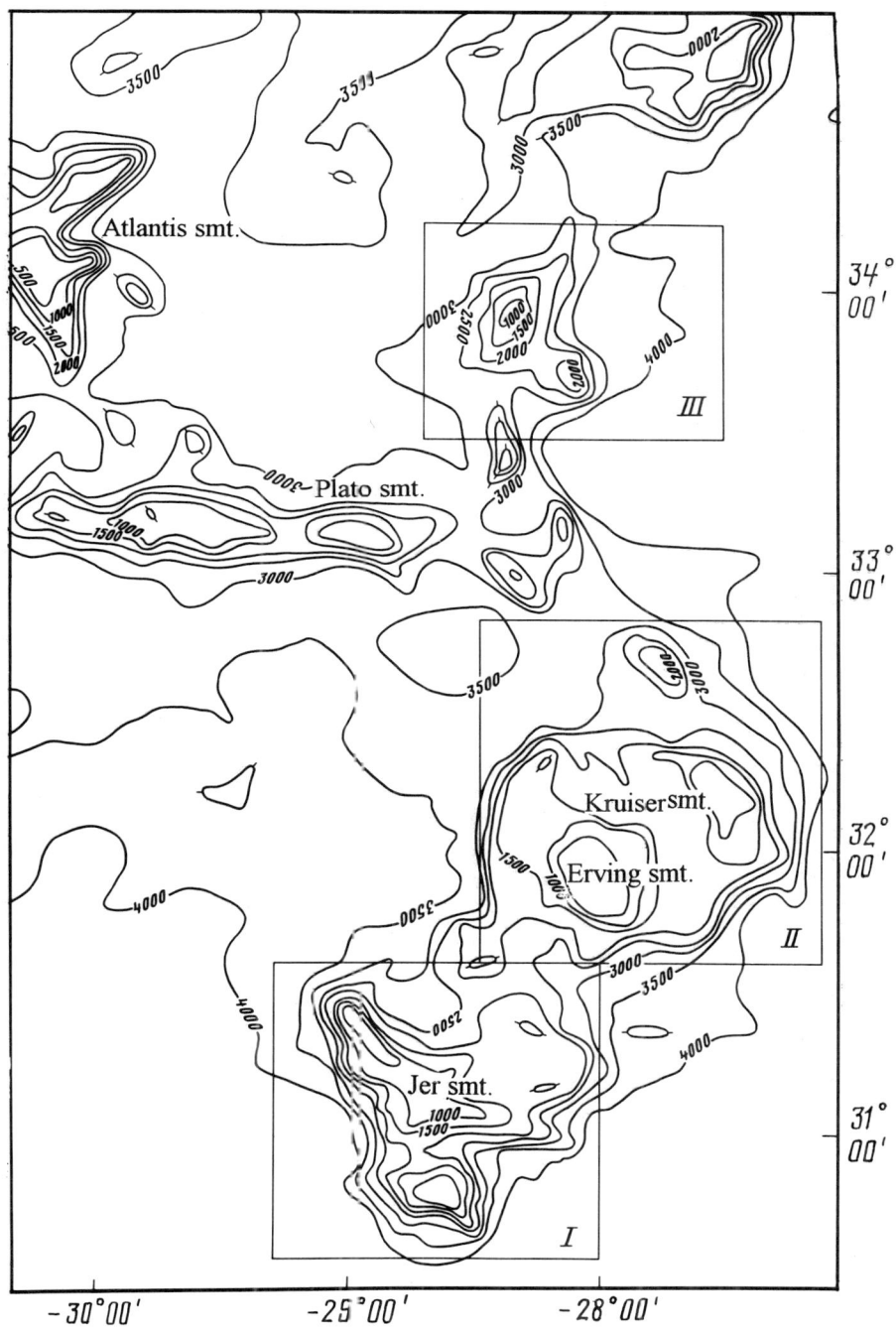

Figure 3 Schematic bathymetric map of the study area. Isobaths are given in meters. I, II, III are numbers of areas for simulation.

Figure 4 Anomalous magnetic field ΔT_a of the study area. Isodynams are given in hundreds of nT. I, II, III are numbers of the areas for simulation.

Comparing these data with the paleoreconstructions (Scotese et al., 1987) and the bearing lithosphere age in the study area (about 89 million years old), deduced from the deep drilling data (borehole 137) and the data on paleomagnetic anomalies, it is possible to suppose that the paleovolcanoes under study were formed in the late Cretaceous. That is also verified by the normal magnetic polarity of the seamounts in general. The approximate age of their formation seems to be no younger than 84 to 88 million years (the Santonian-Coniacian), as at the Campanian/Santonian boundary, frequent reversals of magnetic polarity occurred.

With this account, a mean rate of the meridional drift of the African plate has been calculated from the late Cretaceous through the present time; it was 2.5 cm per year, approximately. The obtained results correlate well with the estimated mean rates of the African plate drift within the mentioned geological time. From paleomagnetic data (Gorodnitsky, 1984), the mean rate of the African plate drift in the mentioned direction was 2.4 cm per year from the late Cretaceous through the present time.

So, magnetic simulation of the Azores Seamounts supposes that the Cruiser-Jer-Erving volcanic range was formed during the late Cretaceous, which is characterized by normal magnetic polarity near the equator (12° N) and then drifted northward at an angular distance of about 20° due to the northward drift of the African plate at a mean rate of about 2.5 cm per year.

II. THE SOMALI BASIN SEAMOUNTS (THE INDIAN OCEAN)

The Indian Ocean is believed to be the most perspective region for paleomagnetic study of seamounts. Individuality of its evolution is favorable for the investigation of absolute and relative motions of the plates framing it. Predominantly submeridional motion of the latter in the Cenozoic and Mesozoic during the Gondwanaland splitting, closure of the eastern Tethys, and opening of the Indian Ocean made the estimation of their absolute kinematics especially pressing.

Until very recent times, determination of plate kinematics of the Indian Ocean region was hardly possible due to scarce information on its magnetometry. That is true for the study of magnetic lineation patterns used for relative estimate of the oceanic floor motion and also for the geomagnetic study of seamounts. Recently, systematic grid geomagnetic surveys have been performed in the western part of the Indian Ocean, within the Arabian and Somali Basins and the mid-oceanic Carlsberg Ridge separating them. The survey was made by the St. Petersburg division of IZMIRAN. The result was that new data on the AMF structure in that part of the Indian Ocean were obtained and were used as a basis for a geohistorical analysis and study of the Indian and Somali plate kinematics (Karasik et al., 1986a).

Recognition of a full sequence of linear anomalies starting from the 29th (66.2 million years old) and completing by the axial one, and compilation of a map of paleomagnetic anomaly axes are the main results of the geochronological interpretation of the AMF (Chapter 3, Figure 14) (Karasik et al., 1986a). The analysis of that map leads to a conclusion about a two-stage opening of the northwestern Indian Ocean, which started relative to the latitudinal Carlsberg protoridge and then relatively to the southeast Recent Carlsberg Ridge. Thus, the relative motion of the Somali and Indian plates has been estimated. However, the problem of the absolute plate motion in that area remained unsolved. Analysis of the geometry of ancient anomalies in the Arabian and Somali Basins corresponding to the period of rapid spreading allowed McKenzie and Sclater (1971) to conclude that they were formed at 10° S and, therefore, the northwestern Indian Ocean bed had moved northward for 2000 to 3000 km during the last 60 to 70 million years after its formation was over.

A systematic geomagnetic survey performed in the southeastern part of the Somali Basin by the St. Petersburg division of IZMIRAN over the grid of profiles made it possible to consider again the problem of absolute and relative plate motions of the region under study. From the map of zero isolines for the investigated part of the Somali Basin, a typical oceanic field presented by a system of magnetic lineations has been recognized between the southwestern Carlsberg Ridge slope and the Seychelles Bank. A regular floor of the oceanic bed is complicated there by a seamount system 3 to 4 km high. However, the real height of seamounts may be greater because the sedimentary thickness of the Somali Basin in that region reaches 0.5 to 1.0 km, according to seismic data. The range of seamounts of presumably volcanic origin starts near the southwestern border of the survey area, that is from the Equator Seamount, and extends northeastward to 57° E (Figure 5). As follows from the map of magnetic field, there are pronounced short-period anomalies over the seamounts, which are isometric in plan and have amplitudes of 200 to 350 nT. They complicate the linear structure of AMF in the system of magnetic lineations 26 to 24 (Figure 5). The mentioned seamount range is orthogonal to the strike of the magnetic lineation system of the Somali Basin. Accounting for the morphostructure of that volcanic ridge, it may be considered a trace of the Somali plate NE-SW motion over the hot spot. The alternative standpoint supposes the occurrence of the deep fault there orthogonal to the system of magnetic lineations along which intraplate basalt eruptions occurred that lead to the formation of the seamount range.

Magnetic simulation has been performed for seven seamounts comprising the range that extends from the southwest to the northeast from 56 to 57° E (Figure 6). When simulating, each mount has been approximated by a set of elementary bodies (from 60 to 200), depending upon the dimensions of the mount and degree of anomaly field complexity. The RMS of the exhaustion varied within a wide range, from 9 to 15% depending on the complexity of the primary magnetic field. Distorting effects of magnetic lineations caused considerable difficulties when simulating.

As follows from the magnetic data analysis, areas of normal and reversed polarity are distinguished by simulations within each seamount. As a rule, the area of reversed polarity coincides in plan with the volcanic construction itself and is confined to that part of the seamount where no distorting effect of magnetic lineations is observed. Average values of paleomagnetic latitude for those areas are 5° S.

Areas of positive magnetic polarity are confined to the areas where anomalies caused by seamounts are distorted greatly by magnetic lineation effect. Therefore, the reliability of the estimate of magnetic inclinations of magnetization vector and paleolatitude may be lower for the areas of positive polarity. Mean paleomagnetic latitude for the areas of positive polarity is about 3° N.

As noted earlier, the formation age of the bearing lithosphere was dated back to the late Cretaceous (magnetic lineations 24 to 26), and the latitude of its formation is about 10° S from the Q method (Karasik et al., 1986a). Arising from the results of magnetic simulation, the formation latitude of volcanic seamounts is about 5° S, on average.

Areas of normal and reversed magnetic polarity revealed by simulation within volcanic constructions may provide evidence that the intraplate volcanism, which resulted in the generation of a seamount range, happened during the epoch of frequent reversals of magnetic polarity. At present, the seamounts are located near the equator, and so the northward drift of the bearing Somali plate, for the time of their existence, was about 5°. The age of these seamounts (about 37 to 40 million years) was calculated from the comparison of the Somali plate drift and paleomagnetic data on the region with due account of the reversals frequency.

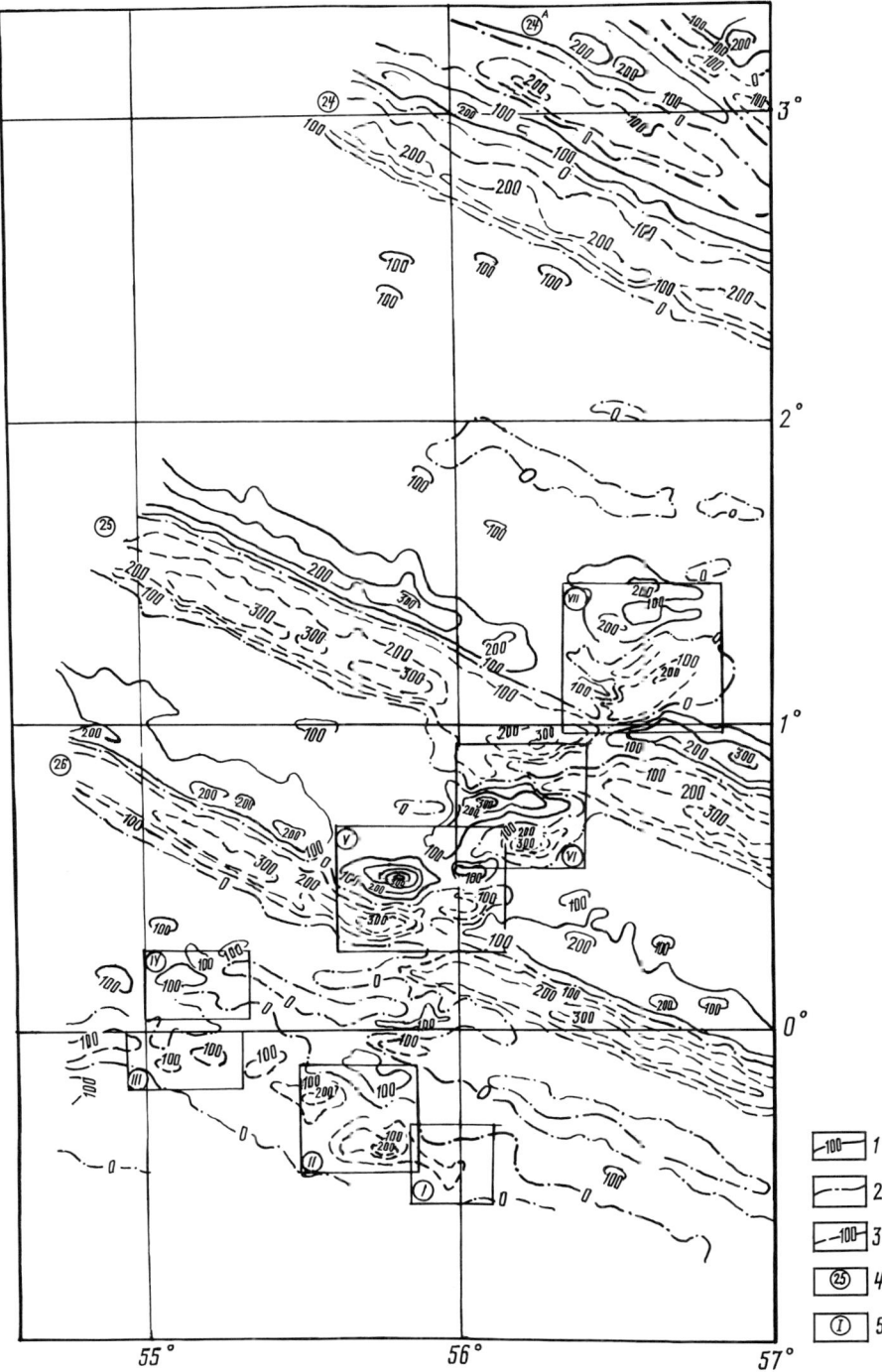

Figure 5 Map of the seamount range AMF: (1) positive isodynams in nT; (2) zero isodynams; (3) negative isodynams; (4) magnetic lineation number; (5) numbers of the simulation areas.

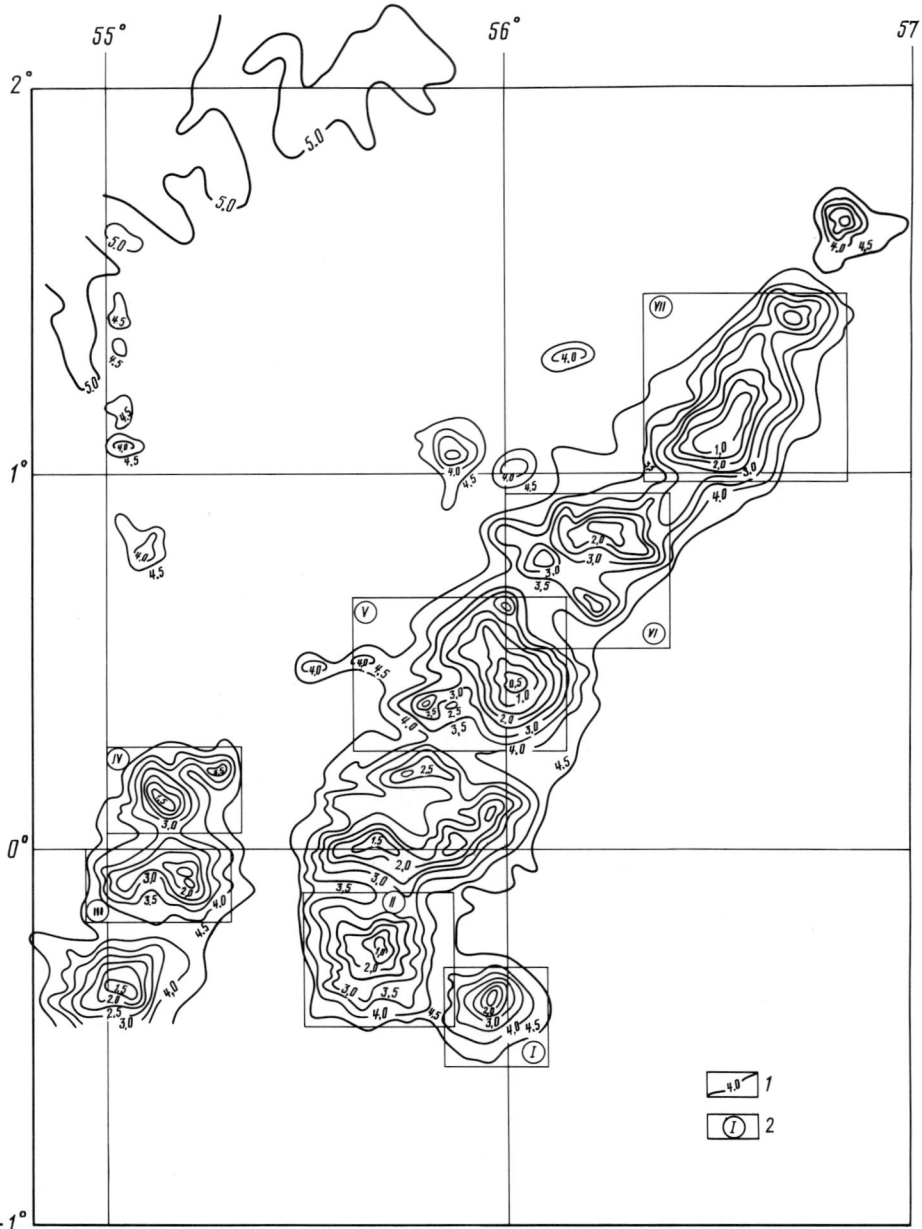

Figure 6 Seamount range for which magnetic simulation is performed: (1) isobaths in km; (2) numbers of simulation areas.

It is worth mentioning that the late Cretaceous-early Paleocene age of the bearing lithosphere, determined from the magnetic lineations, is verified by the deep drilling data obtained near the seamounts analyzed (borehole 236, with coordinates of 01°40.62 S and 57°38.85 E). The age of basalts exposed by that borehole was detemined to be about 58 million years using the absolute age-determination technique.

When comparing the age of the bearing lithosphere and that of the seamounts, it is possible to deduce the mean rate of the northward Somali plate drift since the late Cretaceous. From these calculations, it was about 2.7 cm per year during the geological time interval of 60 to 38 million years, then it decreased sharply at the time of seamount formation (38 to 40 million years), and at the time of 38 through 0 million years, it decreased to 1.5 cm per year.

Comparison of the obtained results with the paleomagnetic reconstructions of the African, Somali, and Indian plates (Gorodnitsky, 1984; Scotese et al., 1987) supposes a sharp fall in the rate of northward drift of the Somali plate at the late Eocene/Oligocene boundary as a result of the largest tectonic collision of that epoch, that is, the closure of the eastern Tethys paleoocean and collision of the Indian and Eurasian plates. That leads to the splitting of the Eurasian plate into a series of microplates at the frontal part of the plate interaction boundary, whereas in the rear part, it reduced sharply the northward drift of the African, Somali, and Indian plates and also intensified intraplate deformations in the Central Indian Basin (Gorodnitsky, 1984). Comparison of the sharp fall of the rate of northward drift of the Somali plate with the minimum rates of the late Mesozoic-Cenozoic continental motion estimated from paleomagnetic data shows good correlation. In fact, the northward drift rate in the interval between the late Cretaceous and late Eocene decreases from 6.8 to 2.5 cm per year for the Indian plate, and from 2.8 to 1.9 cm per year for the African plate (Gorodnitsky, 1984).

The mentioned rate jump of the Somali plate motion correlates well with the corresponding sharp decrease of a linear spreading rate of the Indian Ocean floor, which reflects the jumps in the motion rate of the framing plates (Gorodnitsky, 1984).

So, for its lifetime, the seamount range under study has experienced an absolute northward drift at an angular distance of about 5°, that is, about 300 miles. At the same time, the seamount range has shifted southwestward relative to the spreading axis, that is, to the mid-oceanic Carlsberg Ridge. According to the formula by Sclater-Sorokhtin, the seamount sinking due to general lateral subsidence of a newly formed oceanic lithosphere was about 2 km for 38 million years (Sorokhtin, 1974).

The performed research, one of the pioneering studies in the paleomagnetic simulation of the Indian Ocean seamounts, proved to be very informative and suitable for a geohistorical examination of the Indian Ocean itself and for analysis of the plate motion.

III. SEAMOUNTS OF THE TYRRHENIAN DEEP-SEA BASIN

The geological structure of the Tyrrhenian Sea, which is a part of the Mediterranean Basin, and its tectonic evolution recently became the object of special geological and geophysical concern. Rich information on the surrounding land permits the correlation of the sea and on-land geology. Besides, a volcanic seamount system is located in the southern Tyrrhenian Sea within the Tyrrhenian deep basin, which is accessible to direct geological-geophysical study. The latter provides a key to the understanding of the principal regularities of the basin origin, tectonic evolution, and geodynamic mechanism, related to the late history of the Tethys paleoocean closure and also to the formation and evolution of the oceanic depressions inside the belt of general compression such as the Alpine fold belt.

The Tyrrhenian Basin floor is characterized by large volcanic seamounts like Maghnaghi, Vavilov, and Marsili, elongated in the meridional direction (Figure 7). The previous study showed that these seamounts are composed of tholeiite oceanic basalt, which provides evidence for an extension regime in that part of the basin. From the southeast, the Eolian island arc is attached to that region. It is characterized by intensive manifestations of calc-alkaline volcanism, which led to the formation of the system of seamounts and islands. The Tyrrhenian Basin itself, within which seismic research revealed a suboceanic crust up to 10 km thick, can be subdivided geomorphologically into two parts, namely the Vavilov Basin, in the west, and the Marsili Basin, in the east.

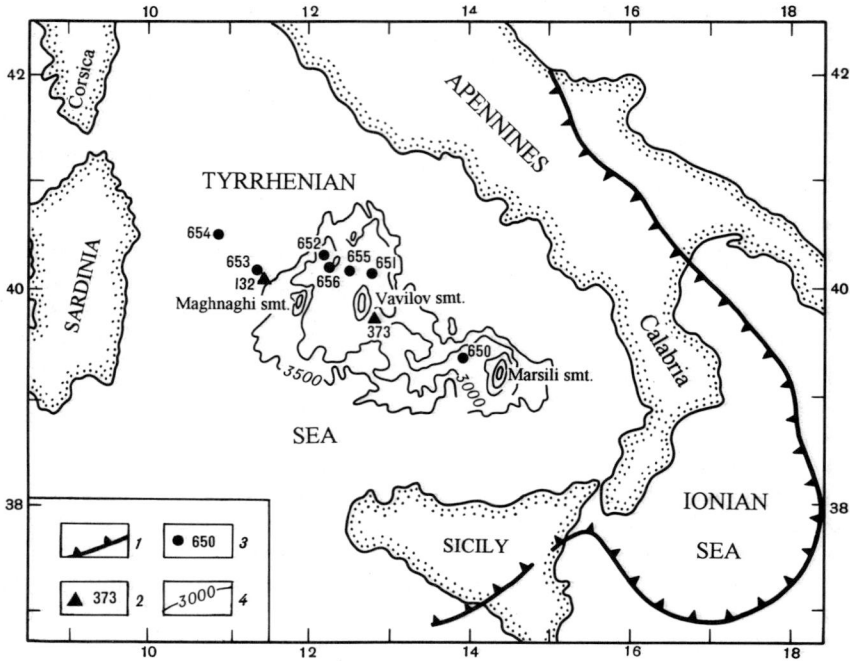

Figure 7 Pattern of the Tyrrhenian Seamounts: (1) subduction zone; (2), (3) deep drilling sites; (4) isobaths in m.

From the recent geomagnetic survey data (Filin et al., 1988), positive and negative anomalies with intensities of 100 to 300 nT and isometric geometry were registered in the region of the Vavilov Basin. In the west of the basin, positive extrema of the field sometimes form elongated sublatitudinal chains. Negative anomalies exceeding 200 to 300 nT are associated with the Maghnaghi Seamount and also with the northern and southern termination of the Vavilov Seamount. The central part of the Vavilov Seamount is characterized by positive anomalous fields up to 300 nT and over. However, in the vicinity of the Vavilov Seamount, some authors noted the tendency of the magnetic field anomalies to have a linear geometry subparallel to the meridional strike of the mount itself (Kogan et al., 1980; Malovitsky et al., 1982).

In the Marsili Basin, the AMF is characterized by sign-reversed anomalies. At the periphery of the basin, their geometry is close to an isometric one at 100 to 400 nT. In the southwest of the basin, some positive and negative extrema have a northwestern strike. In the center of the basin, there is an intensive northeast anomaly up to 800 nT and over, corresponding to the Marsili Seamount topography strike. A negative extremum of the same strike up to −300 nT is attached to it from the northwest, whereas a negative isometric extremum is recorded in the southeast.

Table 2 Magnetic properties of basalt from the Marsili Seamount sampled by the "Argus" submersible

Coordinates	No.	I_n, A/m	$\kappa \times 10^{-3}$ SI unit	Q	$C = 5\kappa/4\pi$ $\times 100\%$
39° 17.0 N	I	2.2	15	4.3	0.6
14° 24.1 E					
39° 17.1	II-2	1.0	15	2	0.6
14° 23.9	II-2	2.2	15	4.3	
39° 17.0	III	13.2	15	25.8	0.6
39° 17.0	IV-1	5.7	15	11.2	0.6
	IV-2	7.1	15	14	0.6
14° 24.0	IV-3	6.5	15	13	0.6
14° 23.0	V-1	1.4	10	4.1	0.4
	VI-1	18.8	10	56.3	0.4
	VI-2	14.5	10	43.5	0.4
39° 16.9	VII	12.6	15	5.2	0.6
39° 16.4	VII-2	2.4	15	4.8	0.6
14° 22.8	VIII-3	2.0	15	4.0	0.6
14° 23.5	VII-4	2.85	15	5.7	0.6
	VIII-1	4.13	15	8.2	0.6
	VIII-2	2.85	15	5.7	0.6
	IX-1	19.2	15	38.4	0.6
	IX-2	19.1	15	38.2	0.6
	IX-3	10.5	15	21	0.6
	X-1	10.1	20	15	0.8
	X-2	9.3	20	14	0.8
39° 12.0	XI	1.7	2	26	0.08
39° 12.0	XII	0.33	2.2	5	0.08
14° 22.0	XIII	0.50	1.9	7.5	0.08
14° 21.1					

The detailed geomagnetic survey within the study areas of the Tyrrhenian Seamounts was made simultaneously with the analysis of the samples uplifted by the submersibles. Tables 2 and 3 present the magnetic properties of rocks sampled from the Maghnaghi, Vavilov and Marsili Seamounts on the 16th cruise of the R/V *Akademik Mstislav Keldysh* by "MIR" submersible.

Table 2 presents magnetic properties of rocks comprising the northern summit of the Marsili Seamount. All of them are presented by fresh calc-alkaline basalt. Mean remnant magnetization is 7.3 A/m, and the Koenigsbergen factor Q is 14. Magnetite concentration in the rock reaches 0.6%. Comparatively large magnetite crystals of several hundred microns were detected when analyzing the basalt thin sections. High value of magnetic susceptibility and total ferromagnetic concentration are associated with those grains.

Table 3 **Magnetic properties of the Marsili Seamount sediments (St. 1669; depth, 3920 m)**

Length of the column, cm	$I_n \times 10^{-2}$ A/m	Inclination $J, °$	Length of the column, cm	$I_n \times 10^{-2}$ A/m	Inclination $J, °$
15	1.8	25	350	1.8	65
50	1.6	59	400	2.3	54
100	2.3	43	450	2.3	−68
210	2.3	64	450	1.6	−55
305	2.7	62			

Small values of I_n and Q are typical of the basalt sampled from the seamount summit and exposed to considerable secondary changes. It should be noted that fresh calc-alkaline basalt contains magnetite as a ferromagnetic mineral, whereas in the toleiite basalt, they are presented usually by titanomagnetite (Table 2).

Relatively small I_n and Q for samples I and II are explained by their high degree of secondary change. These samples were taken directly from the seamount summit in a zone of high hydrothermal activity (Belyaev et al., 1989).

On the 16th cruise of the R/V *Akademik Mstislav Keldysh*, the basalt was sampled from the Marsili Seamount summit (503 m deep) and its northwestern slope in the course of two divings of MIR submersibles: 5/16 and 5/21 (Table 4). Four samples were taken on each dive. The measurement results show the difference in magnetic properties of rock sampled in two divings and also their difference from magnetic properties of basalt from the Maghnaghi and Vavilov Seamounts. So, magnetic susceptibility of the basalt taken from the Marsili summit on dive 5/16 reaches 22×10^{-3} SI unit and exceeds by almost three times that of the basalt taken from its northern slope (7.5×10^{-3} SI unit). That provides evidence of a considerable difference in percent concentration of ferromagnetic minerals, which in turn may be a function of the lava melting conditions. The value of Q for the basalt from the slope is 106, on average, with a maximum of 245, evidencing high magnetic stability of the rock. It should be noted that the measured remnant magnetization correlates well with the geomagnetic survey data obtained on the 12th cruise of the R/V *Vityaz* (Table 2). So, samples of dive 5/16, with mean remnant magnetization of 5 A/m, were taken from the area of magnetic anomaly with an amplitude of 500 nT, whereas samples of dive 5/21, with mean remnant magnetization of 15 A/m, were taken from the zone of positive anomaly with an amplitude of 1450 nT (Belyaev et al., 1991).

Table 3 shows the results of paleomagnetic study of the sedimentary columns uplifted from the very rise of the Marsili Seamount. As follows from the table, the age of the lowermost sediments probably exceeds 700,000 years since both samples had negative magnetization, that is, they seemed to have been magnetized during the Matuyama epoch.

At the Vavilov Seamount, the samples were taken on dives 4/15 and 4/20 (Table 4). Points of sampling are given in Figure 8. On dive 4/15, an oriented core was drilled at a depth of 1825 in the basalt bedrock outcrop. That helped to determine the inclination and declination of natural remnant magnetization. Besides, magnetic parameters were determined of greatly zeolitized basalt sampled from bedrock outcrops on dive 4/20.

According to petrographic determination, basalt samples are presented by porphyriticolivine-plagioclase differences with similar mineralogical composition. They are characterized by relatively low susceptibility (no more than 2×10^{-3} SI unit), which indicates a small concentration of ferromagnetic minerals in the rock (0.04%). Remnant magnetization of three samples analyzed is similar and is 2 A/m, on average. The Koenigsbergen factor Q reaches 40 despite secondary changes in the rock. That supports the assumption that the Vavilov Seamount basalts have great magnetic stability free of the concentration of ferromagnetic minerals. Such a high stability allows a reliable estimate of the natural remnant magnetization declination and inclination in the oriented cores. From the measurement results, remnant magnetization inclination J equals $-5°$ and corresponds to the area of negative magnetic field values from the geomagnetic survey data (Figure 8). The recent magnetic inclination in the vicinity of the Vavilov Seamount is $56°$.

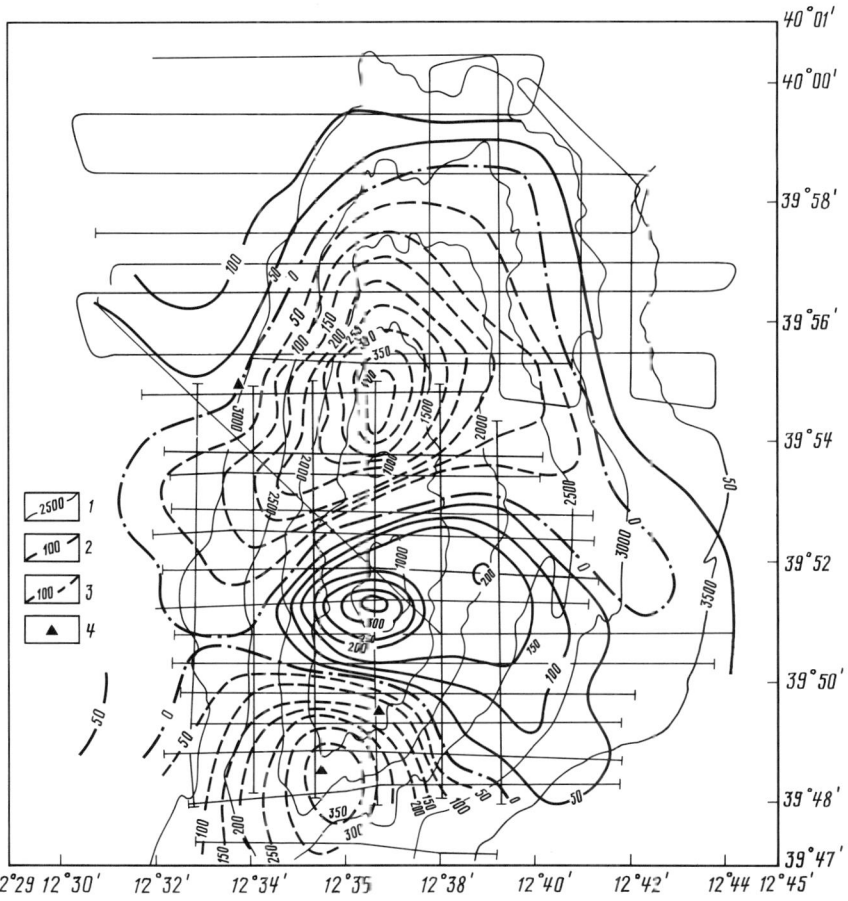

Figure 8 Map of the Vavilov seamount AMF: (1) isobaths in m; (2) positive; (3) negative, in nT; (4) sites sampled by MIR submersible.

Table 4 **Magnetic properties of rock sampled from the Tyrrhenian Basin Seamounts**

Coordinates	Nos.	Short petrographic characteristics	I_n, A/m	$\kappa \times 10^{-3}$ SI unit	Q	$C = 5\kappa/4\pi \times 100\%$
		The Vavilov Seamount				
39° 46.4 N	4/15-1	Basalt porphyric, olivine plagioclase, poorly zeolitized	2.5	2.0	31	0.08
39° 48.9						
12° 36.0 E						
12° 35.6						
39° 49.5	4/20-1	Basalt olivine-plagioclase	2.2	0.9	41	0.04
39° 51.3						
12° 36.9	2/20-2	Same	2.2	1.0	54	0.04
12° 36.5						
		The Marsili Seamount				
39° 19.7	5/16-1	Basalt porphyric	4.2	22	5	0.87
39° 16.8	5/16-2	Same	2.1	18.0	3	0.72
14° 22.8	5/16-3	Lava-breccia, greatly porous	1.4	10.0	3	0.4
14° 24.2	5/16-4	Basalt porphyric	14.0	20.0	18	0.8
39° 20.3	5/21-1	Basalt greatly porous	17.9	4.5	100	0.18
39° 17.5	5/21-2	Same	10.0	7.5	33	0.3
14° 27.7	5/21-3	Basalt porphyric	27.3	2.8	24	0.11
14° 24.5	5/21-4	Basalt polyphyric	7.4	3.8	49	0.15
		The Maghnaghi Seamount				
39° 48.4	6/17-1	Basalt polyphyric	4.9	4.3	28	0.17
39° 51.3	6/17-2	Same	28.9	2.7	269	0.11
11° 47.3	6/17-3	Same	14.9	2.3	163	0.09
11° 45.3	6/17-7	Same	4.4	1.3	84	0.05
	6/17-8	Same	3.3	1.3	63	0.05
	6/17-8A	Same	2.3	1.4	42	0.06
	6/17-8B	Same	4.3	1.5	73	0.06
	6/17-8C	Same	2.6	1.3	50	0.05

Magnetic properties of a plate of carbonate sediments uplifted from the Vavilov Seamount from a depth of 3300 m on the MIR submersible dive 3/19 were determined together with the study of the basalt magnetic properties. Sediments are presented by micrite limestone composed of a limestone nannoplankton with a small admixture of the foraminiferas. Knowing the plate supper and lower boundaries, it is possible to estimate the remnant magnetization inclination vector. As a result of measurements of the 15 cubes cut out of the plate low in the section, the following data were obtained: upper layer, $I_n = 0.003$ A/m, $J = 45°$; middle part, $I_n = 0.015$ A/m, $J = 47°$; lower part, $I_n = 0.007$ A/m, $J = 52°$. As seen from the given results, the J increases regularly downward in the section. That may be probably explained by the gravity differentiation of heavier ferromagnetic minerals in the process of sedimentation. Positive changes of magnetization vector of 48° corresponds to the Brunhes epoch of normal polarity (no older than 0.7 million years).

Thus, the analysis of magnetic properties of the erupted and sedimentary rocks from the Vavilov Seamount supposes a two-stage volcanic activity that resulted in the formation of the seamount under study.

Table 4 also presents the magnetic properties of the basalt sampled from the Maghnaghi Seamount.

The rocks were sampled from a depth range of 2817 to 1888 m, from bedrock outcrops at steep slopes composed of the typical pillow lava. According to mineral composition, the investigated volcanite can be related to the basalt of the subalkaline or enriched tholeiite series. The analyzed basalt has high magnetic susceptibility ($\kappa = 1.9 \times 10^{-3}$ SI unit), high remnant magnetization ($I_n = 7.8$ A/m), and a high value of the Koenigsbergen factor ($Q = 96$).

A. THE MAGHNAGHI SEAMOUNT

The Maghnaghi Seamount is a meridional elongated structure located in the western part of the Tyrrhenian Basin. Its elevation over the seafloor is 1500 m (Figure 9). The minimum depth at its summit is 1470 m. The seamount topography is characterized by wide (to 2.5 km) steps inclined slightly toward the rise. The central part of the western slope forms a large sublatitudinal uplift. The Maghnaghi Seamount is the third-largest volcanic structure (after the Marsili and Vavilov Seamounts) of the Tyrrhenian Basin. Its length is about 29 km, and its maximum width is 18 km. The seamount is elongated in the NNE direction as a ridge with a narrow upper part (2800 to 2100 m and less) and a wide basement. Maximum inclination in the upper part of the slope is 25 to 35°. In the lower part of the seamount, its slopes are flattened to 15°. The western and eastern slopes are complicated by steps. The seamount summit is bounded by a 1900-m isobath. A range of volcanic cones 350 m high is located on it. In its southern part, the range is traced along the mount down to a depth of 3000 m. In the northern part, the mount bifurcates into two offsets.

A detailed geomagnetic survey at the Maghnaghi Seamount was carried out on the 16th cruise of the R/V *Akademik Mstislav Keldysh* together with sounding and continuous seismic profiling. The profiles were spaced 1.5 to 2.0 miles apart. The RMS was ±9 nT. A map of AMF isodynams has been compiled from the survey results (Figure 9).

Analyzing the compiled maps, we can see that the AMF of the Maghnaghi Seamount has a complex structure, which does not seem to have any direct correlation with its topography. Areas of negative magnetic field are confined to the northern and southern parts of the seamount. The maximum negative anomaly of 200 nT has been registered at the northeastern slope of the mount, where depth to the floor reaches 2200 m. The positive maximum anomaly of 150 nT is confined to the western slope (Figure 9), which seems to be the younger superimposed volcano to which positive isometric anomaly is related. A wide belt of slightly positive values of magnetic field is traced from that anomaly. It is orthogonal to the general strike of the seamount and seems to be related to the superimposed volcanism controlled by the fault of the corresponding strike. On the whole, the bulk of the Maghnaghi volcanic massive is characterized by negative values of AMF.

On the basis of the geomagnetic and geomorphological data, a magnetic simulation of that seamount has been performed to study its genesis and the nature of magnetic lineations. A standard deviation of the anomaly differences is about 10%.

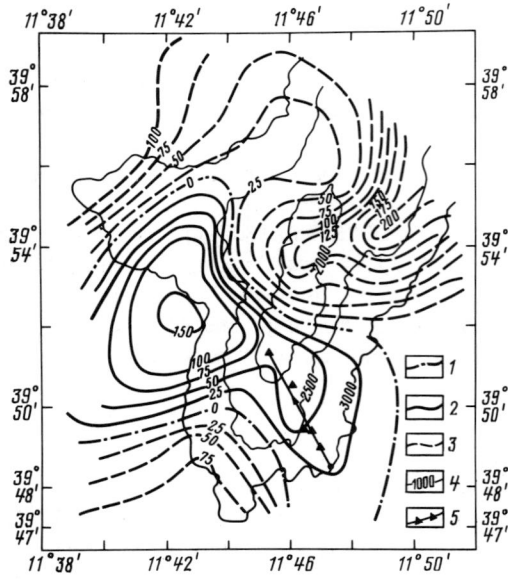

Figure 9 The Maghnaghi Seamount AMF: (1) positive isolines in nT; (2) zero isolines; (3) negative isolines; (4) isobaths in m; (5) sites sampled by MIR submersible.

According to the simulation, a geomagnetic model of the Maghnaghi Seamount is an assemblage of ten areas of normal and reversed magnetic polarity (Figure 10A). The mean effective magnetization for the areas with reversed polarity is 1 to 2 A/m. Area 8, characterized by small dimensions, is not representative of a stable solution. On the whole, two areas of normal polarity, namely, 2 and 10, can be distinguished against the general reversed polarity of the seamount. Both areas are located on the western slope. Effective magnetization of 6.6 A/m, which is close to experimental data, characterizes area 2, to which an epicenter of positive magnetic lineation is attributed (Figure 10A). The analysis of magnetic field transformations showed that its character practically does not vary and that its amplitudes vary with the height of recalculation but slightly. That may suppose a deep locus of the field source, which is also proved by the absence of any correlation between the anomaly and topography. Area 10 coincides spatially with the belt of small positive anomalies orthogonal to the principal strike of the seamount (Figure 10A). It is probably related to the superimposed volcanism controlled by the fault, the strike of which coincides with that of area 10. We may also suppose that area 2 is associated with the volcano channel where the basalt of the latest eruption was crystallized.

Performed magnetic simulation supports a two-stage formation of the Maghnaghi volcano. The formation of the main volcanic construction seemed to occur during the epoch of reversed magnetic polarity in the zone of extension that determined the elongated submeridional geometry of the volcanic seamount. During the epoch of normal magnetic polarity, the processes of secondary volcanism occurred there that resulted in the formation of smaller volcanic bodies controlled by the faults that cut the main volcanic construction. Comparing the results of geomagnetic study with the data of continuous seismic profiling and biostratigraphic analysis of sediments, and also with the deep drilling data (Malovitsky et al., 1978; Anderson and Jackson, 1987; Mascle et al., 1988), we can suppose that the main volcanic construction was formed during the Matuyama epoch and prior to 0.7 million years, whereas the formation of the superimposed volcanoes, related to the normal magnetic polarity, corresponds to the Brunhes epoch. Using the data of continuous seismic profiling (Malovitsky et al., 1982), we can see that the horizontally laminated Recent deposits are attached to the eastern and northern slopes of the mount, whereas at the southwestern flank of the seamount, in the area of positive AMF, seismic profile records a block uplifted at about 200 m over the abyssal basin floor. In many places, the upper laminated horizon, 100 to 150 m thick, is cut by the structures of the lower horizon. The key reflector "M", resting at a depth of 200 to 300 m, also has a rough sharply rugged topography. These data as well as the AMF structure support the idea of young tectonic motions in the Maghnaghi Seamount region and associated volcanism presumably dated back to the Brunhes epoch. So, the data of geomagnetic study show that the Maghnaghi Seamount is a two-stage volcano complicated by superimposed tectonic activity.

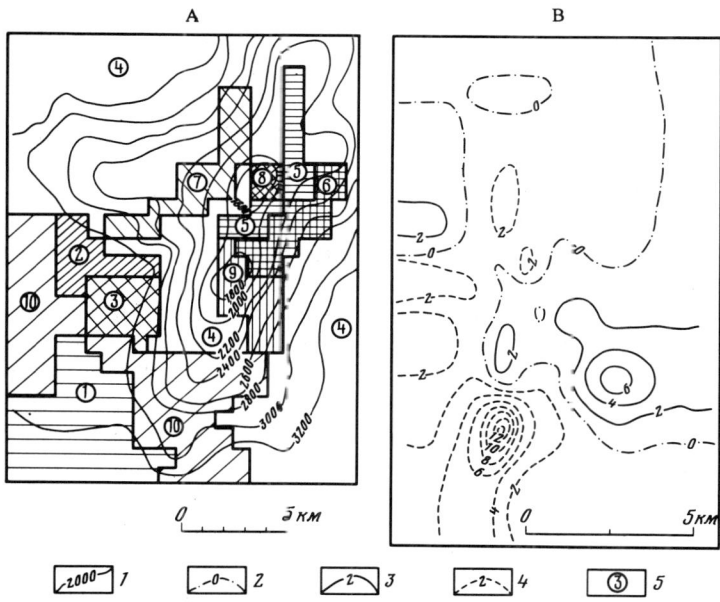

Figure 10 (A) Magnetic simulation of the Maghnaghi Seamount; (B) magnetization pattern of the Vavilov Seamount. Isolines are spaced in A/m; (1) isobath spacing in m; (2) zero isolines; (3) positive isolines; (4) negative isolines; (5) numbers of areas recognized by magnetic simulation.

B. THE VAVILOV SEAMOUNT

The Vavilov Seamount was recognized in 1959 by V. P. Goncharov on the first cruise of the R/V *Akademik Vavilov*. It is a large volcanic construction with a 2840-m elevation over the seafloor and with a minimum depth of 739 m over the southern summit. The seamount occupies an area of 32 × 17 km and is elongated submeridionally along the azimuth of about 15° subparallel to the Maghnaghi Seamount. Geomorphologically, the Vavilov Seamount is asymmetric in the east-west direction. Its western slope is steeper than the eastern and has a tilt of about 24°. Tilt of the more gentle eastern slope is only 16°, on average. The seamount slopes are complicated and dissected by deep troughs from the summit to the base, and as follows from the TV underwater survey by "Zvuk" and "MIR" submersibles, they are spaced between the ranges formed by lava flows.

Analyzing the CDP reflection data obtained along the latitudinal profile crossing the Vavilov Seamount, and the materials of geological-structural study (Kogan et al., 1980), it is possible to conclude that the volcano was set at the crustal block forming one of the scarps tilted eastwardly. That explains the longitudinal symmetry of the volcano stressed by the irregular eruption of lava flows at the western and eastern slopes of the seamount. The regional geomagnetic survey performed there earlier (Robin et al., 1987) showed that an intensive magnetic anomaly of a dipole character was registered over the Vavilov Seamount. The area taken up by that anomaly exceeds the area of the volcano itself. That allows a conclusion that the bulk of the Vavilov volcano massive is covered by a sedimentary layer that is verified by CSP data. An intensive magnetic anomaly with an amplitude of 850 nT was recognized over the seamount by detailed geomagnetic survey (Kogan et al., 1980). Its maximum area (to 468 nT) is correlated with the most uplifted southern part of the crest, and the conjugated minima (to –380 nT) are located north and south along the crest axis.

When dredging the steep western slope of the Vavilov Seamount, volcanic lava fragments with fresh facets were uplifted from a depth of 1600 m, which showed at the bedrock composition of samples (Malovitskiy et al., 1978). According to their structural-mineralogical composition, the samples are related to basalt with low concentration of siltstone and high content of volatiles. The given parameters characterize the Vavilov Seamount as an extinct volcano of basalt composition. A submarine character of eruption is stressed by some peculiar features of the lava texture and structure, first, by a high concentration of volatiles due not only to hydrostatic water layer pressure, but also to rapid cooling of the melt. At the southern summit of the mount, abundant accumulation of corals, forming a needle-top construction, were detected by dredging.

From the data by Morelli (1970), a free-air gravity anomaly was registered over the Vavilov Seamount, its amplitude being over 75 mgal. The Bouguer anomaly over its summit is 230 mgal at a density of 2.67 g/cm^3, whereas it reaches 250 mgal at the seamount rise. So, relative decrease of the Bouguer anomaly over the mount summit is 20 mgal. Meantime, a residual gravity anomaly to –30 mgal at a density of 2.67 g/cm^3 is registered at the radius of 30 km over the Vavilov Seamount. Density of the basalt from the Vavilov Seamount, determined by a single sample, is 2.66 to 0.3 g/cm^3 (Malovitsky et al., 1978).

Four divings of the "Siana" submersible (Robin et al., 1987) showed that the Vavilov volcanic massive may be subdivided structurally into two parts. The lower one, below 1500 m from sea level, is composed of thick lava basalt flows with a submarine character of eruption; the upper one, above 1500 m, is composed of lava with a smaller scale of eruption which does form the central construction of the Vavilov volcano elongated meridionally. From the radiometric determinations, the age of the upper basalt series is 0.1 to 0.4 million years and corresponds to the Brunhes epoch. It has been supposed from the observations that the last volcanic stage might have occurred in the subaerial environment of the Pleistocene. Analyzing the composition of the sampled basalts, we can say that they differ considerably from the basalts of the underlying crust (Robin et al., 1987). They are related to the intermediate alkaline series and are enriched by volatiles.

As for the underlying crust, at which the Vavilov volcano was set, it is regarded as an oceanic one according to its seismic properties, depth to the M boundary, and deep drilling data (Malovitsky et al., 1978).

A map of AMF ΔT_a (Figure 8) was compiled from the data of the module geomagnetic survey performed over a part of the Vavilov Seamount on the seventh cruise of the *Vityaz* and on the 16th cruise of the *Akademik Mstislav Keldysh* along the grid of latitudinal and meridional profiles. As follows from the map analysis, an anomaly maximum with an amplitude of 370 nT is registered over the center of the seamount summit with a minimum of 725 m. One more isometric extremum with an amplitude of 205 nT is registered at the eastern slope of the mount, at a depth of about 2500 m. North of the mentioned maximum, an intensive isometric minimum of AMF with an amplitude of −427 nT and corresponding to a 1400-m isobath is registered over a submeridional elongated crest at the mount summit. One more local isometric minimum of the field was revealed over the southern slope of the mount, which corresponds to a 2100-m isobath. The amplitude there reaches −385 nT.

Two variants of magnetic simulation were performed for the Vavilov Seamount on the basis of geomagnetic survey data and bathymetric materials. The "Mount" program was used as a basis for the first variant (Filin et al., 1988).

Fitting of equivalent models on the basis of the optimal conformity of the estimated and observed anomalies allowed us to localize the areas of maximum magnetization within the study area and also to specify the boundaries of the areas with normal and reversed magnetic polarity. When fitting the models, the azimuth and inclination of magnetization vector in the rocks from the Vavilov Seamount were assumed to be equal to their present value. As the fitting of the equivalent model showed, the Vavilov Seamount consists of three areas, namely, the central, which has a normal polarity, and the southern and the northern ones, which have reversed magnetic polarity. In the northern area, the greatest effective magnetization of 3 A/m is registered within the zone confined to the topography uplift with depths of 1500 to 2000 m. The maximum effective magnetization (6 A/m) of the southern area is located at the southern slope and is confined to a depth range of 2 to 3 km. The central area, with normal polarity, is characterized by effective magnetization of 1.2 A/m, as follows from the simulation. A zone with normal polarity and maximum magnetization of 5 A/m is registered near the summit, with a minimum depth of 725 m.

The second variant of magnetic simulation was carried out by the technique of A. N. Ivanenko. The standard deviation of anomaly differences for the Vavilov Seamount is 29 nT at a general amplitude of 800 nT. Similar to the first variant, the simulation made it possible to conclude that the mount is subdivided into three main areas of normal and reversed polarity (Figure 10B). Its mean effective magnetization is 3 to 4 A/m, and its maximum is up to 14 A/m near the summit. That is considerably higher than in the first variant. The position of the virtual paleomagnetic pole for the Vavilov Seamount is close to the geographic one. Zones of magnetization maxima are elongated in a submeridional direction, and they agree well with the general meridional strike of the mount (Figure 10B).

Thus, analysis of the magnetic simulations showed that the Vavilov Seamount is a volcano, the evolution of which occurred by several stages. The main volcanic construction, expressed in the recent topography seems to have occurred during the period of reversed magnetic polarity, whereas the summit of the seamount associated with the most recent eruptions was formed during the period of normal magnetic polarity. Intrusive formations with high effective magnetization registered at the eastern slope may be associated with the same epoch.

As was stated earlier, examination of the Vavilov Seamount by submersibles, and analysis of the rock samples showed that the seamount is composed of massive basalt lava at a depth of 1500 m and greater, at shallower depths, it changes for the basaltoid forming the summit of the volcano. From the data of radiometric determination (Robin et al., 1987), the age of the upper volcanic construction, to which positive anomaly of magnetic field is confined, is 0.1 to 0.4 million years and coincides with the positive magnetic polarity of the Brunhes epoch. The age of sediments deposited between lava flows at the western slope of the mount at a depth of 2780 m is determined from the foraminifera as the late Pleistocene. The results of the analysis of magnetic properties of rocks sampled from the Vavilov Seamount correspond to the data of age determination and supports the supposition of two-stage volcanic activity that led to the formation of the Vavilov Seamount. All of these data allow a supposition that the main part of the lower volcanic construction was formed during the Matuyama epoch of reversed magnetic polarity, and that the seamount was then built up at the next stage of active volcanism during the Brunhes epoch. That verifies the supposition that the bulk of the Vavilov Seamount was formed not earlier than 2.4 million years ago (Robin et al., 1987).

C. THE MARSILI SEAMOUNT

The Marsili Seamount is located in the southeastern part of the Tyrrhenian Sea within the Central-Tyrrhenian Basin in the direct vicinity of the Calabria island arc seamounts and islands. It presents one of the largest mounts of the Tyrrhenian Basin. The mount is elongated in a submeridional direction and is elevated by 2.5 to 3.0 km over the seafloor. Its minimum depth of 488 m is registered at its northern summit. The dredging performed earlier showed that the mount has a volcanic origin and is composed of alkaline and calc-alkaline basalt. Geomagnetic survey performed in the mount region by the oceanographic vessel Bannock (Morelli, 1970) has registered there an intensive magnetic anomaly with two positive extrema. The greatest positive anomaly value (amplitude of 1500 nT and over) is confined to the northern mount summit with the minimum depth values. The second one, with an amplitude of 700 nT, is confined to the southern slope. The mount rise beyond a 3000-m isobath is surrounded by negative anomalies with altitudes of 250 to 300 m.

A detailed geomagnetic survey was performed at the part of the Marsili Seamount on the 12th cruise of the R/V *Vityaz* together with sounding and CSP along the grid of sublatitudinal and submeridional profiles spaced two to three miles apart and at an RMS of about 30 nT. A map of the AMF isodynams was compiled as a result of the survey (Figure 11).

It follows from the map that a system of intensive magnetic field anomalies with amplitudes up to 2000 nT is attributed to the Marsili Seamount. A positive anomaly corresponds to the upper near-crest part of the mount, whereas the lower parts of the eastern and western slopes are characterized by negative anomalies. The maximum positive anomaly with an amplitude of 1500 nT is confined not to the summit of the mount with minimum depth but to the northern slope, where the depth range is 1000 to 1500 m. At the southern slope, the maximum positive anomaly (over 1000 nT) is associated with depths of about 2500 to 3000 m. The entire zone of positive magnetic anomalies has a NNE strike corresponding to that of the seamount. It appears that in the course of volcanic activity, the inflow of fresh basalt with high magnetization occurred along the seamount axis at its northern and southern slopes, which may provide evidence for the extension regime there. In the northeastern part of the seamount, an intensive positive anomaly with an amplitude of 600 nT and probably related to active young volcanism was registered at a depth of 3000 m at the seamount rise. Horizontal gradients of the field at that part of the seamount are over 500 nT.

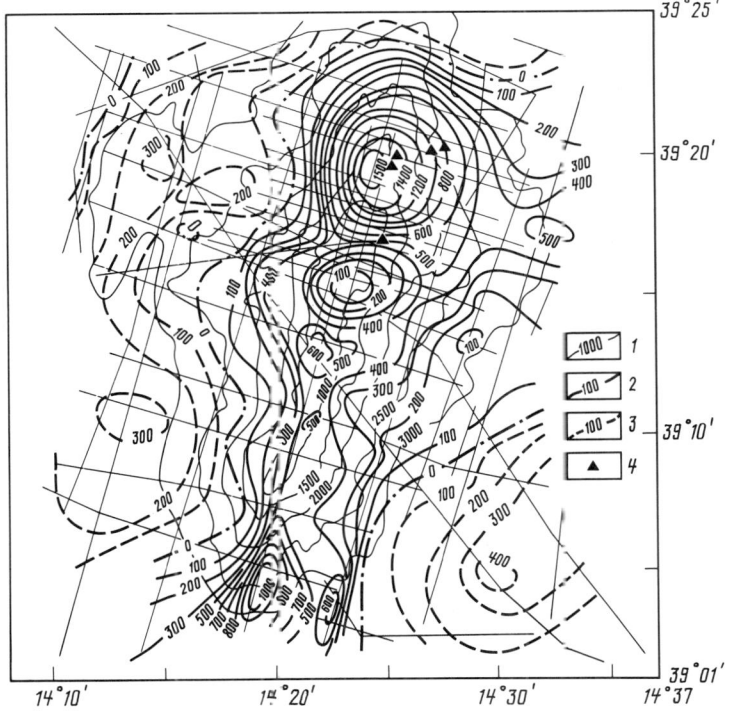

Figure 11 Map of isodynams of the Marsili Seamount AMF. For notations see Figure 8.

Geomagnetic survey results, together with the geomorphological data and study of magnetic properties of basalt, were used for magnetic simulation of the Marsili Seamount. A standard deviation of the estimated anomaly of the Marsili Seamount from the observed one was 103 nT at an amplitude of 2000 nT. Figure 12 presents a magnetic model of the Marsili Seamount.

As follows from the simulation results, blocks of various trends and values of effective magnetization can be recognized within the Marsili Seamount, which has a complex structure. The upper part of the mount, its southern and northeastern slopes, have normal magnetic polarity, whereas its basement and the lower slopes are characterized by reversed polarity. An individual volcanic construction with reversed magnetic polarity and an altitude of about 800 m is recognized in the northwestern part of the seamount. All zones with normal and reversed magnetic polarity have a submeridional strike similar to that of the mount itself. Meantime, at the field downward transformation and calculation of horizontal gradients, submeridional anomaly zones with the strike of 120° and crossing the northern seamount summit can be delineated besides the meridional ones. Blocks with different effective magnetization are recognized within the areas of normal polarity. Maximum effective magnetization (over 10 A/m) is recorded in the northern block confined to the northeastern slope of the seamount at the extension of its long axis. Approximately the same high effective magnetization is typical of the block at the southern slope. The central and southeastern parts of the volcanic construction have lower magnetization: 2 to 3 A/m and 3 to

4 A/m, respectively. A local area of considerably low magnetization is recognized within the central block, it is confined to the seamount summit with a depth minimum of 481 m. It follows, from the comparison of the above-described results of magnetic properties of the rocks sampled there on the divings of the "Argus" and "MIR" submersibles, that part is composed of volcanites with low magnetization considerably changed and destroyed due to hydrothermal activity.

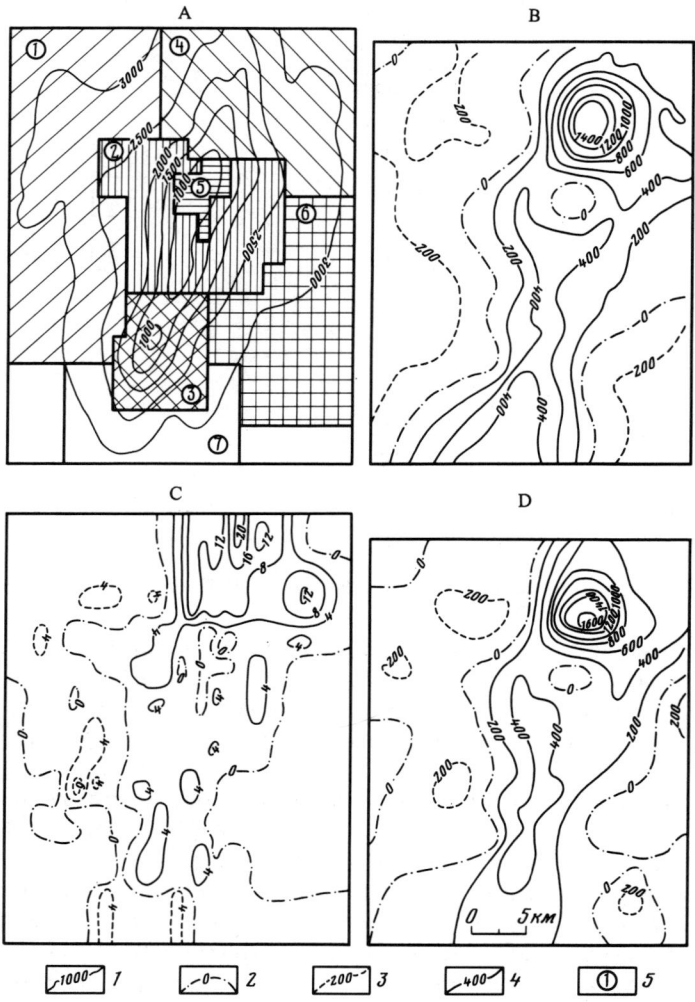

Figure 12 Simulation of the Marsili Seamount: (A) simulated, isobaths are given in m; (B) observed ΔT_a, isolines are spaced in 200 nT; (C) magnetization in A/m, isolines are spaced in 4 A/m; (D) simulated field, isolines are spaced in nT; (1) isobaths in m; (2) zero isolines; (3) negative isolines; (4) positive isolines; (5) numbers of areas recognized by simulation.

In the vicinity of the Marsili Seamount, borehole 650 was drilled down to 600 m and deeper in the deep basin (Figure 11). From the drilling data, the upper layers down to 600 m are composed of sediments, which lie over the crystalline basement. From the magnetostratigraphic data, the basalt comprising the basement is 1.7 to 1.9 million years old, which corresponds to the Olduvai event during the Matuyama epoch of reversed magnetic polarity. Therefore, the beginning of the volcanic activity in the region may be dated back to the epoch with the time range 1.9 to 0.7 million years. That is verified by analysis of the sedimentary columns sampled near the seamount rise. Lower sedimentary layers have negative magnetic polarity which seems to have been acquired during the Matuyama epoch.

As follows from the results of magnetic simulation, the bulk of the Marsili volcano has normal magnetic polarity. This fact supports the idea that the main volcanic construction was formed during the Brunhes epoch, which means that it is younger than 0.7 million years. That is verified by the data that the last manifestations of volcanic activity at the Marsili are dated back to 0.1 to 0.2 million years (Selli et al., 1977).

D. GEOLOGICAL INTERPRETATION OF THE MAGNETIC SIMULATION

A comparison of the results of geomagnetic study of the Maghnaghi, Vavilov, and Marsili volcanic seamounts of the Tyrrhenian deep basin and, first of all, of the magnetic simulation data supports the conclusion that they were formed in at least two stages. Most apparently, the main volcanic activity that resulted in the formation of the main volcanic construction of each of them was confined to the Matuyama and Brunhes epochs. According to the ratio of the regions with normal and reversed polarity and the regular increase of effective magnetization from west to east, from the Maghnaghi Seamount to the Marsili Seamount, the axis of volcanic activity was shifted subsequently from west to east, probably reflecting the shift of the extension axis.

Considering the geomorphological structure of the three seamounts analyzed, it is evident that they have submeridional strike and they have feathering at the southern and northern slopes, which supports the existence of the axial fracture zone in all of them. Besides, they are accompanied by the satellite mounts and have almost flat summits with volcanic cones. These facts may show at the formation phases of the volcanoes. At the first stage, apparently in the Pleistocene, the extension regime and accompanying fracture volcanism governed the region. Further on, an eruption of the central type with the formation of scoria cones was superimposed onto the fracture volcanism.

Analyzing magnetic anomalies of three seamounts under consideration, it is worth mentioning that the magnetic anomaly over the Maghnaghi Seamount is one half or one third of that over the Vavilov Seamount and is one order less than that over the Marsili Seamount. The effective magnetization has the same distribution pattern that has been revealed by magnetic simulation.

If we compare the depth values of the three seamounts, we find that the Maghnaghi Seamount has the maximum depth over its summit, whereas the Marsili Seamount has the minimum depth over its summit and the Vavilov Seamount has an intermediate depth. Such a regularity seems to be explained by the process of subsidence of seamounts. Then, if we compare the depth with the AMF values, we can see the correlation between them, which shows the age increase of seamounts from the Marsili Seamount to the Maghnaghi Seamount. That supposition is verified by the drilling results of 11 boreholes drilled in the 107 leg according to ODP program. So, three boreholes drilled within two subbasins, namely Vavilov and Marsili, in the central Tyrrhenian deep basin, showed that both are underlain by basalt basement and that the Marsili basement is at least 1 million years younger than the Vavilov basement (Boillot et al., 1987; Kastens et al., 1986). The value of the lithosphere thickness calculated from geothermal data agrees well with these data. They show that the minimum thickness of 20 km corresponds to the area between the Vavilov and Marsili Seamounts in the Tyrrhenian Basin.

It should be noted that from the geochronological study, magmatic activity of the Tyrrhenian Sea, since the late Miocene (8.5 to 6.5 million years), has been migrating from west to east due to the change of the geodynamic regime (from compression to extension) in that area. Two stages of intensive spreading and associated volcanism can be distinguished within the Tyrrhenian deep basin: from 5 to 1.8 to 1.3 million years, and from 1.3 to the Recent epoch. Geodynamic reconstructions allow us to associate that process with the counterclockwise rotation of the Apennines Range and with the synchronous opposite drift of the Sicilian block of the Magrebi system. The spreading axis displacement from west to east and volcanic activity within the mentioned time interval is verified by the seismic and geothermal studies in the Central Tyrrhenian Basin.

IV. THE PACIFIC SEAMOUNTS

Magnetic simulation of seamounts is of a chief importance for the Pacific plate which does not bear any continents. Therefore, paleomagnetic study of seamounts is the main source of information on that plate's kinematics. In this section, we give the results of geomagnetic study of seamounts in various regions of the Pacific.

A. THE MARCUS WAKE SEAMOUNT RANGE

The Marcus Wake Seamount Range is attached to the Mid-Pacific Seamount province from the west. It comprises abundant assemblages and individual seamounts and guyots with altitudes of 1 to 4.5 km and rock composition typical of oceanic areas with intraplate volcanism, namely, tholeiite to differentiated alkaline basaltoids (Rudnik and Matveenkov, 1978; Tuezov et al., 1979). The available data provide evidence of a long-time many-stage evolution of the intraplate volcanism in the region. An absolute age of rocks uplifted from various seamounts of the range varies within the interval of 34 to 59 million years (Matveenkov and Marova, 1975) to 78 to 98 million years (Ozima et al., 1983). From the data of magnetic anomaly extrapolation of M sequence, the lithosphere age in that region is dated back to the late Jurassic (165 to 155 million years); that is, the area under consideration is the most ancient part of the Pacific plate.

The V13-1 Seamount, with summit coordinates of 23°45 N and 148°40 E, was studied on the 13th cruise of the R/V *Vulkanolog*. It is located at the western flank of the range near the junction zone of the Izu-Bonin and Marianas deep trench. The seamount is a guyot 4600 m high elongated in the NNW direction (Figure 13D). Its flat summit of 9×4.5 km^2 rests at a depth of 1090 to 1100 m and is complicated by local uplifts 20 to 50 m high. The dimensions of the seamount; typical (in particular, in its lower part) cross-like geometry of its topography; the existence of small positive forms, such as domes, on the slopes; and the accumulative apron at the foot and lower parts of the slopes provide evidence of a complex evolution of that paleovolcano.

As the result of dredging, a fragment of an alkaline olivine basalt was uplifted from the northwestern slope from a depth of about 1500 m, and organogenic limestones were uplifted from the seamount summit. Geological sampling verifies the assumption that the summit of the volcanic construction did reach the ocean surface and that the seamount underwent the stage of a submarine volcano, a volcanic island, an atoll, and subsequent sinking down at the subsidence of the cooling lithosphere plate (Gorodnitsky, 1985).

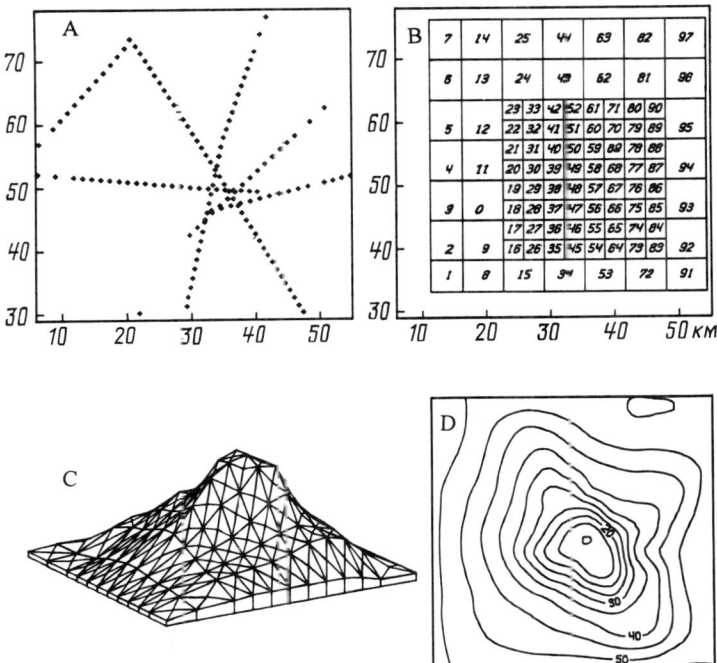

Figure 13 Primary data for magnetic simulation of the V13-1 Seamount: (A) sites of field measurement; (B) grid for topography approximation and number of body; (C) approximated topography; (D) mount bathymetry in hundreds of meters.

A magnetic anomaly over the seamount has an amplitude of 960 nT and geometry typical of the objects with magnetization close to a horizontal one.

Transformations of magnetic field were carried out together with the mutual spectral analysis of magnetic field and topography to find out the primary approximation of the magnetization vector. The results are given in Figure 14 (Ivanenko, 1990). According to the Fourier transformation for the field ΔT_a, we have the magnetization trend with angles $D = -21°$, and $J = 7.5°$. Then, they were used for the inverse problem solution with respect to the magnetization intensity distribution. A somewhat irregular shape of the phase of the transmitting function (Figure 14E) evidences some inhomogeneity of the seamount magnetization, probably the superposition of several components.

As the primary field points, we used the real measurements along profiles of the vessel; that prevents the distorting effect in the observed field due to the subjective interpretation of a specialist compiling a map of magnetic field.

The results of the inverse problem solution and regressive analysis of the simulated model are given in Figure 15.

The following conclusions are possible from the analysis of the obtained results:

- The mount has a heterogeneous magnetization, with mean magnetization being 2 to 3 A/m; at least 85% of its body has magnetization below 4 A/m.
- No less than three quarters of the total seamount body has normal magnetic polarity (Figure 15D)

- The main effect is produced by the central magnetic core and the area of intensive reversed magnetization located within the range at the NNW slope of the seamount
- The lower part of the slopes (deeper than 4 to 4.5 km) has a slight magnetization

A magnetic structure of the seamount (Figure 15D) agrees well with the topography itself. Two orthogonal trends are distinguished in the magnetization direction: the principal one, which coincides approximately with the long axis of the mount (azimuth is about 320°), and a subordinated one (azimuth is 40 to 60°). Arising from the map of the Mesozoic magnetic anomaly axes compiled for the northern part of the Pacific floor (Nakanishi et al., 1989), the main trend of magnetization of the V13-1 Seamount is orthogonal to the trend of the ancient spreading axis that separated the Pacific and Izanagi lithosphere plates (Hilde et al., 1977) (the western branch of the Hawaiian M sequence), whereas the subordinated trend follows approximately the ancient spreading axis (Larson, 1976).

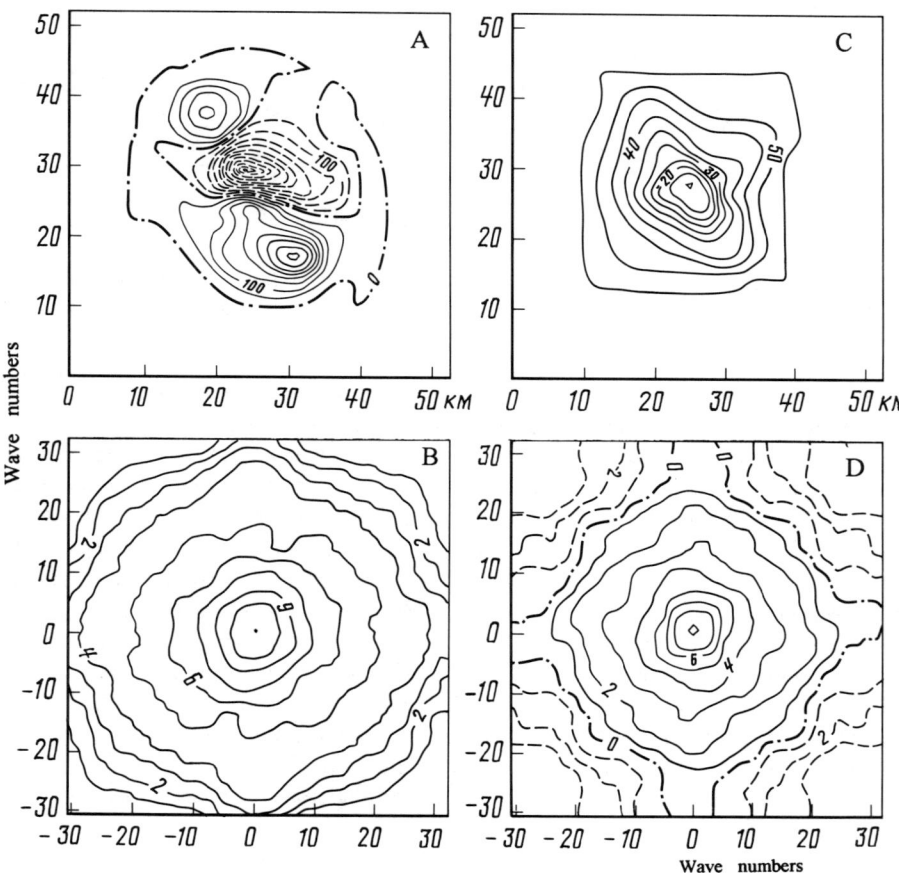

Figure 14 Mutual spectral analysis of magnetic field and topography of the V13-1 Seamount: (A) primary field ΔT_a nT; (B) energy spectrum ΔZ; (C) mount topography in hundreds of meters; (D) energy spectrum of topography; (E) logarithm of the transmittance function amplitude averaged by the circles of constant frequency in spectral plane; (F) phase of transmittance function averaged by frequencies along the radii in spectral plane.

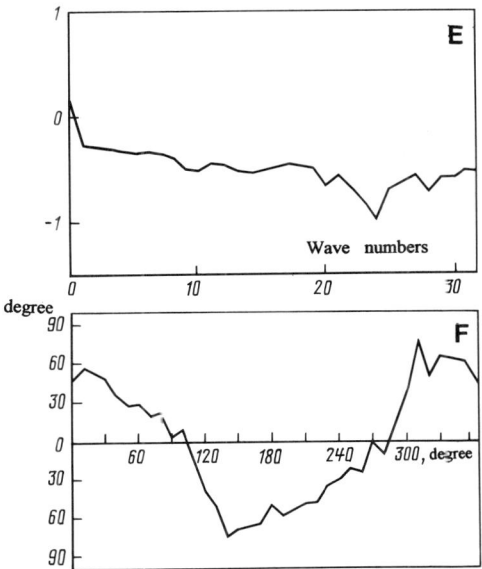

Figure 14 (continued). Mutual spectra analysis of magnetic field and topography of the V13-1 Seamount: (A) primary field ΔT_a nT; (B) energy spectrum ΔZ; (C) mount topography in hundreds of meters; (D) energy spectrum of topography; (E) logarithm of the transmittance function amplitude averaged by the circles of constant frequency in spectral plane; (F) phase of transmittance function averaged by frequencies along the radii in spectral plane.

Such a distribution of magnetic parameters may reflect the geometry of the evolved feeding system of the volcano and the spatial position of both the central magmatic channel and subhorizontal branches of the feeding system related to the evolution of flank rift zones. The existence of such rift zones was detected for the majority of intraplate volcanoes (Ryan, 1988) and is verified by the data of multi-beam bathymetric survey of these seamounts (Vogt and Smoot, 1984). It is quite possible that the spatial position of the magmatic channel system of that volcano is confined to the weak areas of the lithosphere plate and was controlled by the orthogonal (transform and parallel) ancient faults. Magnetic properties of the rocks of the dyke complex feathering the magmatic channels of the flank rift zones (Fornari et al., 1978) are very conservative as compared to the high- and low-temperature destruction of remnant magnetization and may preserve high stability of magnetization for a long time (Smith and Banerjee, 1986). At the same time, magnetic minerals in the erupted pillow lava can be affected more easily by secondary changes and partly lose their magnetic properties (Nazarova, 1981; Bleil and Petersen, 1983). Besides, less magnetization of the seamount slopes may be explained by the occurrence of considerable volume of clastic rock and hyaloclastite accumulated at the slopes and foot of volcanoes (Batiza et al., 1984) and characterized by low integral magnetization due to random orientation of fragments and slight remnant magnetization of hyaloclastites (Gee et al., 1988).

The coordinates of virtual paleomagnetic pole of that mount are $\Phi = 61°8$ N and $\Lambda = 21°1$ W. It should be noted that a 95% confidence ellipse around that paleopole position is overlapped by a 95% confidence area for a virtual paleomagnetic pole of the late Cretaceous seamounts, the age of which is 80 to 90 million years old (Parker, et al., 1987) (Figure 16). Paleolatitude of that volcano formation is estimated as $5.5 + 2°$ N.

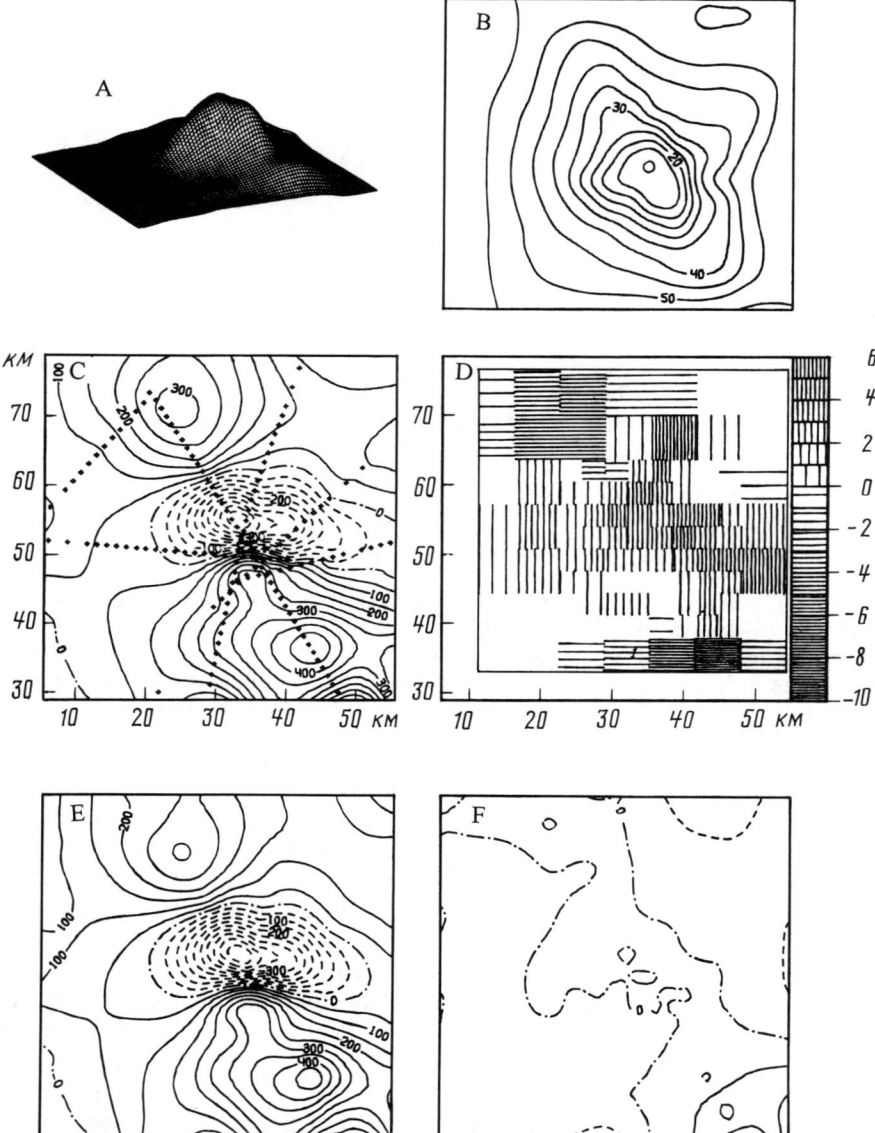

Figure 15 Simulation and analysis of the magnetic model of the V13-1 Seamount: (A) isometric plotting of topography; (B) mount bathymetry; (C) observed field in nT; (D) magnetization pattern in A/m, obtained after solution of inverse problem; (E) field of the finite version of the model; (F) differential field; (G) observed and simulated fields along profiles; (H) differential field and dimension of a 95% confidential area around the simulated field along profiles.

In the central part of the Marcus Wake Range, a seamount with coordinates 23°15 N and 155°50 E was chosen for a simulation. We used the published materials of the bathymetric and magnetic research performed in the expedition of the R/V *Pegas* (1975 to 1976) (Tuezov et al., 1979).

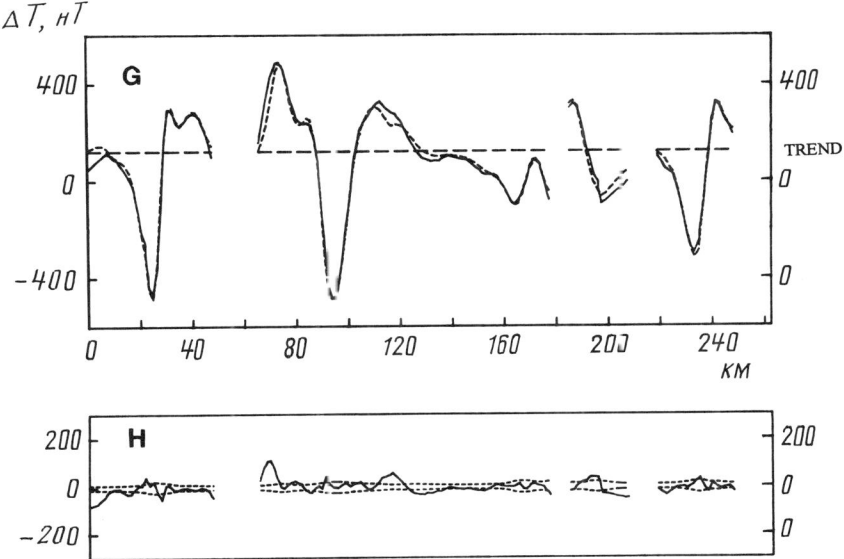

Figure 15 (continued). Simulation and analysis of the magnetic model of the V13-1 Seamount: (A) isometric plotting of topography; (B) mount bathymetry; (C) observed field in nT; (D) magnetization pattern in A/m, obtained after solution of inverse problem; (E) field of the finite version of the model; (F) differential field; (G) observed and simulated fields along profiles; (H) differential field and dimension of a 95% confidential area around the simulated field along profiles.

The mount rises upward from a depth of 5400 to 1500 m and is a cone of 32 × 24 km in the base, elongated slightly in the latitudinal direction (Figure 17B). Magnetic anomaly over the mount has a complex structure (Figure 17C). Two maxima (over the summit and within the northern slope) of 350 nT and 250 nT, respectively, and a narrow sublatitudinal area of negative values to −150 nT (over the southern and southeastern slopes) are registered against the general sublatitudinal isodynam strike.

In the first version, the simulation has stated that the lowermost part of the slope deeper than 5 km is practically nonmagnetized. Therefore, at the further specification of the solution, that part of the mount was excluded from the simulation.

As a result of several cycles of interpretation, we have got that:

- About two thirds of the mount body has reversed polarity of magnetization
- Half of the mount body has a magnetization of 1 to 3 A/m, whereas 20% of the body has less than 1 A/m (Figure 18)
- A radial five-branch system of intensive magnetized zones was traced in the magnetic structure of the mount (Figure 17D). Radial areas in the SE (azimuth of 145°), S (170°), and NE (20°) slopes affect the anomaly most

The virtual paleomagnetic pole of the mount is located in the point with coordinates of $\Phi = 67°5$ N and $\Lambda = 4°5$ W, and its confidence ellipse is overlapped by the confidence area of the Pacific Seamounts of 80 to 90 million years old and also by the area for the mounts of 72 million years old (Sager, 1987) (Figure 16). Paleolatitude of the mount formation is estimated as 7° N.

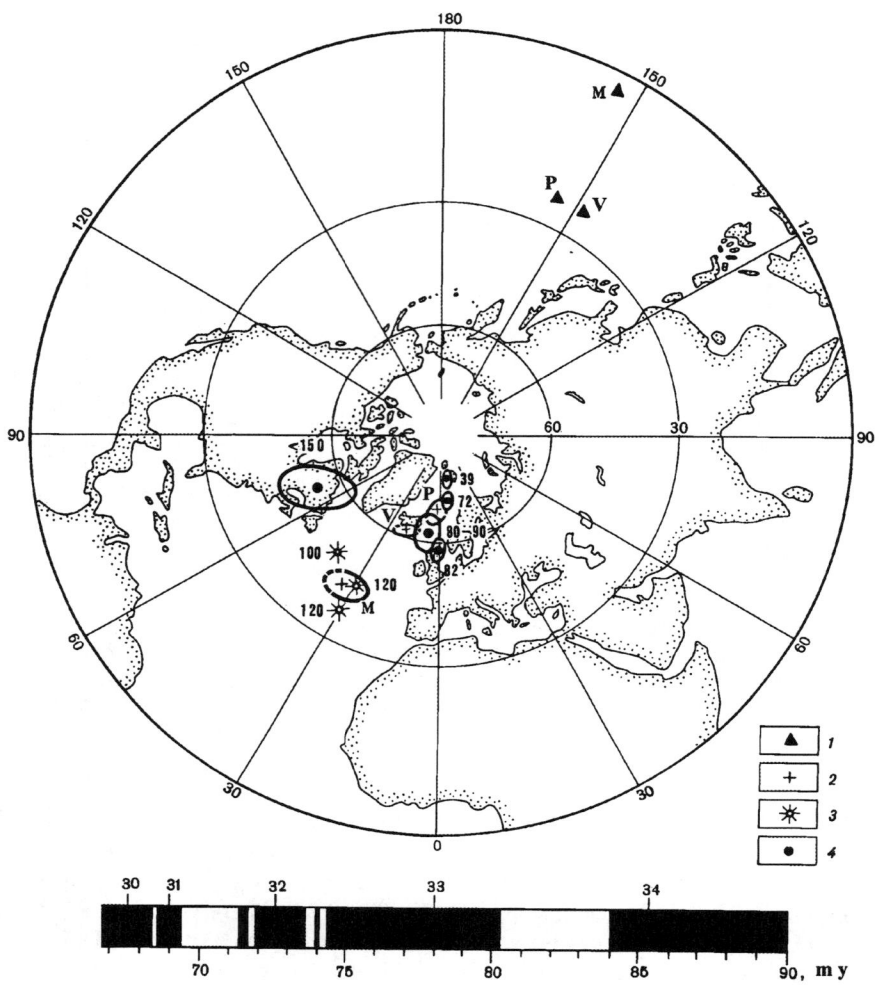

Figure 16 Paleomagnetic poles of seamounts V13-1 (V), Pegas (P), and Mendeleev (M). Triangles indicate the positions of virtual paleomagnetic poles of seamounts with 95% confidential areas; asterisks and numbers show the positions of virtual paleomagnetic poles for the Pacific plate for the epoch of 39, 72, 80-90, 82, 100, 120, and early Cretaceous seamounts over the crust younger than 150 million years (Sager, 1987; Hildebrand and Parker, 1987; Sager and Pringl, 1988; Kononov, 1984; Larson and Chase, 1972) with 95% confidential areas. Below the map is a magnetic chronological scale for the late Cretaceous with numbers of magnetic lineations (Kent and Gradstein, 1985).

Comparing the coordinates of paleopoles and the magnetization pattern of these two mounts of the Marcus Wake Range with the latest data on the paleomagnetism of the Cretaceous Pacific Seamounts (Cox and Gordon, 1984; Parker et al., 1987; Sager and Pringl, 1988) and also with the magnetochronological scale (Kent and Gradstein, 1985), we can suppose that they were set and formed in the late Cretaceous at the time of 86 to 70 million years. The V13-1 Seamount was formed at the end of the Cretaceous superevent, that is, of the normal magnetic polarity, the end of which corresponds to 84 million years. The latest eruption at the NNW slope seems to have occurred later during the reversed polarity epoch. Arising from a more complex magnetization pattern, predominance of reversed magnetization, and position of paleopoles, the mount of the central part of the range seems to have been formed in the time interval of 75 to 73 million years, since four reversals of the Earth's magnetic field polarity did occur there during that period. That age estimate is verified indirectly by the existence of interlayers of the oceanic volcanic sediments of about the same age found in the cores from boreholes 198 and 199 (Rea and Villier, 1983), drilled in the adjacent regions of the Northwestern and Marianas Basins. It is the seamounts of the Marcus Wake range that can be the source of these materials.

Magnetic structure of both seamounts reflects the geometry of the feeding system, which in turn is controlled by the deep lithosphere faults. In both cases, some of the axes of the high magnetization zones coincide with the strike of the transform faults at the bearing lithosphere, and some of them are orthogonal to the latter and are parallel to the strike of ancient spreading axes. A number of researchers have already noted that such orthogonal fault systems are expressed in the morphology and in gravity anomalies of seamounts set at a young lithosphere near the mid-oceanic ridge axes (Batiza and Vanko, 1983; Kellog and Ogujiofor, 1985).

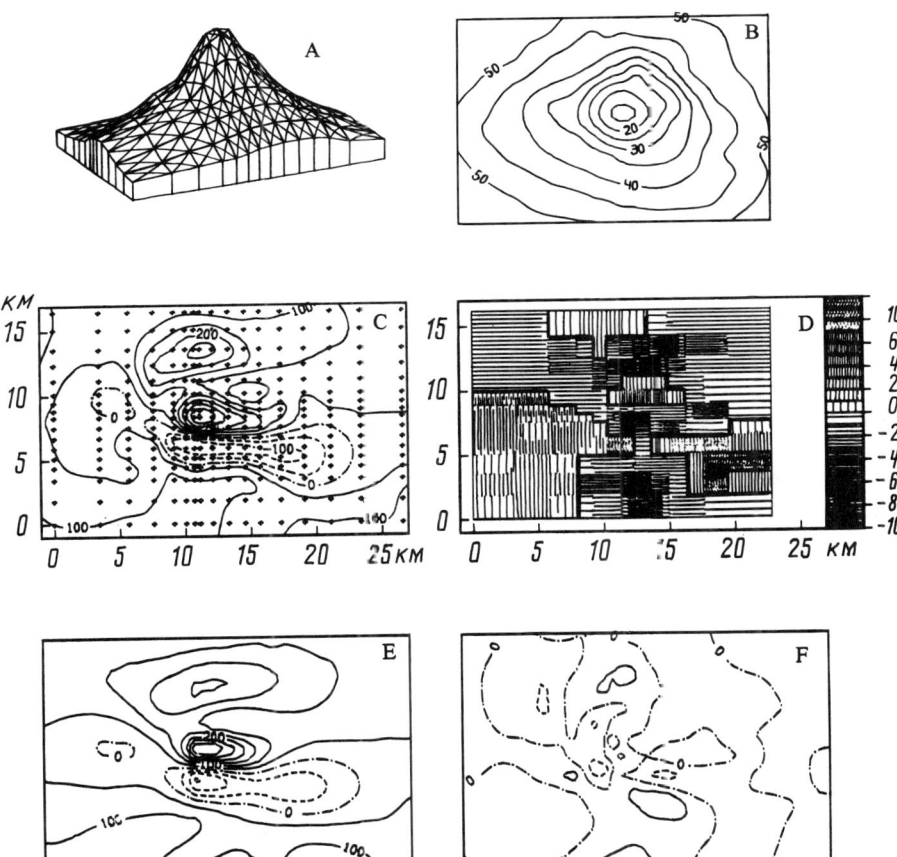

Figure 17 Simulation and analysis of magnetic model of the Pegas Seamount: (A) approximated topography; (B) mount bathymetry; (C) observed field in nT; (D) magnetization pattern in A/m obtained after the inverse problem solution; (E) field of the finite version of the model; (F) differential field.

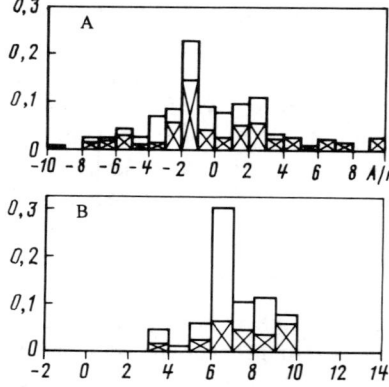

Figure 18 Solution of inverse problem by generalized linear inversion for the Pegas Seamount. Histograms of relatively body magnetization pattern: (A) magnetization pattern; (B) distribution of log 2 ($I_n \times 100$) of magnetization.

The obtained data allowed a conclusion that the older lithosphere (70 to 80 million years is the difference between the underlying lithosphere and the formation age of the Marcus Wake Seamounts calculated from geomagnetic data) "remembers" the structure of the west zones, which remain more permeable to the melt injection when magmatic activity is resumed.

The obtained data, namely:

- Age determination of the seamount formation and the difference between these calculations and the age of the underlying lithosphere
- Paleomagnetic latitudes of seamount formation
- Generally high magnetization of rocks comprising these seamounts
- Their typical magnetic structure, which made it possible to consider the analyzed seamounts of the Marcus Wake Range to the oceanic intraplate basalt paleovolcanoes (Avdeiko, 1979), formed near the equator at the end of the Cretaceous during the general considerable activation of volcanism in the Western Pacific (Rea and Villier, 1983)

B. THE CAROLINE GUYOTS

The Mendeleev Seamount is located in the zone of transition from the Ontong-Java to the Caroline Basin at the periphery of the East Caroline Seamount (Gorodnitsky, 1985). The detailed complex study, by the "Pisces" submersible including (the 21st cruise of the R/V *Dmitry Mendeleev*), showed that the seamount is a cone structure with eroded top covered by reef limestones and that is has a volcanic origin.

The mount base is an ellipse of about 28×22 km resting at a depth of 3500 m. Its flat summit at a depth of 1000 m is crowned by several peaks with minimum depths of 610, 880, and 779 m (Sato and Mogi, 1965) (Figure 19).

Geomagnetic surveys were performed over the mount in the module and component versions. The field anomaly ΔT_a is shown in Figure 20A. According to these data, magnetic simulation was performed by the Vacquier technique. It showed that the mount has reversed magnetization of about 2.9 to 3.2 A/m and that the magnetization vector corresponds to the angles of $D = 220°$, $J = 50°$ (Gorodnitsky and Luk'yanov, 1980). An estimated paleomagnetic latitude showed that the mount was formed in the southern hemisphere at a latitude of 40° S.

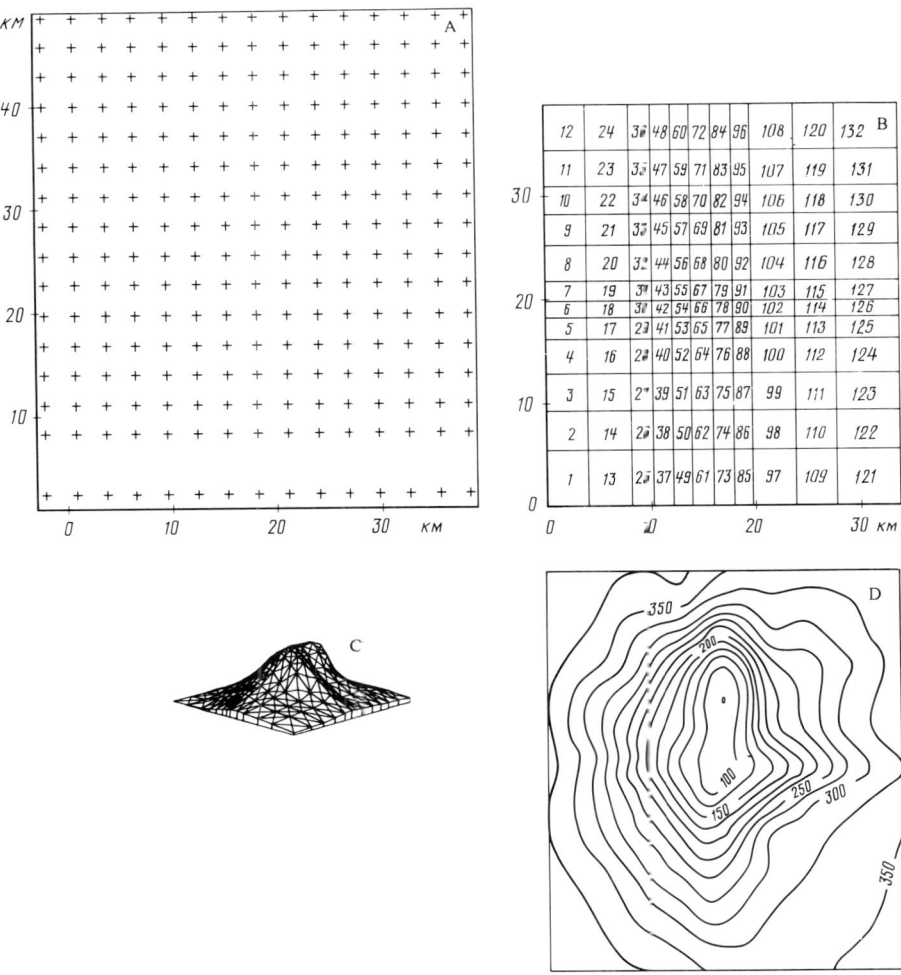

Figure 19 Primary data for magnetic simulation of the Mendeleev Seamount: (A) points of the given primary field; (B) dissection of the mount into elementary bodies; (C) approximated topography; (D) bathymetry in tens of meters.

To check the validity of and specify these conclusions, the model of the Mendeleev Seamount has been resimulated in accordance with the technique described in Chapter 2. The initial data and basic results of the simulation are given in Figure 20. The paleomagnetic pole of the mount is shown, together with the poles of the seamounts of the Marcus Wake Range in Figure 16.

As the result of that simulation it was stated that:

- Magnetization of the mount is almost uniform, with more than two thirds of its body having I_n = 1.5 to 4 A/m with a slight predominance of values 2 to 4 A/m in the western part of the mount. The mount has a reversed polarity with $D = 187 \pm 3°$ and $J = 56 \pm 1.5°$

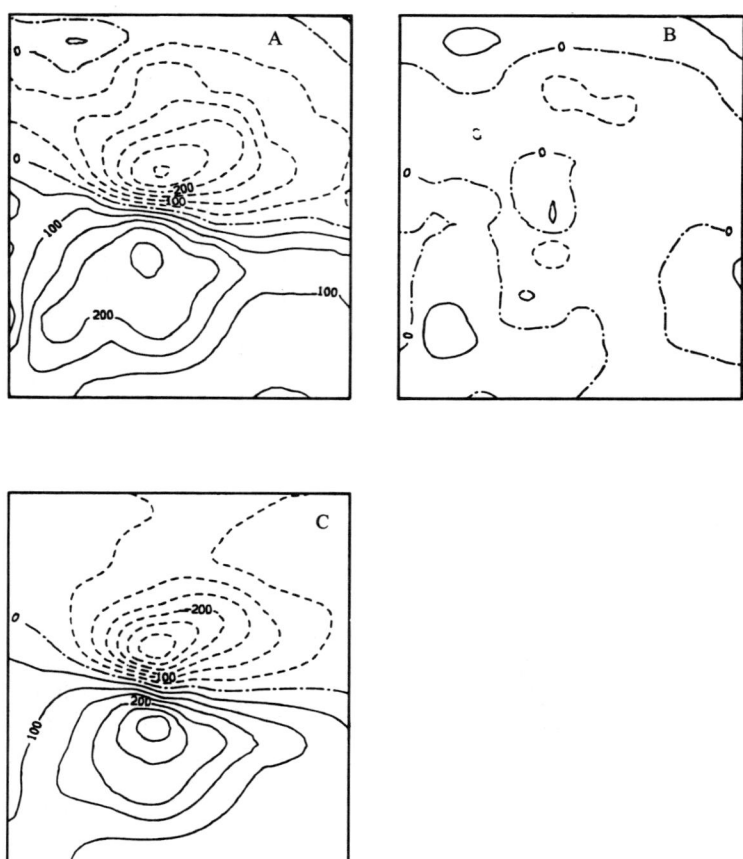

Figure 20 Magnetic fields of the simulated Mendeleev Seamount: (A) primary field; (B) differential field; (C) simulated field.

- The central area of the mount adjacent to its summit has the principal effect upon the magnetic anomaly, whereas the summit itself is less magnetized. That agrees well with the findings of rift formations at the summit
- Paleomagnetic latitude of the mount is estimated as 36.5 ± 1.5° S. Its paleomagnetic pole has coordinates of 48° N and 33° W. The confidence ellipse around the pole is overlapped by the confidence area of the Pacific plate paleomagnetic poles at the time of 120 and 100 million years (Kononov, 1984; Larson and Chase, 1972) (Figure 16)

The obtained estimates of the paleomagnetic pole position with the account of the fact that the first epoch of the reversed geomagnetic polarity in the middle Cretaceous corresponds to anomaly M0, allows a supposition that the Mendeleev Seamount was formed not earlier than 118 to 115 million years ago (the Barremian-Aptian). That conclusion agrees well with the data of deep drilling (boreholes 288 and 289) within the Ontong-Java plateau, which allows us to relate the evolution of the seamount volcanic activity with the general volcanic early Cretaceous history of the region. From the geodynamic reconstructions (Zonenshain and Savostin, 1979; Hilde et al., 1977) and the trajectory of the Cretaceous displacements of the Pacific Seamounts (Zonenshain et al., 1987b), and with due account of the relative uniformity of magnetization and morphological parameters of the mounts, it is possible to conclude that the stage of its active volcanism took up the flank of the mid-oceanic ridge and continued for a relatively short time.

C. THE MAGELLAN SEAMOUNTS

The Magellan Seamounts Range crosses the Marianas Basin of the Pacific from the northwest to the southeast, from the Marianas Trench to the Marshall Islands. It is a part of the world's largest assemblage of intraplate paleovolcanoes at the oceanic lithosphere. North of the Magellan Seamounts and subparallel to them, there runs the Marcus Wake range described earlier. The latter separates the Marianas Basin from the Northwestern Basin. Further eastward, it continues as a large volcanic range of the Mid-Pacific Seamounts. Such a unique assemblage of volcanic seamounts (over 1000 paleovolcanoes) within a relatively small area of oceanic floor made it most promising for study of the nature of an intraplate basalt volcanism in the ocean and its relations to a tectonic evolution of the oceanic lithosphere, and also for exploration of ore minerals associated with submarine volcanoes.

The Magellan Seamounts are spaced in the area of the most ancient oceanic crust of the middle and late Jurassic, east of the Marianas Trench (Joides J. 1990). The Magellan Seamount Range is a boundary between the East Marianas Basin in the south and the Pigafetta Basin in the north (Figure 21). Until recent times, that area was regarded as the Jurassic area of magnetic quiet field (Bogdanov et al., 1990). However, the recent survey performed by Handschumacher and others (1989) allowed them to recognize and delineate there two Mesozoic sequences of magnetic lineations related to the Japanese system, namely from M-21 (154 million years) to M-38 (171 million years).

As follows from Figure 21, the Mesozoic magnetic lineation sequence in the East Marianas Basin is shifted southeastward for approximately 450 to 500 km relative to the anomalies of the Pigafetta Basin. The Magellan Seamount Range is located at the axis of the dextral transform fault along which the crustal blocks are moved in the mentioned basins and which seemed to control the formation of basalt paleovolcanoes.

The results of geological-geomorphological study of the Magellan Seamounts showed that all of them are of volcanic origin and are guyots. It should be noted that until recently, the Magellan Seamounts were not adequately explored. Detailed bathymetric maps on individual guyots do not exist in principle. North of the Magellan Seamounts, between 19 and 20° N, there exists a sublatitudinal volcanic seamount — the Datton Seamount — which comprises five guyots studied thoroughly by a bathymetric survey with the use of a seabeam.

In 1984, a longitudinal transect was performed across the Magellan Seamounts on the ninth cruise of the R/V *Akademik Mstislav Keldysh* (Bogdanov et al., 1990). Altogether, eight guyots were crossed (Figure 21). At the place of the transect, the basement of the guyots of the Magellan group varies from 15 to 54 miles within a 5000-m isobath and is 41 miles, on average. The mean diameter of the flat-topped summit is 22 miles. The basement of the guyots rests at depths of 5600 to 6100 m, and their flat-topped summit starts from depths of 1830 to 1500 m. Since the minimum depth over the Magellan guyots is about 1400 m, on average, their visible altitude over the sea bottom of the Marianas Basin reaches 4700 m. The sedimentary thickness in that region is about 1200 m, according to the deep-drilling data and data of seismic profiling. Therefore, the real altitude of the volcanic constructions of the Magellan Seamounts over the sea bottom is apparently more than 6 km, with due account of the

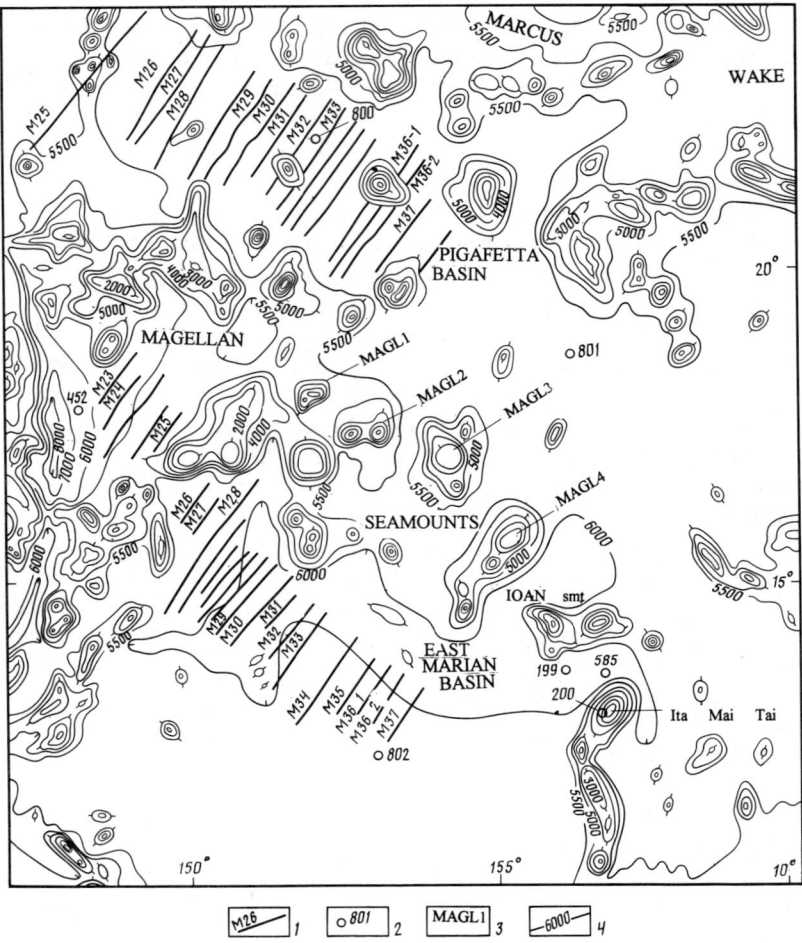

Figure 21 Review scheme: (1) axis of reversed magnetic anomaly, its number; (2) deep drilling borehole, its number; (3) seamount under study; (4) isobaths in m.

erosion of their summit. Some guyots have a pedestal, which starts from their basement and is traced to depths of 5500 to 5000 m, where the tilt of their surface increases twofold.

The analysis of the available bathymetric data (Figure 21) shows that the Magellan Seamounts comprise both isolated paleovolcanoes and complex systems of several intergrown constructions, bounded conventionally by a 5000-m isobath. According to the character of their strike, the Magellan Seamounts may be subdivided into two groups. The first is of the northwestern strike, extending from 10 to 17° N; it is composed of nine isolated mounts. One of them, the Ita Mai Tai guyot, was studied in detail on the ninth cruise of the R/V *Akademik Mstislav Keldysh*. This group also includes three multi-summit complex volcanic systems. The second group, which comprises the intergrown complex volcanic constructions, is traced from 154° N to the Marianas Trench and has a sublatitudinal strike.

On the 20th cruise of the *Glomar Challenger* drilling vessel, three boreholes were drilled at the flat summit of the Ita Mai Tai guyot, namely, 200, 201, and 202 (Heezen et al., 1973) (Figure 21). According to the obtained data, the acoustic basement exposed under the early Eocene globigerina sandstone, is an Eolithic limestone formed during the former island sinking. Below, detrite limestone lies with the remnants of corals and basalt fragments. The latter indicates that the erupted rocks came onto the surface and the Ita Mai Tai guyot was not a coral atoll. During the same cruise, one of the boreholes (199) was drilled in the Marianas Basin bed between the Ita Mai Tai guyot and the IOAN guyot (Figure 21). The borehole did not reach the bed of the basin. At the borehole bottom, the tuff, related presumably to the Magellan Seamount volcanism, was exposed under the foraminifera limestones of the Campanian age. The successive seismic study showed that the crystalline basement lies 500 m below the borehole bottom, the depth of which was 456 m. That allowed the mean sedimentary thickness in the region to be estimated as 1000 m. On the 62rd cruise of the *Glomar Challenger*, 50 miles east of borehole 199, borehole 585 was drilled, and it did not reach the basement either. It did expose the tuff and hyaloclastite of the Campanian-Maestrichtian age (83 to 65 million years), which have a turbidite structure. Titanoaugite was found in the tuff composition, which revealed their relation to the intraplate volcanism of the Hawaiian type probably associated with the Marshall Island group. In the lower borehole part, a thick Aptian-Albian (118 to 98 million years) turbidite layer, composed of the pebble of alkaline basalt and volcanic sand, was exposed at depths of 590 to 983 m. The occurrence of a thick turbidite layer at the borehole bottom allowed a supposition that the considered time was characterized by active volcanism at the Magellan Seamounts. The Albian-Aptian age of the Magellan range guyots was also discussed when detrite limestones were found at the flat summit of the Ita Mai Tai guyot on the ninth cruise of the R/V *Akademik Mstislav Keldysh*. Seismic data in borehole 585 allowed the sedimentary thickness to be estimated as 1200 m. That correlates well with the curve of the Marianas Basin bottom subsidence (Bogdanov et al., 1990). The material obtained in borehole 585 showed that the main stage of the Magellan Seamounts erosion seemed to occur in the Aptian-Albian time.

Deep drilling performed on the 17th and 61st cruises of the *Glomar Challenger* south of the Magellan Seamounts in the Nauru Basin and farther eastward in the Central Pacific Basin, boreholes 167, 169, and 170, has revealed the occurrence there of younger (by approximately 30 to 40 years) basalt cover up to 1 km thick superimposed over the oceanic crust of spreading origin. That type of intraplate basalt volcanism probably associated with the fracture, not central volcanism, seemed to be abundant in the late Mesozoic Pacific.

In 1989, three more boreholes (300, 801, and 802) were drilled from the *Glomar Explorer* drilling vessel near the Magellan Seamounts in accordance with the program of deep oceanic drilling. Borehole 800, drilled within magnetic anomaly M-33, exposed volcanoclastite 125 to 144 million years old. Borehole 801 went through the radiolarian limestone 163 million years old, which covers pillow basalt and basalt flows over 170 million years old, which corresponds approximately to anomaly M-37. Borehole 802 exposed limestone 91 to 116 million years old and penetrated into basalt of unknown age.

In 1988 to 1989, detailed geomagnetic and bathymetric surveys were performed near the Magellan Seamounts from the vessels of the Ministry of Geology of the former U.S.S.R. As a result, a bathymetric map (Figure 22) and a map of the AMF of the analyzed region (Figure 23) were compiled. Figure 22 shows that the survey covers the northern segment of the Magellan Seamounts north of 15° N. The RMS of the geomagnetic survey is 15 nT, which allowed a compilation of the map of the AMF isodynams in each 50 nT (Figure 23).

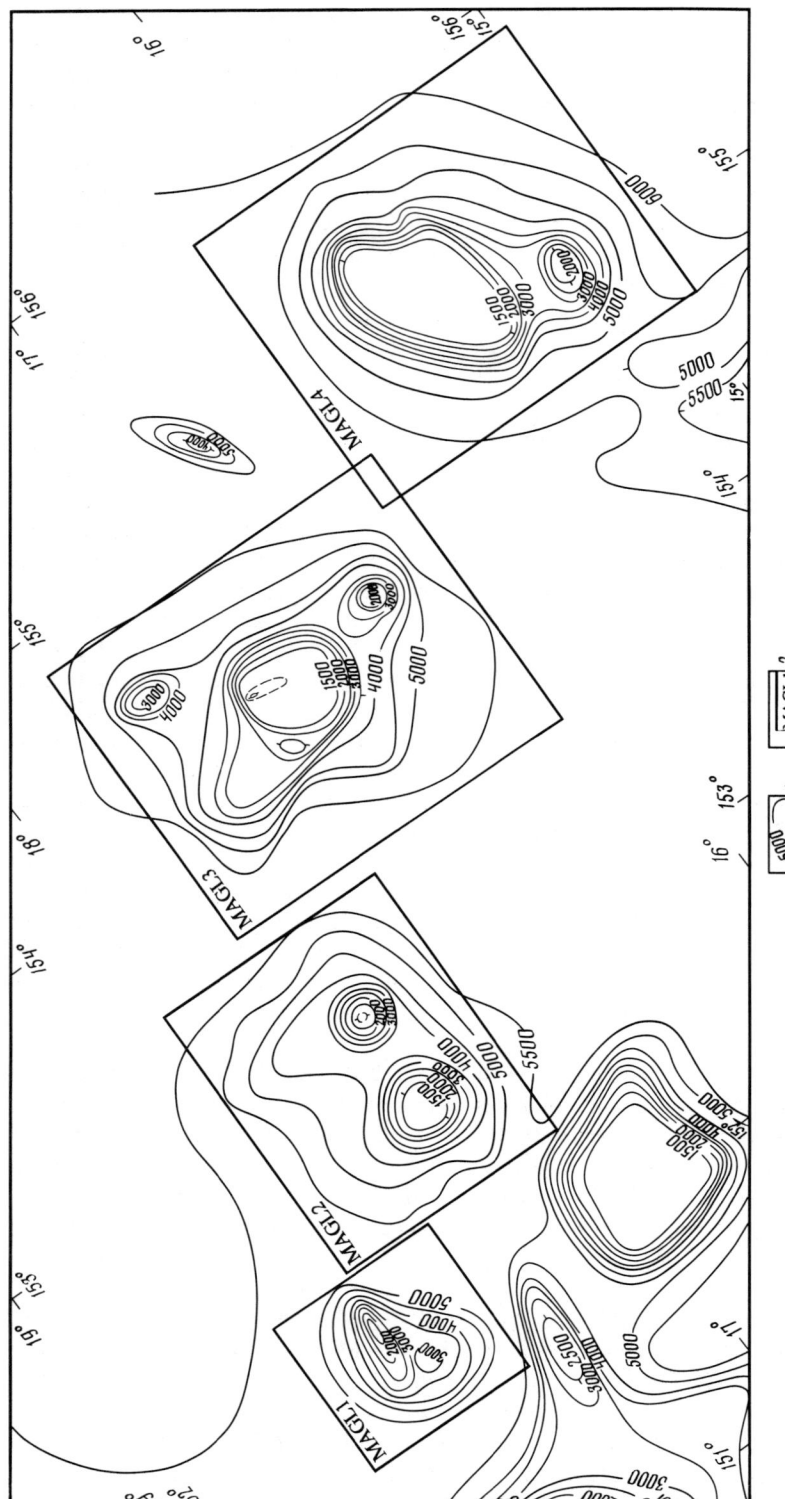

Figure 22 Bathymetric scheme: (1) isobaths in m; (2) simulated area, conventional name of a seamount.

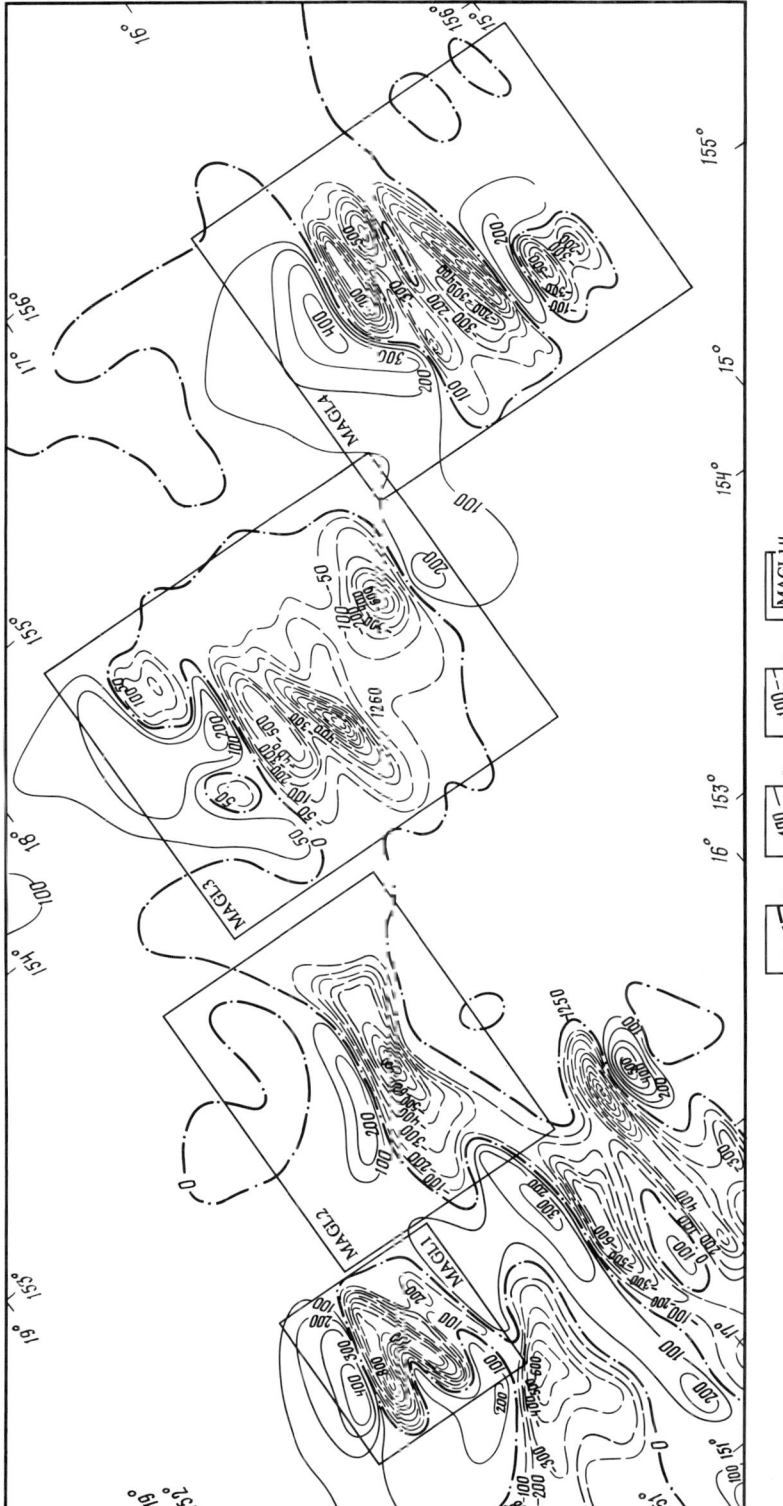

Figure 23 Scheme of AMF ΔT_a isolines: (1) zero in nT; (2) positive; (3) negative; (4) simulated area, conventional name of a seamount.

Analyzing the AMF map, we can see that intensive magnetic anomalies with reversed polarity and amplitude to 1000 nT and more are clearly seen over the Magellan Seamount system. Most intensive negative anomalies with the amplitude of 1500 nT are registered over the guyots MAGL-3, MAGL-4, and M-7. Meantime, the AMF of the northwestern Magellan Range has a complex structure probably and first of all associated with the complex morphology of the intergrown volcanic constructions in that part of the system, which are placed at the same volcanic basement. The latter, despite the detailed character of the survey, greatly hinders the choice of equivalent models for magnetic simulation of seamounts. Using the geomagnetic and bathymetric survey data, magnetic simulation was performed for four mounts, namely, MAGL-1, MAGL-2, MAGL-3, and MAGL-4 (Figure 22). The technique worked out by Ivanenko (1988) was used for that simulation, the results of which are given in Figures 24 and 25.

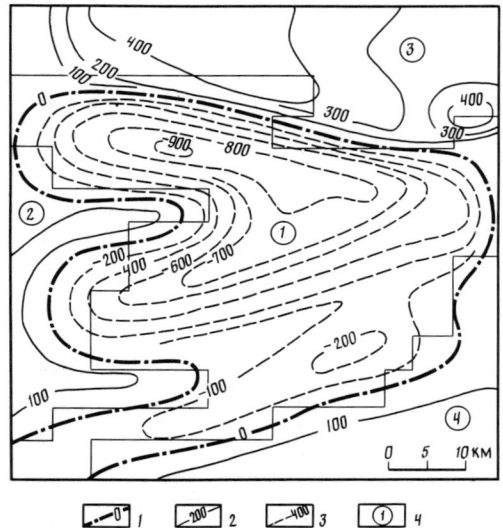

Figure 24 Magnetic simulation of MAGL-1 seamount: (1) zero, isodynams are given in nT; (2) positive; (3) negative; (4) blocks with different magnetization recognized during magnetic simulation.

It follows from the simulation results that an equivalent model for each mount analyzed is an assemblage of areas with different effective magnetization. So, for the MAGL-1 Seamount, the optimal correlation of the observed and estimated fields is provided by a model comprising four areas of different effective magnetization (Figure 24). The MAGL-2 Seamount model (Figure 25) is more complex and is described by seven areas. Magnetic simulation provides evidence that all the seamounts analyzed have reversed magnetic polarity. Table 5 shows the results of the estimates of effective magnetization and paleolatitude of a seamount formation, and also estimated coordinates of the virtual paleomagnetic poles for each seamount.

Table 5 **Results of the estimates of effective magnetization and paleolatitude of a seamount formation** [a]

Seamount name	$\varphi°$	$\lambda°$	$\Phi°$	$\Lambda°$	$\varphi°_c$	I_c, A/m
MAGL-1	17 40	151 42	46	311	−18	−6
MAGL-2	17 05	152 26	51	311	−18	−2
MAGL-3	16 13	153 41	55	338	−17	−6
MAGL-4	15 42	155 13	55	300	−15	−5

[a] Here, $\varphi°$ and $\lambda°$ are modern coordinates of a seamount, $\Phi°$ and $\Lambda°$ are coordinates of the virtual pole, $\varphi°_c$ is a paleolatitude of a seamount formation, and I_c is a mean effective magnetization of a seamount.

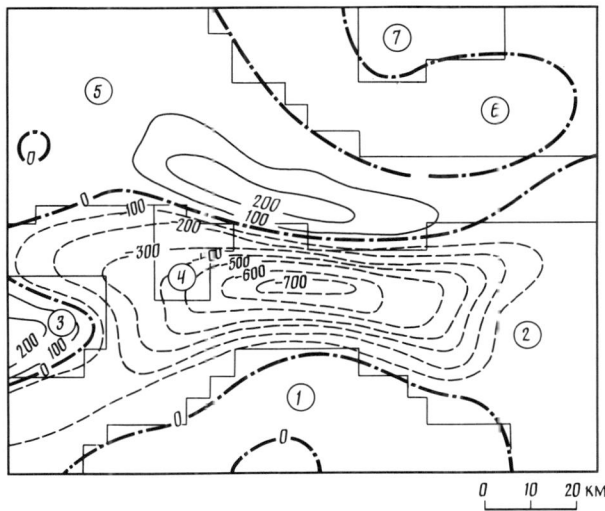

Figure 25 Magnetic simulation of MAGL-2 seamount. For notation see Figure 24.

As follows from Table 5, all of the seamounts analyzed were formed at approximately 17 to 18° S, which agrees with the Pacific plate early Cretaceous reconstructions (Scotese et al., 1987). An average position of the paleomagnetic pole for the Magellan Seamounts (Figure 26) also agrees with the curve of the paleomagnetic pole migration of the Pacific plate for the age of 125 million years (Cox and Gordon, 1984). These data and the fact of reversed magnetic polarity of the Magellan Seamounts allow a supposition that their age corresponds to the interval of reversed magnetic polarity at the Barremian/Hauterivian boundary (123 to 127 million years).

That allows an estimation of the northern component of the absolute Pacific plate drift for the recent 125 million years as about 3.5 cm per year, on average.

So, the independent quantitative estimate from the data of magnetic simulation of the Magellan Seamounts verifies in general the order of the Pacific plate drift rate calculated earlier by Kononov (1989). Meantime, the obtained data introduce considerable corrections in the paleolatitude position of the Magellan Seamount system and the bearing lithosphere during the period of their formation. The average paleolatitude of their formation in the early Cretaceous seemed to be 20° S but not 30° S, as was supposed earlier (Bogdanov et al., 1990).

The study of isotope characteristics of the 12 Cretaceous seamounts of the Magellan, Marshall, and Wake groups by Standigel et al. (1991) allowed them to suppose that the Magellan Seamounts are genetically related to the long-time isotope and thermal anomaly in the mantle under the Pacific plate, the trace of which has been controlling the northwestward drift of the Pacific plate since the early Cretaceous. The performed research showed that the order of isotope concentration in the Magellan Seamounts is the same as in the Tahiti, Samoa, Cook, and Marquisian volcanic island systems.

Table 5 shows that the Recent latitudinal distance between the terminal seamounts of the Magellan system analyzed, namely MAGL-1 and MAGL-4, is 2°. The paleolatitude distance with the account of the error in the paleolatitude estimates is the same (about 3°). However, a small latitudinal distance at which the seamounts are spaced apart and low accuracy of the estimate do not allow us to unambiguously accept a "hot-spot" model of their formation assumed there from the geothermal study (Standiqel et al., 1991). To prove the validity of that model by geomagnetic materials, more detailed geomagnetic survey and magnetic simulation of the southeastern Magellan Range, namely, IOAN and Ita Mai Tai guyots, is necessary. At the same time, from the available geomagnetic data it is possible to assume that the basalt volcanism of the central type, which caused the formation of the Magellan Seamount volcanic range, was controlled by the transform fault between the Pifagetta and East Marianas Basins and was associated with its early Cretaceous tectonic activation. The results of magnetic simulation of the northwestern part of the Magellan Seamount range thus showed that the age of their formation seems to be the Hauterivian-Barremian but not the Aptian-Albian, as was assumed earlier, and is about 125 million years, on average.

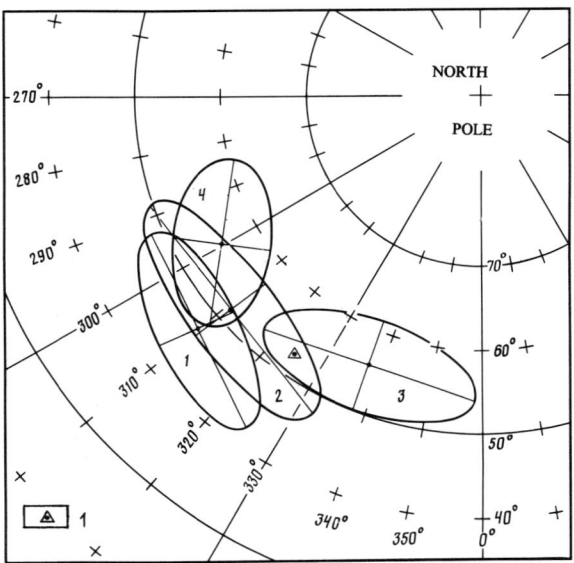

Figure 26 Paleopoles position calculated for each seamount, with the ellipse standing for a confidential area: (1) position of virtual early Cretaceous magnetic pole (Cox and Gordon, 1984).

That correlates well with the results of the geological study performed on the ninth cruise of the R/V *Akademik Mstislav Keldysh* (Bogdanov et al., 1990). According to the materials of that cruise, four of the nine guyots of the Magellan Range studied on that cruise (IOAN guyot including) were formed about 120 million years ago as volcanic islands due to active early Barremian volcanism.

Sorokhtin proposed the formula to estimate the age of the volcanic island transformation into a guyot (Bogdanov et al., 1990). The age of the Magellan guyots estimated by that formula showed that the southeastern Ita Mai Tai atoll became extinct about 105 million years ago, that is, at the very end of the Aptian age and beginning of the Albian age, whereas the erosion of the volcanic basement of the former island happened by the beginning of the Aptian time, that is, about 115 million years ago. As for the IOAN guyot, arising from the depth over the top of consolidated sediments found at its summit (about 1.65 km), it sank down about 108 million years ago, and the erosion of its volcanic basement with a depth of 1.85 km occurred earlier, that is, about 116 million years ago. We calculated approximately the same age for the MAGL-4-MAGL-1 guyots. An extinction of reef biocenocies at the top of islands atolls or banks of that group and the transformation of the latter to guyots occurred 108 to 106 million years ago and coincided with the Aptian transgression. It is possible to assume that the Ita Mai Tai guyot and others of the southeastern part of the Magellan Seamounts was formed a bit later: at the end of the Barremian-Early Aptian (109 to 108 million years ago) their summits were already eroded, and biotherm deposits started to accumulate there. The reefs at the islands of that group seemed to have been extinct already in the Albian time. Meantime, as was noted earlier, to specify the formation age of these paleovolcanoes, we need data of detailed geomagnetic survey.

The recent reconstructions of the Pacific plate position (Scotese et al., 1987) and the data we obtained provide evidence that the area of the Cretaceous intraplate volcanism was located in the central part of the Pacific south of the equator, near the critical latitude 30 to 20° S, approximately in the present region of active volcanism (the Cook, Tahiti, Society, Tuamotu, Markisian, Easter Islands, etc.). Such a coincidence is not accidental. On the one hand, we may suppose that south of the equator an upwelling mantle convective flow of the Earth occurred from the early Cretaceous till the Recent time, to which the abovementioned thermal and isotope anomaly SOPJTA is associated. On the other hand, it may verify the model of membrane tectonics proposed by Turcott (1974). According to that model, the curvature radius of the lithosphere plates varies as they move along the ellipsoid of the Earth's rotation, which caused their splitting. The plate motion from the high latitudes toward the low ones is accompanied by their compression and splitting along the system of orthogonal lateral faults. The faults generated due to lithosphere compression then provide for a central type of volcanism in the places of the fault cross sections that result in the formation of the intraplate volcanoes. The occurrence within plates of transform faults with relatively big shear constituents, like the fault separating the Pigafetta and East Marianas Basins, along which lithosphere blocks of different ages and thicknesses contact, results in the additional stresses in these zones. As a consequence of that, the areas of intraplate volcano formation on the oceanic lithosphere moving from the high latitudes toward the low ones should be attracted to zones of transform faults with great shear displacement. Such regularity is typical of the young volcanic constructions and islands of the Central and Southern Pacific, in particular, the Luisville Seamounts, located at the extension of the Eltanin fault.

In the light of this, it is interesting to note that the Cretaceous guyots at the Pacific floor extend as two sublatitudinal parallel groups, namely, the Marcus-Necker (or the Mid-Pacific) seamounts, in the north, and the Magellan Seamounts, in the East Marianas Basin. Farther eastward, the northern group changes for the Line Islands, whereas the Magellan Seamounts change for the Marshall Islands and Gilbert Islands. The change of the sublatitudinal strike of the guyot groups for the submeridional strike of the ranges of the atoll islands is associated with the great reorganization of the spreading in the Pacific about 110 to 120 million years ago.

As was noted earlier, the age of the bearing lithosphere in the East Marianas Basin near the Magellan Seamount range locus is dated as 165 million years old from the Mesozoic magnetic lineations. If the Magellan Seamounts were formed 125 million years ago, the age of the oceanic lithosphere at the moment of their formation was about 40 million years old. According to the theoretical estimates (Gorodnitsky, 1985), thickness of the lithosphere reached 60 to 65 km. The altitude estimates of the volcanic islands set at the lithosphere of the mentioned thickness (Gorodnitsky, 1985) showed that it should be 6 km and over. These estimates correlate well with the real altitude of the Magellan Islands (calculated from the geomorphological data) during the period of their formation.

The effective magnetization of basalt, comprising the Magellan Seamounts (2 to 6 A/m), was obtained when estimating their age. It corresponds well to the experimental data on the mean remnant thermomagnetization of basalt of intraplate volcanoes. From petrographic study made by Rudnik, all basalts sampled from the slopes of the Ita Mai Tai, IOAN guyots, and the eastern satellite of the IOAN guyot, belong to a single genetic series, that is, from the subalkaline olivine to alkaline, and are close in composition to the basalt of the Hawaiian Islands.

So, the magnetic simulation of seamounts in various parts of the World Ocean allows a solution of a wide circle of geological problems: it is possible to determine the age of their formation, to estimate the kinematics of the bearing lithosphere, and also to find out the sequence and nature of the intraplate basalt volcanism.

Chapter 6

Genesis of Primary Magmatic Titanomagnetite of Tholeiite Basalt from Mid-Oceanic Ridges

A. G. Gorshkov

Anomalous magnetic field of the oceanic crust is generated by magnetized rocks of the crust, mainly by basalt. Magnetization of the latter is related to ferromagnetics comprising them. In their turn, concentration and composition of ferromagnetic phase of tholeiite basalt, namely, titanomagnetite, are determined by the primary magma composition, by conditions of basalt crystallization, and by processes of post-magmatic oxidation.

The target of this work is to examine the correlation between the chemical composition of basalt and the concentration and composition of the primary magmatic titanomagnetite in them.

Titanomagnetite is a solid solution of magnetite Fe_3O_4 (Mt) and ulvospinel Fe_2TiO_4 (Ulv). We estimated the content of these minerals by the modified technique of a standard conversion. The magnetite/ulvospinel ratio determines the composition of titanomagnetite, whereas their summary content defines the titanomagnetite concentration in the rock. It was admitted in the standard conversion that the entire Ti content in the melt is used for ulvospinel formation, whereas the entire Fe^{3+} content is taken up by magnetite formation.

Let us express the molecular weight of a component of an X system through M_x; the percent concentration of component X by (X), for instance (TiO_2) is a concentration of TiO_2 in the system in weight percent; and the molecular concentration of component X by $[X]$.

Thus, it is easy to determine the molecular content of components in system $[X]$:

$$TiO_2 = \frac{(TiO_2)}{M_{TiO_2}} \qquad Fe_2O_3 = \frac{(Fe_2O_3)}{M_{Fe_2O_3}}$$

but $[Ulv] = [TiO_2]$; $[Mt] = [Fe_2O_3]$, that is, the content of ulvospinel molecules is the same as that of TiO_2 molecules, and the content of magnetite molecules in the rock is the same as the content of Fe_2O_3 molecules.

So, the weight content of Mt and Ulv in the rock is

$$(Ulv) = [Ulv]M_{Ulv} = M_{Ulv}/M_{TiO_2} (TiO_2) \qquad (1)$$

$$(Mt) = [Mt]M_{Mt} = M_{Mt}/M_{Fe_2O_3} (Fe_2O_3) \qquad (2)$$

Since the primary data on TiO_2 and Fe_2O_3 concentration, obtained from chemical analysis, are brought to 100%, the estimated weight content of magnetite and ulvospinel given is percentage.

To estimate Fe_2O_3 concentration, we must know what part of the summary iron is presented in a three-valent form. A concentration of a three-valent iron in the system is determined from the oxidation-reduction potential of the latter. Let $Fe^{3+}/\Sigma Fe = K$. Then, Equation 2 has the form

$$(Mt) = M_{Mt}/M_{Fe_2O_3} K(Fe) \qquad (3)$$

From the recent data, tholeiite basalt of the mid-oceanic ridges is generated under conditions of small pressure ($P \leq 1.5$ kbar) (RISE, 1981), the potential being close to the FMO-buffer (quartz-fayalite-magnetite) (Byers et al., 1986; Carmichael and Chiorso, 1986; Christie et al., 1986). In this case, 15% of FeO is presented in a three-valent form. So, parameter K in Equation 3 is 0.15. Substituting content values of molecular weights and parameter K in Equations 1 and 3, we have

$$(Mt) = 0.22 (FeO) \qquad (4)$$

$$(Ulv) = 2.80(TiO_2) \qquad (5)$$

$$(Ti\,Mt) = 0.22(FeO) + 2.80(TiO)_2 \qquad (6)$$

Equations 4 to 6 allow us to estimate the composition and concentration of titanomagnetite, generating from crystallization of tholeiite basalt, from the total chemical composition of the rock.

Equation 4 is a function of the oxidation-reduction regime in the system, which is considerably stable. Besides, even if it is changed by two orders, the K parameter does not response significantly to such a change. So, when the potential of the system is 2 lg units higher than the FMQ buffer, K is 0.10 in Equation 3, and Equation 4 is

$$(Mt) = 0.15(FeO) \qquad (7)$$

which decreases the summary content of titanomagnetite in the system by only 10 to 12%. Besides, such strong deviation of the system potential is extremely rare.

Titanomagnetite, being a solid solution of magnetite Fe_3O_4 (Mt) and ulvospinel Fe_2TiO_4 (Ulv), seems to be formed at the mature stage of basalt melt crystallization at a temperature of 1050 to 1040°C. Composition and concentration of the primary magmatic titanomagnetite depends greatly on many things, first, on the concentration of titane and three-valent iron in the primary basalt magma, on the concentration of the same components in the differentiated basalt melt erupted onto the surface, on oxidation-reduction potential of the system, and on crystallization dynamics of the basalt melt. We shall consider the effect of them all.

Tholeiite basalt, comprising the oceanic crust, is generated in magma chambers of rift zones of the mid-oceanic ridges. Chemical composition of oceanic tholeiite basalt is known sufficiently well by now. In particular, we know that its variations over the World Ocean cannot be explained exclusively by the fractionating of the uniform magma, but also by compositional differences of the primary magma ascending to the upper magma chamber (Dmitriev et al., 1979; Lukanin et al., 1983).

The mid-oceanic ridge system of the World Ocean may be subdivided into isolated segments with a stable composition of the primary magma ascending to the chambers and similar processes of crystallizational differentiation (Gorshkov and Lukashevich, 1989). As a rule, such homogeneous segments are located between large transform faults or uplifts.

We have illustrated that these homogeneous segments are fragments of the mid-oceanic ridges enclosed between large transform faults or uplifts; therefore we may consider them as segments with homogeneous tectonics (Ariskin et al., 1985). Thus, we have homogeneous elements, which can be compared among themselves. We have analyzed the data on chemical composition of 1200 samples of the tholeiite basalt glass collected over the World Ocean. Alongside the data on axial ridge zones, we analyzed the data on hardening glass of the rift tholeiite basalt sampled on ridge slopes and in abyssal basins. Sixty four segments have been distinguished altogether.

It was shown in the paper by Lukashevich and Gorshkov (1989) that chemical composition of glass, related to the same tectonic homogeneous segment, varies within a certain range; that is taken as one of the criteria for the basalt classification. Since the differentiation index FeO*/MgO of basalt glass from various segments varies from 0.86 to 4.80, we would consider the interval from 1.0 to 1.5, which can be found within each segment. This is necessary if we want to explain the difference in glass composition by the different composition of the primary magma, thus leveling the effect of the fractionating in the upper magma chamber.

Three basalt types have been distinguished (Lukashevich and Gorshkov, 1989). Type I is characterized by low concentrations of titane and alkali and by high concentration of CaO. Type III is characterized by high concentration of alkali and low concentration of CaO. Chemical composition of basalt of type II is intermediate between types I and III. Classification of tholeiite basalt of the mid-oceanic ridges by chemical composition of hardening glass is given below:

Type I	Type II	Type III
$TiO_2 \leq 1.2\%$	$1.2\% < TiO_2 \leq 1.6\%$	$TiO_2 > 1.6\%$
$CaO > 12.0\%$	$CaO < 12.0\%$	$CaO < 12.0\%$
$Na_2O \leq 2.4\%$	$2.4\% < Na_2O \leq 2.7\%$	$Na_2O > 2.7\%$
$K_2O \leq 0.1\%$	$0.1\% < K_2O \leq 0.2\%$	$K_2O > 0.2\%$

Note: $1.0 \leq FeO^*/MgO \leq 1.5$.

Figure 1 illustrates the distribution of basalt types over the tectonic homogeneous segments located at the mid-oceanic ridge crests, slopes, abyssal basins, and transform faults. The analysis of their spatial distribution allowed us to correlate the character of tholeiite magmatism with the formation history of the mid-oceanic ridges.

Figure 2 shows how the ratio of TiO_2 concentration in tholeiite basalt glass to the differentiation index depends upon the age of basalt over the Atlantic, Indian, and Pacific Oceans, beyond the zones of geochemical anomaly effect, in places where the age was determined.

The age of basalt formation is assumed to be an interval between the moment the rifting commenced in the region and the moment the basalt erupted, that is, the age of the mid-oceanic ridge at the moment of basalt formation. Despite considering TiO_2, as was done when classifying basalts, we used the ratio of TiO_2 to $I = FeO^*/MgO$, which allowed us to utilize the entire set of data irrespective of the differentiation index and also to demonstrate that only one type of basalt can control the segment with homogeneous tectonics.

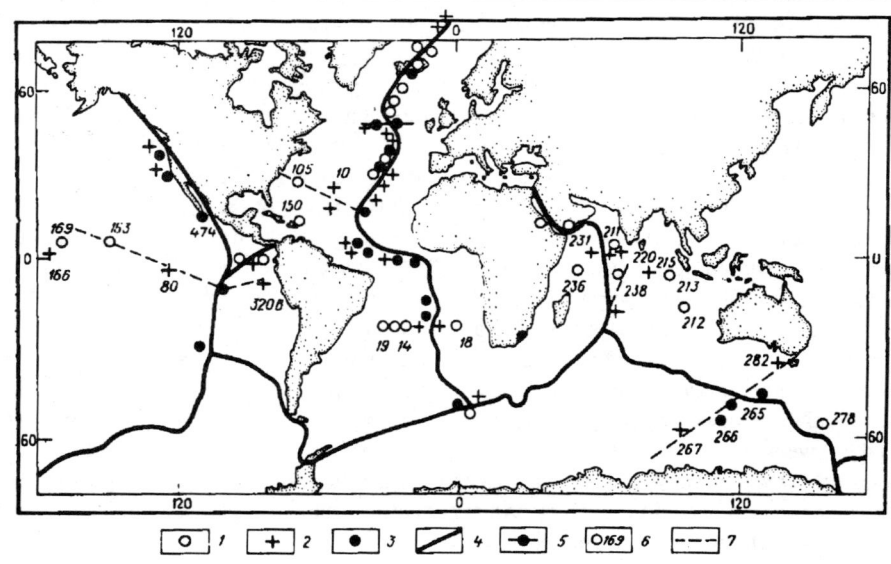

Figure 1 Spatial distribution of basalt types over the World Ocean: (1-3) basalt types I, II, and III, respectively; (4) world pattern of mid-oceanic ridges; (5) basalt glass dredged from transform faults; (6) numbers of boreholes; (7) conventional lines of spreading trend with points of basalt sampling.

When analyzing the data given in Figures 1 and 2, we can see that basalts that have a high concentration of titane and are relatively rich in alkali and depleted of calcium (type III) are practically always confined to rift systems, the age of which was over 80 million years old by the moment of basalt formation. Among these structures we can list, in particular, the central and southern parts of the Mid-Atlantic Ridge, the East Pacific Rise (EPR), and the Australian-Antarctic Rise in the places where it is complicated by superimposed tectonic structures. By contrast, basalt with a low concentration of titane, enriched by CaO (type I), occurs in young rift systems in the abyssal basin areas formed shortly after commencement of the rifting (during the first 30 million years of the rifting), and also at the ridges that had experienced large tectonic reorganization (for instance, Reykjanes Ridge due to Iceland formation). And finally, magmatism of the Arabian-Indian Ridge, the age of which does not exceed 50 million years, is characterized by basalt of type II.

Basalt of type I may thus be associated with the initial stage of rifting, basalt of type III marks the late stage of the ridge evolution, and basalt of type II seems to characterize the transition stages.

Segments with basalt types I and III are never placed next to each other either on ridges or on the lines perpendicular to the spreading axes. They are always separated by the segments with basalt of type II. Therefore, the occurrence of the second intermediate type of basalt is proved not only by its chemical composition but also by its spatial position.

Besides, we shall note that segments with the same basalt type are located symmetrically, both on opposite sides of the ridge (EPR, 13° S, borehole 80 and 320 B; Southern Atlantic Ridge, boreholes 14 and 18; Australian-Antarctic Rise, boreholes

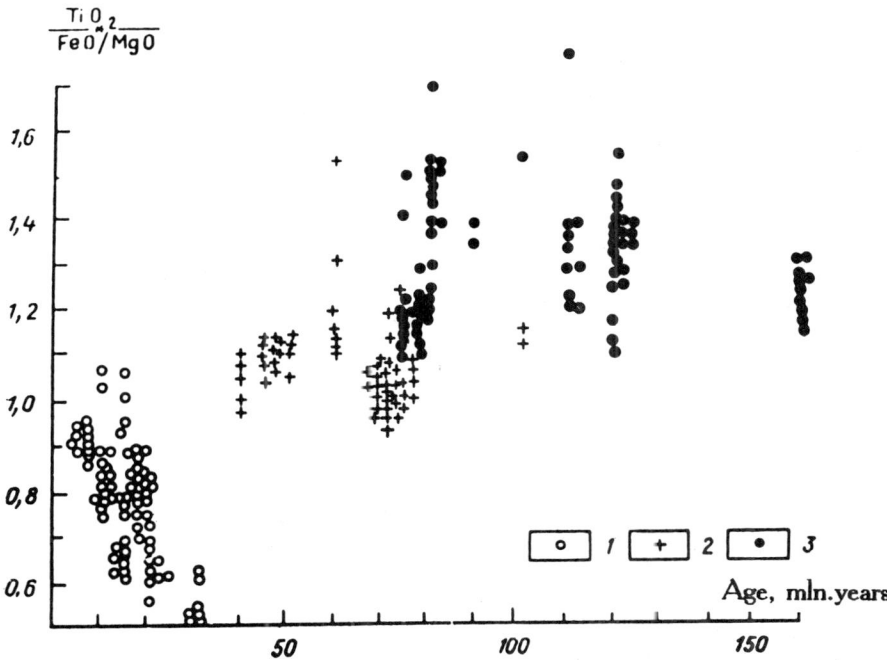

Figure 2 Correlation between chemical composition of basalts and ridge age at the moment of basalt eruption. For notations see Figure 1.

267 and 282) and along it, north and south of Iceland. So, at the Iceland Ridge and Reykjanes Ridge, glass of type I has been sampled, which turnout to have identical principal components. At the Mona Ridge and the Mid-Atlantic Ridge segment of 48 to 43° N, glass of type II, with similar components, was identified (Figure 1).

In a tectonic complex regions like Iceland, Azores-Gibraltar Rise, Galapagos spreading center, Bouvet triple junction, basalts of type I always occur in a combination with basalt of types II and III.

As for the Reykjanes Ridge, the presence of basalt low in titane cannot be explained by the initial rifting stage, because its age is 60 million years old. Seismic data provide evidence for an anomalous structure of the lithosphere under the Reykjanes Ridge, which may be associated with the spreading peculiarities in that region, and in particular with the Iceland formation.

The same is true for the Northern Atlantic Ridge segments of 48 to 43° N and 35 to 25° N. The occurrence of basalt of type II in these regions is explained by their close position to the Azores-Gibraltar Rise characterized by a complex tectonic history, by the anomalous structure of the crust, and by the occurrence there of three basalt types.

Similar situation is described in the Southern Atlantic. Basalt of type I is sampled from boreholes 14, 18, and 19, drilled on opposite sides of the ridge at a latitude of 28° S, whereas at the same latitude but on the ridge proper, basalt belongs to type II. The Walvis Ridge and the Rio Grande Ridge approach the Southern Atlantic Ridge in the region between 27 and 36° S. Remarkable geological events seemed to occur in this part of the ridge prior to 53 million years ago (53 million years is the age of basalt from borehole 19), which resulted in the formation of two large uplifts. At present, magma formation in the region is at the transitional stage and is characterized by basalt of type II (Figure 1).

As for transform faults, they are characterized by basalt of types II and III.

Figure 2 illustrates that regular changes in chemical composition of tholeiite basalt occurs as the mid-oceanic ridge evolves. Basalt of type I, low in Ti_2O and Na_2O and rich in CaO, erupted for the first 30 million years of the system life. Next, was basalt of type II, which erupted for 50 million years, that is, in the systems 30 to 80 million years old, and at last, it was the turn of basalt of type III, rich in titane and alkali and low in calcium, which erupted in the mature systems over 80 million years old.

In other words, the duration of the process of mid-oceanic ridge evolution is associated with the regular changes of chemical composition of the primary magma ascending to the upper chambers. During the first 80 to 90 million years after the mid-oceanic ridge had been set, magma was enriching in titane and alkali and depleting of calcium, but since the time of 90 million years, this process stopped and chemical composition of magma did not change; so, magma rich in titane and alkali continued to arrive in the rift system.

We ascribe these changes in chemical composition of the primary magma to a regular change of the mid-oceanic ridge recharge zones as they evolve. According to the paper by Ryabchikov (1984), the higher the lherzolite melting degree and the lower is pressure, the lower the concentration of TiO_2, Na_2O, K_2O and the higher the concentration of CaO is observed. The situation is the opposite when lherzolite melting degree decreases and pressure increases.

Estimates from the seismic and electromagnetic sounding data, however, provide evidence that in the upper asthenosphere layers down to a depth of about 30 km, the share of the liquid phase under the mid-oceanic ridges is 10 to 15%, and it reduces to 1 to 3% at a depth of 60 to 80 km. So, we can conclude that at the early stages, the recharge area of the mid-oceanic ridges was located in the upper asthenosphere layers, where the share of the liquid phase is relatively great. As the rift system evolved, the area of primary magma generation migrated down to the deeper asthenosphere layers, where the share of the liquid phase is less.

Since the rate of this migration is about 1 km per million years (for 80 million years it sank by 80 km), which is one to two orders of magnetude higher than the rate of convective flow, the input of the matter in the area of magma generation due to convection cannot have much effect on the composition of the liquid phase.

As a result, we can conclude the following. Chemical composition of tholeiite basalt erupting in the mid-oceanic ridges changes with geological time. Basalt of three types erupts in the following turn: for the first 30 million years, it is basalt of type I, rich in calcium and with low concentrations of titanium and alkali; then, for the successive 50 to 65 million years, it is basalt with intermediate concentrations of these components (type II); and only then is it basalt of type III, with low concentrations of CaO and rich in TiO_2, Na_2O, and K_2O. So, in the process of evolution, the prime magma is getting richer in TiO_2, Na_2O, and K_2O and depleted of CaO.

As the mid-oceanic ridge is getting more mature, its magma-generating system is evolving toward the formation of deeper primary melts. When a certain level of depth is reached, it stops.

Regions with complex tectonics have all three types of basalt. That means that titanomagnetite composition varies alongside the changes in chemical composition of the primary magma. So, at the early stages of rifting, basalt of type I, with FeO/MgO $\approx$ 1, FeO $\approx$ 8%, $TiO_2 \approx 1\%$ when being crystallized, forms, according to equations 4 to 6, 1.76% of Mt and 2.8% of Ulv, that is, basalt contains 4.56% of TiMt in 60 Ulv; for mature rifts, basalt of type III, with the same differentiation index FeO/MgO ≈ 1 and the same concentration of iron FeO 8%, would already contain 1.6 to 1.7% of TiO_2. That is, at the same concentration of Mt = 1.76%, the concentration of Ulv increases to 4.5%, and basalt would already contain 6.26% of TiMt of 70 Ulv. So, as the mid-oceanic ridge is evolving, titanomagnetite concentration in the nondifferentiated basalt increases by 1.5% due to changes in the primary magma. Ulvospinel concentration in titanomagnetite increases as well.

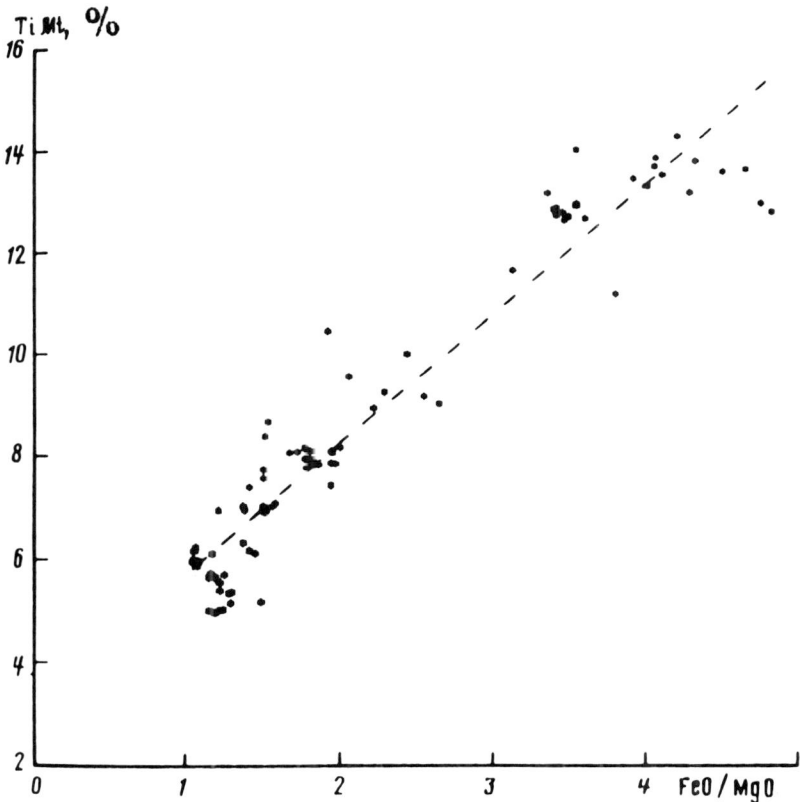

Figure 3 TiMt concentration in basalt as a function of FeO/MgO ratio in the rock: figurative points of composition and lines of regression.

Concentration and composition of titanomagnetite depend even more upon the processes of crystallizational differentiation. In the course of a shallow differentiation of tholeiite basalt, high-magnesite minerals, namely, olivine or eutectic mixture of high-magnesite olivine and plagioclase, are first to be crystallized out of the melt. So, the melt is enriched by iron and titane, which leads, in accordance with Equations 4 to 6 to increase of TiMt in basalt generating at the eruption of such high-differentiated melts. Considerably higher concentrations of TiMt in the rock may be obtained due to crystallizational differentiation contrast to changes of the primary magma.

Figure 3 shows the variations of TiMt in the basalt of the Galapagos spreading center. Changes in TiMt concentration are given as a function of the FeO/MgO ratio, which reflects the degree of the basalt differentiation. Concentration of TiMt increases proportionally to the increase in this ratio, and may reach 13 to 14% if the ratio is 4.5

to 5. From our estimates, partly presented in Figure 3, it follows that there is a close correlation between the FeO/MgO ratio in the rock and concentration of TiMt in it. It can be written as an equation of linear regression. Coefficients of the equation would vary slightly when we consider various segments.

It follows from these data that enrichment of the basalt melt by TiMt in the process of crystallizational differentiation is a more powerful process than enrichment of the melt by Fe and Ti due to evolution of rift zones.

Usually, 1 < FeO/MgO < 2, that is, 5% < TiMt < 9%, but in many cases the index of FeO/MgO differentiation reaches 5, which corresponds to 13 to 14% of TiMt.

This approach gains interesting results when applied to the zones of high strength of magnetic field, a so-called H zones (Vogt, 1979), in particular, the Galapagos spreading center, the Juan de Fuca Ridge, and some segments of other mid-oceanic ridges. According to petromagnetic data, ferromagnetics both in H zones and in zones of normal strength of magnetic field (N zones) are presented mainly by pseudo-single-domain crystals of titanomagnetite that have 50 to 65% concentration of ulvospinel in them. So, there are no sharp difference between these zones, according to the domain structure or composition of titanomagnetite.

From the data on chemical composition of basalt in the axial zones (Melson et al., 1977; Perfit et al., 1983; Sigurdsson, 1981), we have compiled histograms of the distribution of the FeO/MgO rock; differentiation degree determined by the linear ratio of TiMt concentration in the rock (Figure 4) over two H zones, namely, the Galapagos spreading center and the Juan de Fuca Ridge, and over several N zones of mid-oceanic ridges. Analyzing the histograms, we can see that basalt with 1 < FeO/MgO < 1.5 (over 80% of all contents) predominates in N zones, where composition with FeO/MgO > 2 is rare, whereas differentiated basalt with a high concentration of TiMt predominates in H zones. This picture is most clear in the Galapagos spreading center, where more than 50% of the rocks have an FeO/MgO ratio over 2 and contain 10 to 15% of TiMt.

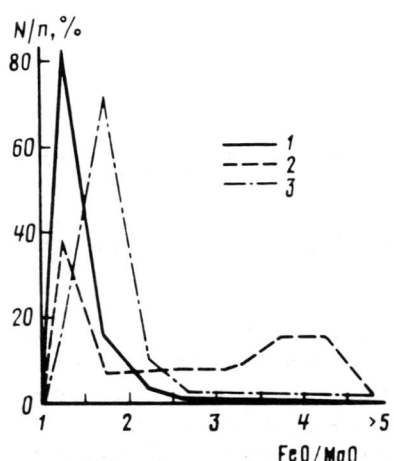

Figure 4 Distribution of FeO/MgO ratio in tholeiite basalt in various regions: (1) N zones; (2) Galapagos spreading center; (3) Juan de Fuca Ridge.

Thus, using the factual data on rock composition in H and N zones, we have concluded that high-differentiated basalts with high concentrations of titanomagnetite are more abundant in H zones, which can explain the high value of natural remnant magnetization of rocks and the high strength of the magnetic field in H zones.

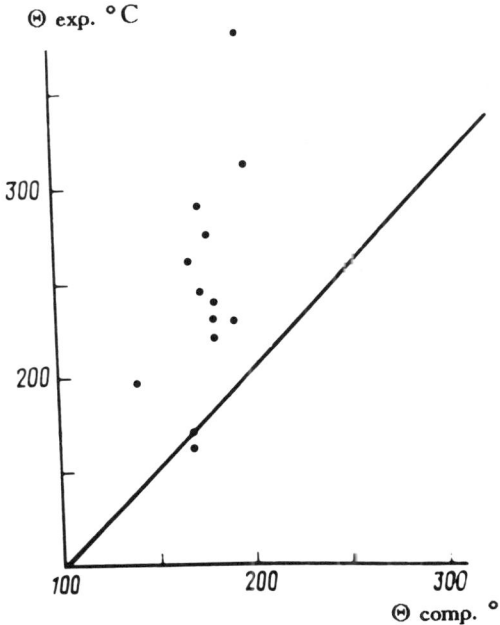

Figure 5 Comparison of the estimated and experimental data on Curie point θ value for the same samples: figurative points of composition and lines of θ and θ'.

Equations 4 to 6 can be used to estimate the composition of titanomagnetite. The crystallization of the primary magmatic titanomagnetite is followed by its oxidation, during which the total chemical composition of the rock remains unchanged and the titanomagnetite composition may experience great changes. The proposed technique allows us to estimate the primary composition of titanomagnetite generated in the process of the rock crystallization. A comparison of this primary composition with the real, estimated for instance from the Curie point θ, allows us to determine the degree of a one-phase oxidation of titanomagnetite.

Figure 5 presents the θ estimated and θ' experimental data used for the same samples (θ estimated was inferred from Equations 4 to 6, and experimental was inferred from curves of thermodemagnetization) (Kashinskaya et al., 1986). In the case of a complete coincidence of both, the figurative points of the rock composition would be spaced at the θ = θ' line. In fact, such coincidence can be achieved only for a few points, and as a rule, θ > θ'. A systematic shift of experimental data toward high values of θ seems to evidence a one-phase oxidation in them, whereas the discrepancy between the experimental and theoretical values of θ is the degree of ferromagnetics oxidation in the given sample. However, this problem remains unsolved.

So, we can make the following conclusions. We can calculate the composition and concentration of titanomagnetite in the tholeiite basalts from their total chemical composition.

In the process of a mid-oceanic ridge evolution, the evolution of the chemical composition of the primary basalt magma also occurs toward its enrichment by titanomagnetite and, in its turn, enrichment of titanomagnetite by an ulvospinel component.

Considerably greater enrichment of the melt by titanomagnetite occurs in the process of crystallizational differentiation of the tholeiite basalt.

High strength of the magnetic field at some segments of the mid-oceanic ridges, at least in some cases, seems to be associated with the occurrence there of high-differentiated basalt with high concentration of titanomagnetite.

Comparison of the estimated and experimental values for the same samples allows us to estimate the degree of titanomagnetite oxidation in these samples.

References

Afanas'ev, Yu. V., Prishepo, V. A., Bakalinskii, S. P., and Tsutskarev, B. M., Marine three-component digital fluxgate magnetometer, in *Proc. 3rd All-Union Geom. Congr.*, Kiev, 1986, 307.

Afonyashin, A. A., Belyaev, I. I., Belyaeva, E. N., Neronov, B. N., Paramonov, A. A., Popov, E. A., and Filin, A. M., Marine proton magnetometer MPM-5M, *Okeanologiya*, 24, 6, 1022, 1984.

Albert, A., *Regression, Pseudoinversion and Recurrent Estimation*, Nauka, Moscow, 1972, 223.

Al'mukhamedov, A. I., Kashintsev, G. L., and Matveenkov, V. V., *Evolution of the Red Sea Basalt Volcanism*, Nauka, Novosibirsk, 1985, 191.

Anderson, H. and Jackson, J., The deep seismicity of the Tyrrhenian Sea, *Geophys. J. R. Astron. Soc.*, 91, 613, 1987.

Andreev, B. A., Estimates of spatial distribution of potential fields for exploration geophysical purpose, *Izv. Akad. Nauk SSSR Ser. Geogr. geofiz.*, part 1, 1, 30, 1947.

Aplonov, S. V. and Popov, E. A., Spatial-temporal unstability of seafloor spreading and its reflection in anomaly magnetic field, *Izv. Akad. Nauk SSSR Fiz. Zemli*, 6, 21, 1991.

Ariskin, A. A., Boulanger, A. Yu., Gorshkov, A. G., et al., Thermal regime of the Mid-Atlantic Ridge and association with the tectonics of the region, *Dokl. Akad. Nauk SSSR*, 280, 6, 1405, 1985.

Aronov, V. I., On calculations of transformation and reduction of gravity anomalies to upper plane in mountain region, *Fiz. Zemli*, 7, 79, 1970a.

Aronov, V. I., On optimal filtering of random errors in gravimetry, *Fiz. Zemli*, 10, 79, 1970b.

Aronov, V. I., *Computer Processing of Gravity Anomalies at a Random Topography Surface*, Nedra, Moscow, 1976, 128.

Atwater, T., Propagating rifts in seafloor spreading patterns, *Nature*, 290, 5803, 185, 1981.

Atwater, T. and Grow, J. G., Mid-Tertiary tectonic transition in the Aleutian arc, *Geol. Soc. Am. Bull.*, 81, 3715, 1970.

Atwater, T. and Mudie, J. A., A detailed near-bottom geophysical study of the Gorda Rise, *J. Geophys. Res.*, 78, 35, 86, 1973.

Aurende, J. M., Charvet, J., Le Lann, A., et al., Le banc de Gorringe un fragment de mantean et de eroute oceanique reconnu par submersible, *C. R. Acad. Sci.*, 285, 16, 1403, 1977.

Avdeiko, G. P., Types of modern oceanic volcanism and their role in geological history, *Vulkanol. Seismol.*, 3, 53, 1979.

Babenko, K. M., Svistunov, Yu. I., and Shlezinger, A. E., Oceanic and continental bottom structures of the eastern Arabian Sea margin, *Dokl. Akad. Nauk SSSR*, 253, 1, 203, 1980.

Backus, G. and Gilbert, F., The resolving power of gross Earth data. *Geophys. J. R. Astron. Soc.*, 16, 169, 1968.

Balyberdin, V. V., Erofeev, Yu. G., Koptyaev, A. O., and Sinitsin, B. I., Automatic marine proton magnetometer, in IZMIRAN, Moscow, 1987, 115.

Barazangi, M. and Dorman, J., World seismicity maps compiled from ESSA Coast and Geodetic Survey epicenter data, *Bull. Seismol. Soc. Am.*, 59, 369, 1969.

Batiza, R., Fornari, D. J., Vanko, D. A., and Lonsdale, P., Craters, calderas and hyaloclastites on young Pacific seamounts, *J. Geophys. Res.*, 89, 10, 8371, 1984.

Batiza, R. and Vanko, D., Volcanic development of small oceanic central volcanoes on the flank of the East Pacific Rise inferred from narrow-beam echo-sounder surveys, *Mar. Geol.*, 54, 53, 1983.

Belyaev, I. I., Belyaeva, E. N., Sviridov, S. A., Tompofol'sky, A. V., and Filin, A. M., Marine proton magnetometer MPM-6, *Okeanologiya*, 27, 3, 532, 1987a.

Belyaev, I. I., Brusilovsky, Yu. V., Gorodnitsky, A. M., Popov, K. V., Shreider, A. A., and Scherbakov, V. P., Detailed geomagnetic study of Marsili Seamount, *Okeanologiya*, 29, 1, 81, 1989a.

Belyaev, I. I., Brusilovsky, Yu. V., Gorodnitsky, A. M., Popov, K. V., Filin, A. M., and Shreider, A. A., Geomagnetic study of the Tyrrhenian seamounts, *MOIP Ser. Geol.*, 66, 3, 45, 1991.

Belyaev, I. I. and Filin, A. M., Marine differential magnetometer DPM-2 and experience in its application, *Okeanologiya*, 30, 6, 1031, 1990.

Belyaev, I. I., Filin, A. M., and Popov, E. A., Apparatus for magnetic investigation, in *Geological, Geophysical and Submersible Research of the Baikal Lake*, IOAN SSSR, Moscow, 1979, 35.

Belyaev, I. I., Gorodnitsky, A. M., Palshin, N. A., and Filin, A. M., New data on anomalous magnetic field structure of the Cape Verde and Brazil basins, *Izv. Akad. Nauk SSSR Fiz. Zemli*, 6, 78, 1989b.

Belyaev, I. I., Valyashko, G. M., and Mirlin, E. G., Seafloor spreading irregularity reflected in magnetic anomaly structure: the Red Sea, in *Stationary Magnetic Field, Petromagnetism and Paleomagnetism*, Proc. 2nd All-Union Geom. Congr., TGU, Tbilisi, 1981, 132.

Belyaev, I. I., Van'yan, L. L., Gorodnitsky, A. M., Palshin, N. A., Filin, A. M., and Shilovsky, P. P., On anomalous magnetic field structure of the Central Basin of the Indian Ocean, *Geomagn. Aeron.*, 27, 2, 293, 1987b.

Bendat, J. S. and Piersol, A. G., *Engineering Applications of Correlation and Spectral Analysis*, John Wiley & Sons, New York, 1980, 312.

Bhattacharrya, B. K., A generalized multibody model for inversion of magnetic anomalies, *Geophysics*, 45, 2, 255, 1980.

Blakely, R. J. and Christiansen, R. L., The magnetization of Mount Shasta and implications for virtual geomagnetic poles determined from seamounts, *J. Geophys. Res.*, 83, B12, 5971, 1978.

Bleil, U. and Petersen, N., Variations in magnetization intensity and low-temperature titanomagnetite oxidation of ocean floor basalts, *Nature*, 301, 384, 1983.

Bocharova, E., Maschenkov, S., Roest, W., Verhoef, J., and Macnab, R., A study of Mid-Atlantic spreading process between the Kane and Atlantis fracture zones as seen in detailed bathymetric and magnetic maps, paper presented at Int. Assoc. Geomagnetism and Aeronomy, *IUGG 20th Gen. Assembly*, Vienna, 1991.

Bogdanov, Yu. A., Sagalevitch, A. M., Kuzin, M. I., and Kuznetsov, A. P., Geological structure of the Reykjanes Ridge rift zone near 58N, *Dokl. Akad. Nauk SSSR*, 273, 6, 1478, 1983.

Bogdanov, Yu. A., Sorokhtin, O. G., Zonenshain, L. P., et al., in *Manganese Nodules of the Pacific*, Nauka, Moscow, 1990, 229.

Boillot, G., Recg, M., Winterer, E. L., et al., Tectonic denudation of the upper mantle along passive margins: a model based on drilling results O.D.P., leg 103, *Tectonophysics*, 132, 4, 335, 1987.

Bonatti, E., Vertical tectonics in oceanic fracture zones, *Earth Planet. Sci. Lett*, 37, 366, 1976.

Bott, M. H. P., Inverse methods in the interpretation of magnetic and gravity anomalies, *Methods Comput. Phys.*, 13, 133, 1973.

Bowin, C. and Vogt, P. R., Magnetic lineation between Carlsberg and Seychelles Bank, Indian Ocean, *J. Geophys. Res.*, 71, 2625, 1966.

Briggs, I. C., Machine contouring using minimum curvature, *Geophysics*, 39, 1, 39, 1974.

Brusilovsky, Yu. V. and Gorodnitsky, A. M., Magnetic simulation of the Gorringe seamount range, in *Electromagnetic Induction in the Ocean*, Nauka, Moscow, 1990, 31.

Byers, C. D., Garcia, M. O., and Muenov, D. W., Volatiles in basaltic glasses from East Pacific Rise at 21 N; implications for MORB sources, *Earth Planet. Sci. Lett.*, 79, 9, 1986.

Cambell, W. H. An analisys of the Spectra of geomagnetic variations having periods from 5 min to 4 hours, *J. Geophys. Res.*, 81, 7, 1369, 1976.

Canadian-American Seamount Expedition, Hydrothermal vents on an axis seamount of the Juan de Fuca ridge, *Nature*, 313, 59, 212, 1985.

Carmichael, I. S. E. and Chiorso, M. S., Oxidation-reduction relations in basaltic magma: a case for homogeneous equilibria, *Earth Planet. Sci. Lett.*, 79, 200, 1986.

Carres, D. W., Menard, H. W., and Hey, R. N., Eocene reorganization of the Pacific-Farallon spreading center north of the Mendocino fracture zone, *J. Geophys. Res.*, 93, 2813, 1988.

Christie, D. M., Carmichael, I. S. F., and Langmuir, C. H., Oxidation statuses of mid-oceanic ridge basalt glasses, *Earth Planet. Sci. Lett.*, 79, 397, 1986.

Cochran, J. R., The Gulf of Aden: structure and evolution of young ocean basin and continental margin, *J. Geophys. Res.*, 86, 1, 263, 1981.

Cochran, J. R., The magnetic quiet zone in the eastern Gulf of Aden: implication for the early development of the continental margin, *Geophys. J.R. Astron. Soc.*, 62, 171, 1982.

Constable, S. C., Parker, R. L., and Constable, C. G., Ocean's inversion: a practical algorithm for generating smooth models from electromagnetic sounding data, *Geophysics*, 52, 3, 289, 1987.

Cox, A., Angular dispersion due to random magnetization, *Geophys. J. R. Astron. Soc.*, 8, 4, 345, 1987.

Cox, A. and Gordon, R., Paleolatitudes determined from paleomagnetic data from vertical cores, *Rev. Geophys. Space Phys.*, 22, 1, 47, 1984.

Cunningham, K. I., Roberte, A. A., and Donovan, T. J., Horizontal gradient magnetic and helium surveys, *U.S. Geol. Surv. Bull.*, 1778, 1, 209, 1987.

Dalrymple, G., Critical tables for conversion of K-Ar ages from old to new constants, *Geology*, 7, 558, 1979.

Den, N., Ludwig, R., and Muranchi, S., Seismic refraction measurements in the North-West Pacific basin, *J. Geophys. Res.*, 74, 6, 1421, 1969.

Dickin, A. P., Fallick, A. E., et al., An isotopic and geochronological investigation of the younger igneous rocks of the Seychelles microcontinent, *Earth Planet. Sci. Lett.*, 81, 1, 46, 1986.

Dmitriev, L. V., Sobolev, A. V., and Suchevskaya, N. M., Formation conditions for primary melt of oceanic tholeiite basalt and variations of this melt, *Geokhimiya*, 2, 163, 1979.

Donets, E. G., Litvinov, E. M., and Shkitin, A. I., Magnetic characteristics of the northwestern Atlantic seamounts and their genesis, in *Marine Geology and Geophysics*, vol.2, VIEMS, Moscow, 1975, 3.

Douglas, R. G. and Mouhade, M., Age of basal sediments on the Shatsky Rise, Northwestern Pacific, *Bull. Geol. Soc. Am.*, 83, 4, 1163, 1972.

Dreiper, N. and Smith, G., Applied regression analysis, *Finansy Statistika*, 1, Moscow, 1986, 366.

Dunlop, D. and Prevot, M., Magnetic properties and opaque mineralogy of drilled submarine intrusive rocks, *Geophys. J. R. Astron. Soc.*, 69, 763, 1982.

Ekinyan, P., Submarine volcanoes, *Am. Sci.*, 9, 12, 1984.

Elvers, D., Srivastava, S. P., Potter, K., Morley, J., and Sdidel, D., Asymmetric spreading across the Juan de Fuca and Gorda Rises as obtained from an oftailed magnetic survey, *Earth Planet. Sci. Lett.*, 20, 211, 1973.

Engebretson, D. C., Cox, A., and Gordem, R. G., Relative motions between oceanic plates of the Pacific Basin, *J. Geophys. Res.*, 89, 10291, 1984.

Engebretson, D. C., Cox, A., and Gorden, R. G., Relative motions between oceanic and continental plates in the Pacific Basin, *Geol. Soc. Am. Spec. Pap.*, 206, 59, 1985.

Fastovsky, U. V., *Marine Geomagnetic Survey Methods*, IZMIRAN, Moscow, 1989, 117.

Filin, A. M., Belyaev, I. I., and Alampieva, E. M., Marine differential magnetometer DPM-1, *Okeanologiya*, 25, 6, 1040, 1985.

Filin, A. M., Gorodnitsky, A. M., Pal'shin, N. A., Popov, K. V., Shishkina, N. A., Shreider, A. A., and Dimitrov, P. S., New data on anomalous field structure over the Verchelli and Vavilov seamounts (Tyrrhenian Sea), *Okeanologiya Bulg. Acad. Sci.*, 17, 58, 1988.

Fornari, D. J., Malahoff, A., and Heezen, B. C., Volcanic structure of the crest of the Puna Ridge, Hawaii: geophysical implications of submarine volcanic terrain, *Geol. Soc. Am. Bull.*, 89, 606, 1978.

Forsight, J., Malcolm, D., and Mowler, K. *Machine Methods of Mathematical Simulations*, Mir, Moscow, 1980, 279.

Francheteau, J., Harrison, C. G. A., Sclater, J. C., and Richards, M. L., Magnetization of Pacific seamounts: a preliminary Polar curve for the northeastern Pacific, *J. Geophys. Res.*, 75, 11, 2035, 1970.

Gardner, J. V., Dean, W. E., and Blakely, R. J., Shimada seamount: an example of recent mid-plate volcanism, *Geol. Soc. Am. Bull.*, 95, 855, 1984.

Gee, J., Tauxe, L., Hildebrand, J., Staudingel, H., and Lonsdale, P., Nonuniform magnetization of Jasper seamount, *J. Geophys. Res.*, 93, B10, 12159, 1988.

Gladkyi, K. B., *Gravity and Magnetic Prospecting*, Nedra, Moscow, 1967, 319.

Gordin, V. M., Registration of geomagnetic field variations by hydromagnetic survey material, in *Marine Geology and Geophysics*, vol.3, VIEMS, Moscow, 1980, 1.

Gordin, V. M., Gorshkova, L. K., Mihailov, V. O., and Strakhov, V. N., Inverse problems of marine magnitometry, in *Electromagnetic Study of Deep Structure of the Earth Crust and Upper Mantle of Sea and Ocean*, IZMIRAN, Moscow, 1981, 53.

Gordin, V. M., Roze, E. N., and Uglov, B. D., *Marine Magnetometry*, Nedra, Moscow, 1986, 232.

Gorodnitsky, A. M., Thickness of Oceanic Lithosphere and altitude limit of submarine volcanoes, in *Plate Tectonics*, IOAN SSSR, Moscow, 1977, 109.

Gorodnitsky, A. M., Demenitskaya, R. M. and Kaminsky, V. D., *Seamounts*, Nedra, Moscow, 1978, 163.

Gorodnitsky, A. M., Phanerozoic continental drift velocity according to paleomagnetic data, in Proc. 3rd Far East Workshop on Paleomagnetic Study, *SVKNII DVNZ Akad. Nauk SSSR*, Magadan, 1984, 65.

Gorodnitsky, A. M., Structure of Oceanic Lithosphere and Seamount Formation, Nauka, Moscow, 1985, 165.

Gorodnitsky, A. M., Geophysical study of transform faults of the World Ocean, in *Marine Geology and Geophysics*, VIEMS, Moscow, 1987, 58.

Gorodnitsky, A. M., Belyaev, I. I., Brusilovsky, Yu. V., Popov, K. V., and Scherbakov, V. P., Geomagnetic characteristics of the Gorringe seamount range (the North Atlantic), *Okeanologiya*, 28, 5, 814, 1988.

Gorodnitsky, A. M. and Koryakin, E. D., *Geophysical Survey of Transform Faults of the World Ocean*, VIEMS, Moscow, 1987, 57.

Gorodnitsky, A. M. and Lin'kova, T. I., Magnetic properties of bed rocks of transform faults, in *Deep Faults of Oceanic Bottom*, Nauka, Moscow, 1984, 75.

Gorodnitsky, A. M. and Lukyanov, S. V., Anomalous magnetic field of the northwestern Pacific correlated with tectonic evolution of the region, in *Marine Electromagnetic Investigation Problems*, IZMIRAN SSSR, Moscow, 1980, 46.

Gorodnitsky, A. M., Lukyanov, S. V., Litvinov, E. M., and Suzyumov, A. E., Anomalous magnetic field structure and magnetoactive layer, in *Deep Faults of Oceanic Bottom*, Nauka, Moscow, 1984, 50.

Gorodnitsky, A. M., Nasarova, E. A., and Shishkina, N. A., On depth limit of active magnetic layer in the oceanic lithosphere, in *Electromagnetic Induction in the World Ocean*, Nauka, Moscow, 1990a, 3.

Gorodnitsky, A. M., Suzyumov, A. E., and Ivanenko, A. N., Paleomagnetic survey of the West Pacific seamounts, in *A New Problem of Marine Geology*, Proc. 4th Int. Marine Geology Symp. IOAN SSSR, Moscow, 3, 1990b, 30.

Gorodnitsky, A. M., Valyashko, G. M., Palshin, N. A., Filin, A. M., and Lukyanov, S. V., Anomalous magnetic field, in *Geophysical Fields and Seafloor Structure of Oceanic Basins*, Nauka, Moscow, 1990c, 48.

Gorshkov, A. G. and Lukashevich, I. P., Computing of magma chamber temperature in rift zones of the World Ocean, *Tectonophysics*, 159, 337, 1989.

Gunn, P. J., Linear transformation of gravity and magnetic fields, *Geophys. Prospect.*, 23, 300, 1975.

Handschumacher, D., Sager, T., Hilde, T., and Bracey, D., Pre-Cretaceous tectonic evolution of the Pacific plate and extention of the geomagnetic polarity reversal time scale with implications for the origin, *Tectonophysics*, 160, 365, 1989.

Harland, W. B., Cox, A. V., Llewellyn, P. G., Pickton, C. A. G., Smith, A. G., and Walters, R. A., *A Geologic Time Scale*, Cambridge University. Press, Cambridge, 1985, 141.

Harrison, C. G. A., A seamount with a nonmagnetic top, *Geophysics*, 36, 2, 349, 1971.

Harrison, C. G. A. and Ball, M. M., Geophysical observations on an exposed seamount in the Afar depression, *Bull. Volcanol.*, 38, 26, 1975.

Hayes, D. E. and Pitman, W. C., Magnetic lineations in the North Pacific, geological investigations of the North Pacific, *Mem. Geol. Soc. Am.*, 126, 291, 1970.

Heezen, B. C., MacGregor, I. D., et al., *Init. Rep. DSDP*, 20, 959, 1973.

Heirtzler, J. R., Dickson, G. O., and Herron, E. M., Marine magnetic anomalies, geomagnetic field reversals and motion of the ocean floor and continents, *J. Geophys. Res.*, 73, 2119, 1968.

Helbig, K., Some integrals of magnetic anomalies and their relation to the parameters of the disturbing body, *Z. Geophysik*, 29, 2, 83, 1963.

Hey, R. A., New class of "pseudofaults" and their bearing on plate tectonics: a propagation rift model, *Earth Planet. Sci. Lett.*, 37, 2, 321, 1977.

Hey, R., Evidence for spreading-center jumps from the scale bathymetry and magnetic anomaly near the Galapagos Islands, *Geology*, 7, 504, 1979.

Hey, R., Duennebier, F. K., and Morgan, N. J., Propagation rifts on mid-oceanic ridges, *J. Geophys. Res.*, 85, 137, 3647, 1980.

Hey, R., Kleinrock, M. G., Miller, S. P., et al., Sea-beam/deep-tow investigation of active oceanic propagating rift system, Galapagos 95,5 W, *J. Geophys. Res.*, 93, B3, 3369, 1986.

Hey, R. and Wilson, D. S., Propagating rift explanation for the tectonic evolution of the northeast Pacific: the pseudoview, *Earth Planet. Sci. Lett.*, 58, 2, 167, 1982.

Hilde, T. W. C., Isezaki, N., and Wageman, J. M., Mesozoic seafloor spreading in the North Pacific, in *The Geophysics of the Pacific Ocean Basin and Its Margin*, Washington, D.C., 1976, 205.

Hilde, T. W. C., Uyeda, S., and Kroenke, L., Evolution of the western Pacific and its margin, *Tectonophysics*, 38, 145, 1977.

Hildebrand, J. A. and Parker, R. L., Paleomagnetism of Cretaceous Pacific Seamounts revisited, *J. Geophys. Res.*, 92, 12, 12695, 1987.

Hoerl, A. E. and Kennard, R. W., Ridge regression: Basel estimation for nonorthogonal problem, *Technometrics*, 12, 55, 1970.

Houtz, R., Seismic properties of layer 2A in the Pacific, *J. Geophys. Res.*, 81, 35, 6321, 1976.

Houtz, R., Windich, C., and Murauchi, S., Changes in the crust and upper mantle near Japan-Bonin trench, *J. Geophys. Res.*, 85, 131, 267, 1980.

IAGA, Division 1, Working Group 1, International Geomagnetic Reference Field Revision, EOS, *Trans. Am. Geophys. Union*, 67, 523, 1986.
Initial Reports of the Deep Sea Drilling Project, Washington, D.C., 6, 1971.
Initial Reports of the Deep Sea Drilling Project, Washington, D.C., 18, 1975.
Initial Reports of the Deep Sea Drilling Project, Washington, D.C., 23, 1180, 1974a.
Initial Reports of the Deep Sea Drilling Project, Washington, D.C., 24, 1180, 1974b.
Initial Reports of the Deep Sea Drilling Project, Washington, D.C., 24, 1977.
Initial Reports of the Deep Sea Drilling Project, Washington, D.C., 66, 1982.
Ivanenko, A. N., On dimension decrease in inverse linear geophysical problems, in *Volcanological Research at Kamchatka*, Volcanology Institute DVO AN SSSR, Petropavlovsk-Kamchatsky, 1988, 79.
Ivanenko, A. N. and Semenets, N. V., Choice formalization of model parameters in geotomography problems, in Proc. 4th All-Union Symp. Computer Tomography, Tashkent, 2, 1989, 36.
Jackson, D. D., Interpretation of inaccurate, insufficient, and inconsistent data, *Geophys. J. R. Astron. Soc.*, 28, 99, 1972.
Joides J., 16, 2, 1990.
Kaminsky, V. D. and Simovsky, I. S., Magnetization vector determination for a group of anomaly-generating bodies from their gross magnetic field, in *Geophysical Methods of Prospecting in Arctic*, NIIGA, Leningrad, 11, 1976, 155.
Karasik, A. M., Merkuriev, S. A., Mitin, L. I., Sochevanova, N. A., and Yanovsky, B. N., The main features of the Arabian Sea opening according to systematic magnetic mapping results, *Dokl. Akad. Nauk SSSR*, 286, 4, 933, 1986a.
Karasik, A. M., Merkuriev, S. A., Parakhin, A. M., Sochevanova, N. A., and Yanovsky, B. N., Magnetic and tectonic inhomogeneities of the Northwestern part of the Indian ocean as a reflection of spreading geometry changes, in Proc. 3rd All-Union Oceanological Congr., 7, 1, 1987, 180.
Karasik, A. M., Merkuriev, S. A., Sochevanova, N. A., Tanichev, A. A., and Yanovsky, B. N., Anomalous magnetic field of Carlsberg ridge from regular survey data, in Proc. 3rd All-Union Geomagnetic Congr., Kiev, 1986b, 58.
Karasik, A. M., Merkuriev, S. A., Sochevanova, N. A., and Yanovsky, B. N., Magnetic anomalies, structure and age of the Arabian Sea bottom, in Proc. 27th Int. Geol. Congr., Moscow, 1984a, 40.
Karasik, A. M., Merkuriev, S. A., Sochevanova, N. A., and Yanovsky, B. N., Zoning and nature of anomalous magnetic field in the northwestern Indian ocean, in *Geophysical Fields of the Pacific and Indian Oceans*, Moscow, 1988, 80.
Karasik, A. M., Merkuriev, S. A., Sochevanova, N. A., and Yanovsky, B. N., Evolution features of the northwestern Indian ocean from regular magnetic survey, in Proc. 4th All-Union Workshop on Marine Geology: Geology of Oceans and Seas, Moscow, 2, 1984b, 54.
Karasik, A. M. and Sochevanova, N. A., Axes of paleomagnetic anomalies in the World Ocean (at 1.01.1981), in *Electromagnetic Study of Deep Structure of the Earth Crust and Upper Mantle of Sea and Ocean*, IZMIRAN, Moscow, 1981, 205.
Karasik, A. M., Sochevanova, N. A., Desimon, A. I., and Shreider, A. A., The map of the paleomagnetic anomaly axes of the World Ocean, in *Electromagnetic Study of Deep Structure of the Earth Crust and Upper Mantle of the Sea and Ocean*, IZMIRAN, Moscow, 1981a, 207.
Karasik. A. M., Zimoglyadov, V. A., Zolotov, I. G., Merkuriev, S. A., Sochevanova, N. A., and Yanovsky, B. N., Magnetic anomalies and chronology of the Arabian Sea bottom, in Proc. 2nd All-Union Oceanology Congr., Sevastopol', 7, 1, 1981b, 21.
Kashinskaya, I. V., Trukhin, V. I., and Shreider, A. A., Petromagnetic study of the Red Sea basalts, in *Geomagnetic Field Fine Structure*, IFZ Press, Moscow, 1986, 152.

Kashintsev, G. L. and Rudnik, G. B., Igneous rocks associated with the Earth's crust faults, in *Deep Faults of the Oceanic Bottom*, Nauka, Moscow, 1984, 174.

Kastens, K., Mascle, J., Auroux, Ch., et al., Young Tyrrhenian Sea evolved very quickly, *Geotimes*, 31(8), 11, 1986.

Kellog, J. N. and Ogujiofor, I. J., Gravity field analysis of Sio Guyot: an isostatically compensated seamount in Mid-Pacific Mountains, *Geol. Mar. Lett.*, 5, 91, 1985.

Kent, D. V. and Gradstein, F. M., A Cretaceous and Jurassic geochronology, *Geol. Soc. Am. Bull.*, 96, 11, 1419, 1985.

Khutorskoi, M. D., Gorodnitsky, A. M., Gol'mshtok, A. Ya., Sochel'nikov, V. V., and Kordyurin, A. V., The Tyrrhenian Sea heat flow, basaltic volcanism and lithosphere structure, *Geotektonika*, 5, 116, 1986.

Klitgord, K. D. and Mammerickx, J., Northern East Pacific Rise: magnetic anomaly and bathymetric framework, *J. Geophys. Res.*, 87, 6725, 1982.

Klitgord, K. D. and Schouten, H., Plate kinematics of the Central Atlantic, in *The Geology of North America: The Western North Atlantic Region (DNAG Ser.)*, Vogt, P. R. and Tucholke, B. E., Eds., Geologied. Society of America, 1986, 351.

Kodama, K. and Uyeda, S., Magnetization of Izu Islands with special reference to Oshima Volcano, *J. Volcanol. Geotherm. Res.*, 6, 353, 1979.

Kodama, K., Uyeda, S., and Isezaki, N., Paleomagnetism of Siko Seamount, Emperor Seamount Chain, *Geophys. Res. Lett.*, 5, 3, 165, 1978.

Kogan, L. I., Baniolessi, V. M., and Evsyukov, Yu. D., Results of detailed bathymetric and magnetic mapping in Tunis-Sardinian channel and Vavilov Seamount in the Mediterranean sea, *Okeanologiya*, 20, 1059, 1980.

Kogan, L. I., Zonenshain, L. P., and Shmidt, O. A., Structure of the Earth's crust at the Shatsky and Hess Rises (Pacific Ocean) from DSS-RW, *Bull. MOIP Ser.Geol.*, 1, 4, 15, 1983.

Kolesova, V. I., *Analytical Methods of Magnetic Cartography*, Nauka, Moscow, 1985, 221.

Kono, M., Paleomagnetism of DSDP leg 55 basalt and implications for the tectonic of the Pacific plate, *Init. Rep. DSDP*, 55, 1980, 737.

Kononov, M. V., Pacific plate motion for the last 130 m.y., *Okeanologiya*, 24, 3, 857, 1984.

Kononov, M. V., *Plate Tectonics of the Northeast Pacific*, Nauka, Moscow, 1989, 168.

Krasil'nikov, A. I., Lyubimov, V. V., Perunov, B. S., et. al., Nuclear magnetometers designed by SCB FP AN SSSR, in *Analysis of Space-Time Structure of Geomagnetic Field*, Nauka, Moscow, 1975, 243.

Kuz'min, M. I., Magmatism, in *Geology of Tadgjura Rift*, Nauka, Moscow, 1987, 150.

La Breque, J. L., Kent, D. V., and Cande, S. C., Revised magnetic polarity time scale for Late Cretaceous and Cenozoic time, *Geology*, 5(6), 330, 1977.

Lanczos, C., *Linear Differential Operators*, van Nostrand, London, 1961, 564.

Lansdale, P., Tectonic and magnetic ridges in the Eltanin fault system, South Pacific, *Mar. Geophys. Res.*, 8, 203, 1986.

Larson, R. L., Late Jurassic and Cretaceous evolution of the western Central Pacific, *J. Geomagn. Geoelectr.*, 28, 219, 1976.

Larson, L. R. and Chase, C. G., Late Mesozoic evolution of the Western Pacific Ocean, *Bull. Geol. Soc. Am.*, 83, 3627, 1972.

Larson, L. R. and Pitman, C., World-wide correlation of Mesozoic magnetic anomalies and its implication, *Bull. Geol. Soc. Am.*, 83, 2, 3646, 1972.

Lashkov, V. V., Mironov, V. S., and Protsaenko, S. V., Gradient magnetic survey experience in auroral zone, *Geophys. Probl.*, 32, 1988, 94.

Laughton, A. S., The Gulf of Aden, *Philos. Trans. R. Soc. London Ser. A*, 259, 1099, 150, 1966.

Laughton, A. S. and Whitmarsh, R. B., The Azores-Gibraltar plate boundary, *Geodynamics of Iceland and the North Atlantic Area*, Reykjavik, 1974, 63.

Laughton, A. S., Whitmarsh, R. B., and Cones, M. T., The evolution of the Gulf of Aden, *Philos. Trans. R. Soc. London Ser. A*, 257, 227, 1970.

Laurenso, P. and Morrison, H. F., Vector magnetic anomalies derived from measurements of a single component of a field, *Geophysics*, 38, 2, 359, 1973.

Leibov, M. B., Melikhov, V. R., Bulychev, A. A., Shamaro, A. M., Uglov, B. D., and Gainanov, A. G., Modern methods to improve accuracy of marine magnetic survey, in *Marine Geology and Geophysics*, VIEMS, Moscow, 1988, 48.

Leibov, M. B. and Mirlin, E. G., Oligocene magnetic time scale according to oceanic linear magnetic anomalies, *Vest. Mosk. Univ.*, Geol., 2, 1979, 68.

Leibov, M. B., Uglov, B. D., Bulychev, A. A., Lygin, V. A., Gainanov, A. G., and Melikhov, V. R., *Practical Questions of Marine Magnetic Survey Accuracy Increasing*, VINITI, Moscow, 9041, B86, 140, 1986.

Le Mouel, J. L., Galdeano, A., and Le Pichon, X., Remanent magnetization vector direction and the statistical properties of magnetic anomalies, *Geophys. J. R. Astron. Soc.*, 30, 353, 1972.

Le Pichon, X., Francheteau, J., and Bonnin, J., *Plate Tectonics*, Elsevier, Amsterdam, 1973, 300.

Lin, J. G., Purdy, G. M., Shouten, H., et al., Evidence from gravity data for focussed magmatic accretion along the Mid-Atlantic Ridge, *Nature*, 344, 627, 1990.

Lin'kova, T. I. and Raikevitch, M. I., *Paleomagnetic Research of Igneous Rocks of the Western Pacific*, SVKNII AN DVO, Magadan, 1989, 41.

Livotov, L. L., Nikolaev, A. A., and Semevskyi, R. B., Marine towed magnetometer MBM, *Geofiz. Appar.*, 69, 1979, 31.

Lobkovsky, L. I., *Geodynamics of Spreading Zone, Subduction and Two-Level Plate Tectonics*, Nauka, Moscow, 1988, 291.

Louson, Ch. and Henson, P., *Numerical Solution of Least Square Problems*, Nauka, Moscow, 1986, 232.

Ludvig, W. S. and Houtz, R. E., Isopach map of sediments in the Pacific Ocean basin and marginal sea basin, *AAPG MAP Ser.*, Catalog 647, N. Y., 1979.

Lukanin, O. A., Kadik, A. A., et al., Subsurface magmatic evolution of tholeiite of the Atlantic, *Geokhimiya*, 3, 382, 1983.

Lukashevich, I. P. and Gorshkov, A. G., Tholeitic magmatism and the evolution of Mid-Oceanic Ridges, *Geotectonica*, 6, 105, 1989.

Lumb, J. T., Hochstein, M. P., and Woodward, D. J., Interpretation of magnetic measurements in the Cook Islands, south-west Pacific Ocean, in *The Western Pacific Islands Arcs, Marginal Seas, Geochemistry*, Crane, Russak and Co., New York, 1973, 79.

Lyubimov, V. V. and Perfilov, V. I., New constructions of nuclear magnetic apparatus in SKB FP, Ac. Sci. USSR, review in *Investigations of Cosmic Plasma*, IZMIRAN, Moscow, 1980, 159.

Macdonald, K. C., Near bottom magnetic anomalies, asymmetric spreading, oblique spreading and tectonics of the Mid-Atlantic Ridge near lat. 37N, *Bull. Geol. Soc. Am.*, 88, 4, 191, 1977.

Macdonald, K. C. and Fox, P. J., Overlapping centers: new accretion geometry on the East Pacific Rise, *Nature*, 302, 3, 55, 1983.

Macdonald, K., Sempere, J. C., and Fox, P. J. E., Pacific Rise from Siqueros to Orozco FZ: along strike continuity of axial neovolcanic zone and structure and evolution of overlapping spreading center, *J. Geophys. Res.*, 89, B7, 6049, 1984.

Macnab, R., Verhoef, J., and Srivastava, S. P., A compilation of magnetic data from the Arctic and North Atlantic oceans, *Current Res., Part II, Geol. Survey Can.*, paper 90-1, 1, 1990.

Malahoff, A. and Handschumacher, D. W., Magnetic anomalies south of the Murrey fracture zone: new evidence for a secondary sea-floor spreading and strike-slip movement, *J. Geophys. Res.*, 76, 26, 6265, 1976.

Malovitsky, Y. P., Baniolessi, V. M., et al., On geological nature of Vavilov and Magnaghi seamounts in the Tyrrhenian Sea, *Okeanologiya*, 18, 6, 1053, 1978.

Malovitsky, Y. P., Chumakov, I. S., Shimkus, K. M., Esina, L. A., and Moskalenko, V. N., *The Earth's Crust and Development History of the Mediterranean Sea*, Nauka, Moscow, 1982, 207.

Markov, I. M., Gradiometric Method for Marine Magnetic Measurement and Its Application in Registration of Geomagnetic Variations, *Ph.D. Thesis*, LGU, Leningrad, 1986, 16.

Markov, I. M., Merkuriev, S. A., and Roze, E. N., Methods of gradiometry data processing, Constant geomagnetic field, rock magnetism and paleomagnetism, in *Proc. 2nd All-Union Assoc. on Geomagnetism*, Tbilissi, 1, 1981, 164.

Marquardt, D. W., Generalized inversion, ridge regression, biased linear estimation, and nonlinear estimation, *Technometrics*, 12, 3, 591, 1970.

Maschenkov, S. P., Magnitometrical information unification problem ingeotravers database, in *Object Subsystem Development Problems*, PGO, "Sevmorgeologiya", Leningrad, 1989, 25.

Maschenkov, S. and Pogrebitsky, Y., Preliminary results of the Canary-Bahamas geotransect project, *EOS, Trans. Am. Geophys. Union*, 73, 37, 393, 397, 1992.

Maschenkov, S., Roest, W., Verhoef, J., Macnab, R., Astafurova, E., Bocharova, E., and Glebovsky, V., Temporal variations in the Atlantic seafloor spreading process between the Kane and Atlantis fracture zones, *J. Geophys. Res.*, in press.

Mascle, J., Kastens, K., and Auroux, C., A land-locked back-arc basin: preliminary results from O.D.P. Leg 107 in Tyrrhenian Sea, Leg 107, Scientific Party, *Tectonophysics*, 146, 10, 1-4, 149, 1988.

Matthews, D. H., The Owen fracture zone and the northern end of the Carlsberg Ridge, *Philos. Trans. R. Soc. London Ser A*, 259, 1099, 172, 1966.

Matthews, D. H. and Loncarevic, B. D., Bathymetric, magnetic and gravity investigation, H.M.S. Owen, 1961-1962, Admiralty Marine Sci. Publ.; as cited in McKenzie, D. and Sclater, J., *Geophys. J. R. Astron. Soc.*, 125, 437, 1971.

Matveenkov, V. V. and Marova, N. A., Formation age of magmatic complexes and related structures of the Marcus-Necker Rise, *Izv. Akad. Nauk SSSR, Ser. Geol.*, 5, 126, 1975.

McKenzie, D. and Sclater, J., Evolution of the Indian Ocean since the Late Cretaceous, *Geophys. J. R. Astron. Soc.*, 25, 437, 1971.

McNutt, M., Nonuniform magnetization of seamounts: a least squares approach, *J. Geophys. Res.*, 91, B3, 3686, 1986.

Melankholina, E. N., Tectonics of the Northwestern Pacific: relations between oceanic structures and continental margins, *GIN Akad. Nauk SSSR Trans.*, 434, 216, 1988.

Melankholina, E. N., Puscharovsky, J. M., and Rudnik, G. B., Oceanic structure and magmatism of the north-west Pacific, in Proc. Int. Workshop Geodinamics of the Western Pacific, Yuzhno-Sakhalinsk, 2, 1981, 8.

Melankholina, E. N., Savelieva, G. N., Kudryavtsev, D.I., et al., Substantial composition of oceanic crust and upper mantle in the Clarion fracture zone (Pacific Ocean) *Dokl. Akad. Nauk SSSR*, 268, 4, 943, 1983.

Melikhov, V. R., Bulychev, A. A., and Shamaro, A. M., Frequency method of resolution of stationary and variable components of geomagnetic field in hycromagnetic gradient survey, in *Deep Electromagnetic Survey*, IZMIRAN, Moscow, 1937, 85.

Melson, W. G., Byerly, G. R., and Nelson, J. A., A catalog of the major element chemistry of abyssal volcanic glasses, Contribut. Earth Sci. 19, 31, 1977.

Merill, R. T. and Burns, R., A detailed magnetic study of Gobb seamount, *Earth Planet. Sci. Lett.*, 14, 413, 1972.

Merkuriev, S. A., Identification of linear magnetic anomalies of slow ridges from regular magnetic survey of the Carlsberg ridge, in *Proc. 8th All-Union Workshop on Marine Geology*, Moscow, 2, 1988, 195.

Merkuriev, S. A., Karasik, A. M., and Sochevanova, N. A., On some process at slow rifts from regular magnetic survey of the Carlsberg ridge, in Proc. 8th All-Union Workshop on Marine Geology, Moscow, 2, 1988, 185.

Mikhno, M. F., Investigation of the structure of the oceanic crust of the Northwestern Pacific by refracted waves method, *Izv. Akad. Nauk SSSR Ser. Geol.*, 3, 43, 1964.

Miller, S. P. and Hey, R. N., Three dimension magnetic modelling of a propagating rift 95° 30W, *J. Geophys. Res.*, B, 91, 3, 3395, 1986.

Mirchick, I. M., Panaev, V. E., and Pogrebitsky, J. E., Angolan-Brazilian geotraverse, in *Geology of Sea and Oceans*, VNIIOkeangeolog., Leningrad, 1988, 141.

Monger, J. W. H., The global geoscience transect project, *Episodes*, 9, 217, 1986.

Monin, A. S., Bogdanov, J. A., Zonenshain, L. P., et al., *Submarine Geological Investigations from Submersible Apparatus*, Nauka, Moscow, 1985, 232.

Monin, A. S., Voitov, V. L, and Yastrebov, V. S., Red Sea expedition of P. P. Shirshov Oceanology Institute, USSR Acad. Sci., Pikar operation, *Okeanologiya*, 20, 4, 743, 1980.

Morelli, C., Physiography, gravity and magnetism of the Tyrrhenian Sea, *Bull. Geophys. Theor. Appl.*, 12, 48, 275, 1970.

Morgan, W. J., Mid-Ocean Ridge dynamics: observation and theory, Rev. Geophys. Suppl., *U.S. Natl. Rep. to IUGG 1987-1990*, 807, April 1991.

Mudie, J. D., Grow, J. A., and Bessey, J. S., A near-bottom survey of lineated abyssal hills in the Equatorial Pacific, *Mar. Geophys. Res.*, 1, 397, 1972.

Muller, R. D. and Roest, W. R., Fracture zones in the North Atlantic from combined Geosat and Seasat data, *J. Geophys. Res.*, 97, 3337, 1992.

Nafikov, V. M., Russak, Yu. S., and Gusev, V. K., Security estimation of hydroacoustic navigation system in various regimes, in *Submarine Technical Devices for Ocean Investigation*, IFAN USSR, Moscow, 1988, 90.

Nakanishi, M., Tamaki, K., and Kobayashi, K., Mesozoic magnetic anomaly lineations and seafloor spreading history of the Northwestern Pacific, *J. Geophys. Res.*, 94, D, 11, 15437, 1989.

Nazarova, E. A., Magnetic properties of oceanic basalts, in *Magnetic Anomalies and New Global Tectonics*, Nauka, Moscow, 1981, 131.

Nazarova, E. A., Plutonic rocks of oceanic lithosphere and anomalous magnetic field of the World Ocean, in *Marine Geology and Geophysics*, VIEMS, Moscow, 1987, 38.

Nazarova, E. A. and Gorodnitsky, A. M., On oceanic crust structure of active magnetic layer, in Geophysical Fields of the Atlantic Ocean, Int. Geol. Congr., Moscow, 1988, 131.

Neprochnov, Yu. P., Complex geophysical fault zone investigations in 24th cruise of R/V "Akademic Kurchyatov" and 23d cruise of R/V "Dmitriy Mendeleev", in *Deep Faults of Oceanic Bottom*, Nauka, Moscow, 1984a, 4.

Neprochnov, Yu. P., Investigation of deep faults by DSS, in *Deep Faults of Oceanic Bottom*, Nauka, Moscow, 1984b, 117.

Neprochnov, Yu. P., El'nikov, I. N., Semionov, G. A., Sedov, V. V., and Neprochnova, A. F., The Earth's crust structure, in *Deep Faults of Oceanic Bottom*, Nauka, Moscow, 1984, 121.

Neprochnov, Yu. P., Gorodnitsky, A. M., Levchenko, O. V., Merklin, L. R., and Sedov, V. V., Anomaly geophysical fields and the Earth's crust structure of the Central Indian ocean, in *Geophysical Fields of the Pacific and Indian Ocean*, MGK, Moscow, 1988, 61.

Northrop, I., Fresch, R. A., and Frassetto, R., Bermuda-New England seamount arc, *Bull. Geol. Soc. Am.*, 73, 5, 47, 1962.

Norton, I. O. and Sclater, J. G., A model for the evolution of the Indian ocean and the break of the Gondwanaland, *J. Geophys. Res.*, B, 84, 312, 6830, 1979.

Novysh, V. V., Belyaev, I. I., Figner, D. L., and Abramov, L. M., Marine magnetometric apparatus, in *Marine Geology and Geophysics*, VIEMS, Moscow, 1974, 58.

Oldenburg, D. W. and Brune, J. W., An explanation for the orthogonality of ocean ridges to transform faults, *J. Geophys. Res.*, 80, B17, 2575, 1975.

Onischenko, E. L., Russak, Yu. O., Ruzkov, N. V., and Filatov, A. V., A ship informative computer complex of real time to ensure work of submarine apparatus, in *Submarine Apparatus and Robots*, IFAN SSSR, Moscow, 1986, 78.

Ozima, M., Kaneika, I., Saito, K., et al., Summary of geochronological studies of submarine rocks from the Western Pacific Ocean, in *Geodynamics of the Western Pacific-Indonesian Region* (Geodyn. Ser. 11), Boulder, CO, 1983, 137.

Parker, R. L., The rapid calculation of potential anomalies, *Geophys. J. R. Astron.Soc.*, 31, 447, 1972.

Parker, R. L., A statistical theory of seamount magnetism, *J. Geophys. Res.*, 93, B4, 3105, 1988.

Parker, R. L. and Huestis, L. P., The inversion of magnetic anomalies in the presence of topography, *J. Geophys. Res.*, 79, 1587, 1974a.

Parker, R. L. and Huestis, L. P., The inversion of magnetic anomalies, *J. Geophys. Res.*, 77, 7089, 1974b.

Parker, R. L., Shure, L., and Hildebrand, J. A., The application of inverse theory to seamount magnetism, *Rev. Geophys.*, 25, 1, 17, 1987.

Parkinson, U., *Introduction to Geomagnetism*, Mir, Moscow, 1986, 525.

Patriat, P., Deplus, C., Rommevaux, C., Sloan, H., Hunter, P., and Brown, H., Evolution of the segmentation of the Mid-Atlantic Ridge between 28 and 29N during the last 10 Ma, Preliminary results from the SARA cruise (R/V *Jean Charcot*, May, 1990), *EOS, Trans. Am.Geophys.Union*, 71, 1629, 1990.

Pechersky, D. M. and Didenko, A. L., Petromagnetic model of the Earth's crust under paleooceans, in Proc.3rd All-Union Geomagnetic Cong., 1986, 61.

Pechersky, D. M. and Tikhonov, L. V., Pertomagnetic features of the Atlantic and Pacific basalts, *Izv. Akad. Nauk SSSR Fiz Zemli*, 4, 79, 1983.

Peddie, N. W., International geomagnetic reference field: the third generation, *J. Geomagn. Geoelectr.*, 34, 309, 1982.

Perfit, M. R., Fornari, D. S., Malahoff, A., and Embley, R. W., Geochemical studies of abyssal lavas recovered by DSRV Alvin from eastern Galapagos rift, Inca transform, and Ecuador rift, *J. Geophys. Res.*, 88, B12, 10551, 1983.

Pitman, W. and Talwani, M., Sea floor spreading in the North Atlantic, *Bull. Geol. Soc. Am.*, 83, 3, 405, 1972.

Plouff, D., Gravity and magnetic field of polygonal Prisms and application to magnetic terrain corrections, *Geophysics*, 41, 727, 1976.

Popov, K. V., Scherbakov, V. P., Gorodnitsky, A. M., and Nazarova, E. A., Magnetic properties of oceanic crust rocks in Clarion transform fault zone, *MOIP Ser. Geol.*, 64, 3, 34, 1989.

Prichard, H. M. and Gann, G. R., Petrology mineralogy of dredged gabro from Gettisberg bank in the Eastern Atlantic, *Contrib. Mineral. Pet.*, 79, 46, 1982.

Purdy, G. M., The eastern end of Azores-Gibraltar plate boundary, *Geophys. J. R. Astron. Soc.*, 43, 973, 1975.

Puscharovsky, J. M. and Melankholina, E. N., Tectonics of the Northwestern Pacific, *Geotektonika*, 1, 5, 1981.

Pylaeva, T. A. and Roze, E. N., On one registration technique of geomagnetic field variations from hydromagnetic survey data, in *Electromagnetic Study of the Earth's Crust and Upper Mantle Structure in Sea and Ocean*, IZMIRAN, Moscow, 1981, 199.

Raff, A. D. and Mason, R. G., Magnetic survey off the west coast of North America, 40 N-52 N latitude, *Bull. Geol. Soc. Am.*, 72, 1267, 1961.

Raitt, R. W., The crustal rocks, in *The Sea*, vol.3 Hill, M. N., Ed., Interscience, New York, 1963, 85.

Rea, D. K., Late Cretaceous and Paleogene tectonic evolution of the North Pacific Ocean, *Earth Planet. Sci. Lett.*, 65, 145, 1983.

Rea, D. K. and Villier, T., Two Cretaceous volcanic epicodes in the Western Pacific Ocean, *Bull. Geol. Soc. Am.*, 94, 1430, 1983.

Renkin, M. L. and Sclater, J. G., Depth and age in the North Pacific, *J. Geophys. Res.*, 93, 4, 2919, 1988.

Richards, M. L., Vacquier, V., and Van Voorhis, G. D., Calculation of the magnetization uplifts from combining topographic and magnetic surveys, *Geophysics*, 32, 4, 678, 1967.

RISE Project Group, Crustal processes of the mid-ocean ridge, *Science*, 213, 4503, 31, 1981.

Robin, C., Colantoni, P., Gennesseaux, M., and Rehault, J. R., Vavilov seamount: a mildly alkine Quaternary volcano in the Tyrrhenian basin, *Mar. Geol.*, 78, 125, 1987.

Roest, W. R., Seafloor spreading pattern of the North Atlantic between 10 and 40N, *Geol. Ultraiectina*, 123, 1984.

Rona, P. A., Asymmetric fracture zones and seafloor spreading, *Earth Planet. Sci. Lett.*, 30, 109, 1976.

Rona, P. A., The Central North Atlantic Basin and Continental Margins: Geology, Geophysics, Geochemistry, and Resources, Including the Trans-Atlantic Geotraverse (TAG), Atlas 3, NOAA, Washington, D.C., 1980.

Rona, P. A. and Gray, D. F., Structural behavior of fracture zones symmetric and asymmetric about a spreading axis: Mid-Atlantic Ridge (latitude 23N to 27N), *Bull. Geol. Soc. Am.*, 91, 485, 1980.

Roze, E. N., On informativity of the gradient method applied to the Earth magnetic field measurements, *Geomagn. Aeron.*, 13, 4, 762, 1973.

Roze, E. N., On data integration in gradiometer, *Geomagn. Aeron.*, 18, 5, 948, 1978.

Roze, E. N. and Markov, I. M., Gradiometric method applied to geomagnetic field measurements in ocean, in *Inclusion of Temporal Variations in Marine Magnetic Survey*, IZMIRAN, Moscow, 1984, 194.

Roze, E. N. and Semevsky, R. B., Gradiometric method applied in marine magnetometry for the electromagnetic field survey, in *Devices and Technique of Gravity and Magnetic Fields Survey in the World Ocean*, PO Yuzhmorgeo., Gelendzhik, 1986, 3.

Rudnik, G. B. and Matveenkov, V. V., Peculiar features of chemism and development stages of volcanic rocks from the Marcus-Nekker rise (the Pacific Ocean), *Okeanologiya*, 17, 3, 489, 1978.

Rudnik, G. B., Melankholina, E. N., and Kudryavtsev, D. I., Rocks of the second oceanic layer at the Shatsky and Hess Rises cross-sections, *Izv. Akad. Nauk SSSR* Ser. Geol., 11, 21, 1981.

Russak, Yu. S., Navigation service in submarine apparatus, in *Submarine Apparatus and Robots*, Izd. IFAN SSSR, Moscow, 1986, 82.

Ryabchikov, I. D., Primary magma generation in primitive and modified mantle, in Proc. 27th Int. Geol. Congr., Moscow, 9, C-09, 1984, 381.

Ryan, M. P., The mechanics and three-dimensional internal structure of active magma system: Kilauea volcano, Hawaii, *J. Geophys. Res.*, 93, B5, 4213, 1988.

Sager, W. W., Paleomagnetism of Abbott Seamount and application for latitudinal drift of the Hawaiian hot spot, *J. Geophys. Res.*, 89, 1984, 6271.

Sager, W. W., Late Eocene and Maestrichtian paleomagnetic poles for the Pacific plate: implications for the validity of seamount paleomagnetic data, *Tectonophysics*, 144, 4, 301, 1987.

Sager, W. W., Davis, G. T., Keating, B. N., and Philpotts, J. A., A geophysical and geological study of Nagata seamount, northern Line Islands, *J. Geoelectr.*, 34, 283, 1982.

Sager, W. W., Handschumacher, D. W., Hilde, T. W. C., and Bracey, D. R., Tectonic evolution of the northern Pacific plate and Pacific-Farallon-Izanagi triple junction in the Late Jurassic and Early Cretaceous (M21-M10), *Tectonophysics*, 155, 345, 1988.

Sager, W. W. and Keating, B. N., Paleomagnetism of Line Islands seamounts: evidence for Late Cretaceous and Early Tertiary volcanism, *J. Geophys. Res.*, 89, B13, 1135, 1984.

Sager, W. W. and Pringl, M. S., Mid-Cretaceous to Early Tertiary apparent polar wander path of the Pacific plate, *J. Geophys. Res.*, 93, B10, 11753, 1988.

Sato, T. and Mogi, A., Guyots found from the Marshall and East Caroline Ridges, *J. Oceanogr. Soc. Jpn.*, 21, 4, 139, 1965.

Schlich, R., Echelle chronologique des inversions champ magnetique terrestre pour l'Ecene, le Paleocene et le Cretace supper, *Phys. Earth Planet. Inter.*, 24, 191, 1981.

Schlich, R., The Indian ocean: aseismic ridges, spreading centers and oceanic basins, in *The Ocean Basins and Margins*, vol. 6, Nairn, A.E.M. and Stehli, F.G., Plenum Press, New York, 1982, 51.

Scotese, C. R., Gahagan, L. M. and Larson, R.L., Plate tectonic reconstructions (1985): map of the age of the ocean basins, May 18, 1987.

Seber, G. A. F., *Linear Regression Analysis*, John Wiley & Sons, New York, 1977, 456.

Selli, R., Lucchini, F., Rossi, P. L., and Savelli, C., Del Monte, Datigeologici petrochimici e radiometrici sui vulcani Centro-Tirrenici, *G. Geol.*, 42(1), 221, 1977.

Semevsky, R. B., The advantages of gradiometer for magnetic survey, *Geofiz. Appar.*, 26, 1976.

Semevsky, R. B., Some aspects of application and investigation of errors in differential magnetometer, *Geomagn. Aeron.*, 21, 6, 1107, 1981.

Semevsky, R. V., Chernoburov, E. I., and Poddubnyi, A. I., Registration of geomagnetic field variation in motion, *Geofiz. Appar.*, 61, 46, 1977.

Sherman, G. F. and Rigch, D. L., Northwest Pacific tectonic evolution in the Middle Mesozoic, *Tectonophysics*, 155, 331, 1988.

Shouten, H. and McCamy, K., Tilting marine magnetic anomalies, *J. Geophys. Res.*, 77, 7089, 1975.

Shreider, A. A., Belyaev, I. I., Popov, E. A., et al., Geomagnetic studies in rift zone of the Reykjanes Ridge, in *Rift Zone of the Reykjanes Ridge*, Nauka, Moscow, 1990, 62.

Shreider, A. A. and Trukhin, V. I., Geomagnetic field, in *Geology and Geophysics of the Eastern Indian Ocean Seafloor*, Nauka, Moscow, 1981, 54.

Sigurdsson, H., First-order major element variation in basalt glasses from the Mid-Atlantic Ridge: 29N to 73N, *J. Geophys. Res.*, 86, B10, 9483, 1981.

Sloan, H. and Patriat, Ph., Kinematics of the North American–African plate boundary during the last 10 Ma: evolution of the axial geometry and spreading rate and direction between 28 and 29N, *Earth Planet. Sci. Lett.*, in press.

Smith, G. M. and Banerjee, S. K., Magnetic structure of the upper kilometer of the marine crust at Deep Drilling Project Hole 504B, Eastern Pacific Ocean, *J. Geophys. Res.*, 91, B10, 10337, 1986.

Soloviev, O. A., Interpretation of magnetic anomalies T with frequent method, *Geol. Geophys.*, 11, 126, 1962.

Sorokhtin, O. G., Mid-oceanic ridge topography as a function of spreading rate of lithosphere plates, *Dokl. Akad. Nauk SSSR*, 208, 6, 1338, 1973.

Sorokhtin, O. G., *The Earth's Global Evolution*, Nauka, Moscow, 1974, 181.

Sorokhtin, O. G. and Ushakov, S. A., *Nature of Earth Tectonic Activity*, VINITI, Moscow, Ser. Earth Physics, 12, 292, 1993.

Spector, S. and Grant, F. S., Statistical models for interpretation of aeromagnetic data, *Geophysics*, 35, 2, 293, 1970.

Spiess, P. N. and Mudie, J. D., Small scale topographic and magnetic features, in *The Sea*, vol. 4, New York, 1970, 205.

Standigel, H., Park, R. N., Pringle, M., et al., The longevity of the South Pacific isotopic and thermal anomaly, *Earth Planet. Sci. Lett.*, 102, 1, 24, 1991.

Steiner, F., M-fitting (fitting according to the most frequent value) and its comparison with the method of least-squares, *Acta Geogaet. Geophys. Montanist.*, 15, 37, 1980.

Strakhov, V. N., To filtration and transformation potential field theory using the a priori information about errors in input data, *Fiz. Zemli*, 3, 87, 1977.

Strakhov, V. N., On common solutions of inverse problems in gravimetry and magnitometry, *Izv. Vuzov, Geol. Razv.*, 4, 104, 1978.

Strakhov, V. N., Some problems of interpretation of marine geomagnetic survey data, in *Magnetic Anomalies and New Global Tectonics*, Nauka, Moscow, 1981, 20.

Strakhov, V. N. and Lapina, M. I., On ambiguity of inverse problem solution in magnetometry, in *Anomaly of the Earth's Interior*, Naukova Dumka, Kiev, 1976, 185.

Strakhov, V. N., Lapina, M. I., and Efimov, A. B., Solution of direct problems in gravimetry and magnitometry based on new analytical representation of field elements from typical approximating bodies, *Izv. Akad. Nauk SSSR Fiz. Zemli*, 6, 55, 1986.

Strakhov, V. N. and Valyashko, G. M., On choice of a regularization parameter in linear ill post problems, *Dokl. Akad. Nauk SSSR*, 228, 1, 128, 1976.

Strakhov, V. N. and Valyashko, G. M., Technique of in-time interpretation of hydromagnetic survey data in the ocean, *Dokl. Akad. Nauk SSSR*, 235, 1, 67, 1977.

Strakhov, V. N. and Valyashko, G. M., Algorithms of adaptive regularization of linear ill post problems, *Dokl. Akad. Nauk SSSR*, 259, 3, 120, 1981.

Strakhov, V. N. and Valyashko, G. M., Adaptive regularization of linear ill post problems and its application in problems of gravimetry and magnetometry, *Izv. Akad. Nauk SSSR Fiz. Zemli*, 11, 55, 1984.

Structure of the floor of the Northwest Pacific (Edit. Gu. M. Pushcharovsky), Nauka, Moscow, 1984, 227.

Talwani, M., Computation with the help of digital computer of magnetic anomalies caused by bodies of arbitrary shape, *Geophysics*, 30, 5, 797, 1965.

Tapscott, C., Patziut, P., Fisher, L., Sclater, J., Haskins, H., and Parsons, B., The Indian Ocean triple junction, *J. Geophys. Res.*, 85, 4723, 1980.

Taylor, P. T., Interpretation of North Arabian Sea aeromagnetic survey, *Earth Planet. Sci. Lett.*, 4, 3, 232, 1968.

Technical recommendations, in *Book of References for Geophysicists*, Nikitsky, V. E. and Glebovsky, J. S., Eds., Nedra, Moscow, 1990, 470.

Tikhonov, A. N. and Arsenin, V. Ya., *Methods for Solution of Incorrect Problems*, Nauka, Moscow, 1986, 288.

Tisseau, J. and Patriat, P., Identification des anomalies magnétiques sur les donsales a faible taux d'expansion: methode des taux fictifs, *Earth Planet. Sci. Lett.*, 52, 2, 381, 1981.

Tomoda, I. and Fuimoto, H., The possibility to measure diurnal geomagnetic variations during the ship's motion, in *Marine Geology and Geophysics*, vol.3, VIEMS, Moscow, 1982,1.

Tretiyak, A. I., Viglyanskaya, L. I., Makarenko, V. N., and Dudkin, V. P., *Fine Structure of Geomagnetic Field in the Late Cenozoic*, Naukova Dumka, Kiev, 1989, 156.

Tsutskarev, B. M., Bakalinsky, S. P., and Rusanova, N. V., Magnetometric complex for component measurements, in *Marine Geomagnetic Survey on Research Vessel "Zarya"*, Nauka, Moscow, 1986, 17.

Tucholke, B. E., Macdonald, K. C., and Fox, P. J., ONR seafloor natural laboratories on slow- and fast-spreading Mid-Oceanic Ridges, *EOS, Trans. Am. Geophys. Union*, 72, 268, 1991.

Tuezov, I. K., Il'ichev, A. Ya., Ostapenko, V. F., et al., Geological structure of Marcus-Necker rise, *Sov. Geol.*, 4, 85, 1979.

Turcotte, D. L., Membrane tectonics, *Geophys. J. R. Astron. Soc.*, 36, 1, 133, 1978.

Turcotte, D. L. and Oxburgh, E. R., Intra-plate volcanism, *Phil. Trans. R. Soc. London Ser.A*, 288, 561, 1978.

Uchupi, E., Phillips, J. D., and Prada, K. E., Origin and structure of the New England seamount chain, *Deep Sea Res.*, 17, 3, 483, 1970.

Ueda, Y., Geomagnetic study on seamounts Daiitikasima and Katori with special reference to a subduction process of Daiitikasima, *J. Geomagn. Geoelectr.*, 37, 602, 1985.
Uglov, B. D., Complex of marine magnetometric apparatus, in *Mineral Resources of World Ocean: Study and Exploration*, Sevmorgeologiya, Leningrad, 1984, 125.
Uglov, B. D., Leibov, M. B., Lygin, V. A., Skripka, A. P., et al., *Technical and Methodical Aspects of Differential Magnetometric Survey: Problems of Their Practical Application*, VINITI, Moscow, 1989, 7720-889.
Uglov, B. D. and Lygin, V. A., On geometrical stability of towed marine gradiometric systems, *Vest. Mosk. Univ. Geol*, 6, 89, 1988.
Uglov, B. D., Skripka, A. P., Leibov, M. B., and Lygin, V. A., High accuracy geomagnetic survey, in *Devices and Methods of Gravity and Magnetic Fields Investigation in World Ocean*, Gelendjik, PO Yuzhmorgeo., 1986, 16.
University of Washington, *RIDGE Initial Science Plan, a Component of the US Global Change Program*, Seattle, 1989, 90.
Uyeda, S. and Richards, M. L., Magnetization of four Pacific seamounts near the Japanese Islands, *Bull. Earthquake Res. Inst. Univ. Tokyo*, 44, 1, 179, 1966.
Vacquier, V., A machine method for computing the magnitude and direction of magnetization of a uniformly magnetized body from its shape and a magnetic survey, in *Proc. Benedum Earth Magnetism Symp.* Nagata, T., Ed., University of Pittsburgh Press, 1962, 123.
Vacquier, V., *Geomagnetism in Marine Geology*, Nauka, Moscow, 1976, 192.
Vacquier, V. and Uyeda, S., Paleomagnetism of nine seamounts in the western Pacific and of three volcanoes in Japan, *Bull. Earthquake Res. Inst.*, 45, 815, 1967.
Valyashko, G. M., Magnetic survey interpretation method, in *Oceanic Magnetic Anomalies and Plate Tectonics*, Nauka, Moscow, 1981, 60.
Valyashko, G. M., Belyaev, I. I., and Popov, E. A., *Detailed Geomagnetic Investigation in the Red Sea*, VINITI, Moscow, 1983, 4207.
Valyashko, G. M., Filin, A. M., Lukyanov, S. V., Gorodnitsky, A. M., and Osipova, I. L., Application of marine gradiometric surveys and data of observatories to study geomagnetic variations and mapping of the top of the active magnetic layer in the Barents Sea, in *Electromagnetic Induction in the World Ocean*, Nauka, Moscow, 1990a, 82.
Valyashko, G. M., Gorodnitsky, A. M., and Lukyanov, S. V., Magnetic modelling of the Shatsky Rise and magnetoactive layer structure, in *Electromagnetic Induction in the World Ocean*, Nauka, Moscow, 1990b, 23.
Valyashko, G. M., Lukyanov, S. V., and Popov, E. A., Comparative analysis of spreading between the Tadjura and Red Sea rifts, in *Proc. 3rd All-Union Congr. Geomagnetism*, Naukova Dumka, Kiev, 1985, 52.
Valyashko, G. M., Lukyanov, S. V., and Popov, E. A., Detailed geomagnetic investigation of the Tadjura rift, in *Geology of the Tadjura Rift*, Nauka, Moscow, 1987a, 76.
Valyashko, G. M., Lukyanov, S. V., and Popov, E. A., Geological-geophysical data processing by onboard computers, in *Geology of the Tadjura Rift*, Nauka, Moscow, 1987b, 28.
Valyashko, G. M. and Strakhov, V. N., Filtration and differentiation of experimental data by adaptive ill post problem method, *Izv. Akad. Nauk SSSR Fiz. Zemli.* 12, 68, 1984.
Vanyein, L. L. and Butkovskaya, A.J., *Magnetotelluric Sounding of layered Media*, Nedra, Moskow, 226,1980.
Verhoef, J. and Macnab, R., Definitive Magnetic Reference Field (DMRF) evaluation based on marine magnetic anomalies, *Phys. Earth Planet. Inter.*, 54, 332, 1989.
Verlan, A. F. and Sisikov, V. S., *Integral Equations: Methods, Algorithms, Programs*, Naukova Dumka, Kiev, 1986, 543.

Verzhbitsky, E. V., Gorodnitsky, A. M., Emel'yanov, E. M., Lobkovsky, L. I., Marova, N. A., et al., New geological and tectonic data on the Gorringe ridge (North Atlantic), *Geotektonika*, 1, 1, 1989a.

Verzhbitsky, E. V., Gorodnitsky, A. M., and Lobkovsky, L. I., On thermal field anomaly nature in the eastern Azores-Gibraltar fault zone (North Atlantic), in *Theoretical and Experimental Study of Marine Geothermal Regime*, Nauka, Moscow, 1989b, 1.

Vine, F., Spreading of the ocean floor: new evidence, *Science*, 154, 1405, 1966.

Vogt, P. R., Amplitudes of oceanic anomalies and the chemistry of oceanic crust: synthesis and review of "magnetic telechemistry", *Can. J. Earth Sci.*, 16, 2236, 1979.

Vogt, P. R. and Byerly, G. R., Magnetic anomalies and basalt composition in the Juan de Fuca-Gorda Ridge area, *Earth Planet. Sci. Lett.*, 33, 185, 1976.

Vogt, P. R. and Smoot, N. C., The Geisha guyots: multibeam bathymetry and morphometric interpretation, *Geophys. Res.*, 89, B13, 11085, 1984.

Whitmarsh, R. B., Some aspects of plate tectonics of the Arabian Sea, *Init. Rep. DSDP*, 23, 527, 1974.

Wiggins, R. A., The general linear inverse problem: implication of surface waves and free oscillations for Earth structures, *Rev. Geophys. Space Phys.*, 10, 1, 251, 1972.

Won, T. J., Determination of thickness and susceptibility of magnetized crust from Magsad data, in *Abstr.* IAGA:IAMAP, 1, Moscow, 1985, 172.

Yastrebov, V. S., Valyashko, G. M., Gorodnitsky, A. M., and Rimsky-Korsakov, N. A., Geomagnetic near bottom mapping with towed device, *Okeanologiya*, 31, 1, 1991.

Zagurcky, A. S., Belyaev, I. I., Shterengarts, E. M., and Kuznetsov, G. A., Marine proton magnetometer MPM-7, *Geofiz. Appar.*, 89, 7, 1988.

Zhivago, A. V., *Fault and Rift Morphostructure of the Pacific and Atlantic Bottom*, Nauka, Moscow, 1984, 23.

Zonenshain, L. P., Tectonics, in *Geology of the Tadjura Rift*, Nauka, Moscow, 1987, 123.

Zonenshain, L. P., Kononov, M. V., and Savostin, L. A., Pacific and Kula Eurasia relative motions during the last 130 m.y. and their bearing on orogenesis in the Northeast Asia, *Amer. Geophys. Union Geodyn. Ser.* 18, 29, 1987a.

Zonenshain, L. P., Kuz'min, M. I., Lisitsin, A. P., Bogdanov, Yu. A., et. al., Tectonics of the rift valley of the Mid-Atlantic Ridge between 26 and 24 N, Evidence of vertical motion, *Geotektonika*, 4, 99, 1989.

Zonenshain, L. P., Matveenkov, V. V., Baranov, B. V., and Khain, V. V., Pacific seamounts movements during last 110 m.y., *Okeanologiya*, 27, 4, 592, 1987b.

Zonenshain, L. P., Monin, A. S., and Sorokhtin, O. G., Tectonics of the Red Sea rift near 18 N, *Geotektonika*, 2, 3, 1981.

Zonenshain, L. P. and Savostin, L. A., *Introduction in Geodynamics*, Nauka, Moscow, 1979, 310.

Zonenshain, L. P., Savostin, L. A., and Sedov, A. P., Global paleodynamic reconstruction for last 160 million years, *Geotektonika*, 3, 3, 1984.

INDEX

A

Accuracy of field reconstruction, 28
Acoustic basement, 83, 139
Active magnetic layer, 47, 149, 162
Adaptive
   algorithm, 24
      of filtering, 127
   families, 11
   regularization, 24, 32, 41
   reparameterization, 45, 46, 145
Admittance, 64
Alarm disconnection system, 3
Alkaline, 165
Amphibole, 177
Amphibolite, 162
Analysis, regressive, 203
Anisotropy, 162
Annihilator, 44, 149, 152, 162
Anomalous
   gradient zone, 4
   magnetic field (AMF), 6-8, 10-19, 51, 145, 152
Atoll, 202, 221
Auroral zone, 12

B

Basalt, 147, 152, 166, 168-178,
   alkaline, 174-178, 180, 198, 215, 222
   calc-alkaline, 189, 190, 198
   intraplate, 184
   metamorphism, 162
   olivine-plagioclase, 191, 192, 202
   porphyritic, 191
   subalkaline, 193
   tholeiite, 177, 178, 180, 187, 190
   zeolitized, 190, 192
Base of measurements, 5, 10, 22, 24
Bathymetric
   data, 77
   information, 79
   map, 124
Bathymetry, 153, 155, 158
Biostratigraphic age, 98
Bottom geomagnetic survey, 29, 30
Bouguer anomaly, 78, 79, 196
Bruness epoch, 118, 120, 124

C

Canary-Bahamas geotraverse (CBGT), 67, 75-80
Carbonate sediments, 192
Charamilio epoch, 125
Coercive force, 131, 150
Common reference plane, 35
Confidential area, 220
Continuous seismic profiling (CSP), 11
Correlation
   discriminator, 32
   radius, 7, 8
Cross-spectral analysis, 79
Curie
   point (Q), 129, 150, 231
   temperature, 147

D

Database
   coherent digitized, 69, 72
   computer, 69, 71
   real-time, 67
Declination, 52, 173, 190, 191
Deep-sea towed system, 2
Derivative of the field, 4, 5
Destructive field, remnant, 150
Detrite, 215
Deviation, 9-12, 24, 25, 27
   difference, 25, 27
Differential channel, 5, 6, 8
Discrete cosine Fourier transform, 31
Discrete Fourier transform (DFT), 60, 62
Dissection of the object, 38
Dolerite, 152, 161, 166, 177
Dredging, 152, 156, 165, 169
Dyke complex, 166

E

Echo-sounding, 11, 79
Electrical conductivity, 4
Electromagnetic sounding, 228
Error
   absolute, 8
   base instability, 24,
   closure, 42
   closure principle, 40

deviational, 8
high frequency, 27
instrumental, 8
instability, 8
interpolation, 32
observations, 45
orientation, 24, 27
random, 8
relative, 22
root mean square (RMS), 10-18, 36, 176-198, 215
survey, 32
Evaporite, 122

F

Ferromagnetic
concentration, 147, 189, 191
grains, 147, 172
material, 190-192
minerals, 168, 192-194
Ferromagnetite, 117
Field
spectrum, 58, 66
transformants, 30, 31
transformation, 58
Filtering of the data, 10
Fine spatial structure of the AMF (FSS AMF), 1, 10, 14, 16, 19
Foraminifera, 175, 198
Fracture zones, 73, 75-77
Frequency filtering, 27
Function, transmitting, 203

G

Gabbro, 83, 166, 174, 175
Gabbroid, 166, 175, 177
Gabbro-dolerite, 174, 175
Generalized linear inversion, 38
Geochemical effect, 225
Geomagnetic
data, 155, 166
measurements, 158
survey, 152, 166
Geometrical mean values, 173
Geothermal studies, 201, 202
Globigerina, 215
Graben, 82
Gradient
horizontal, 7, 14, 16-19, 84, 149
longitudinal, 6, 10-16, 19, 25

magnetic potential, 51
measured, 25
superposed, 24, 25
vertical, 8, 24
Gradiometer, 5-8, 12, 18, 19
base, 9
survey, 1, 8-10, 13-14, 17
Gradiometry technique, 4, 6
Gravimagnetic survey, 68
Gravimetric (gravity)
anomalies, 78, 80
data, 77
information, 79
survey, 79
Gravity anomaly, 194
Green-stone metamorphism, 161
Guyot, 202, 213-215, 218- 221
Gyro-vertical SU-2, 2

H

H zone, 230
Hawaiite, 177
Heading of the vessel, 9
Heat flow measurement, 68
Hemisphere, 210
Heterophase changes, 152
Hot-spot, 95, 184, 220
Hyaloclastite, 172, 205
Hyperbasite, 174

I

Inclination, 9, 52, 173, 190, 191, 197
Instrumental noise, 8
Integrated field, 28, 29
Integrated latitudinal profile, 158, 161
Integration process, 28
Intraplate, 112, 113
trough, 168
volcanism, 168
Intrusion, 18
Inverse filtering procedure, 13
Island arc, 187, 198
Isotope
anomaly, 219, 221
characteristic, 219
Isotopic data, 10

J

Junction, Tadjura and Cheba Ridges, 116

K

Key reflector "M", 195
Koenigsbergen factor, 114, 150-161, 171-193

L

Leucite, 177
Lherzolite, 228
Limestone
  foraminifera, 215
  micrite, 192
  organogenic, 202
  radiolarian, 215
  reef, 210
Listric fault, 116
Lithological analysis, 167
Low-frequency filter, 7

M

Maggemite, 161
Magma chamber, 132
Magnetic
  anomaly, 8, 50, 65, 71, 76, 147, 158-168
  basement, 172
  concentration, 191
  declination, 2, 3
  geochronological scale, 14, 47
  inclination, 83, 85
  induction, 71
  intensity, 54, 176
  layer, 74, 83, 158, 166, 168
  lineation, 13, 22, 47, 71-85, 123-127, 146, 213, 221
  mapping, 94
  moment, 58, 61, 88
  polarity, 83, 173, 183
    direct, 172
    normal, 178-199, 208
    reversed, 179-180, 184, 194-210, 218
  potential, 50, 61
  quiet zone, 88
  resistance, 129
  resonance, 129
  reversals, 74
  simulation, 48, 83, 121, 145, 168, 171
  source, 2, 14, 76
  stability, 190, 191
  storm, 3, 11-13
  survey, 4, 11, 68-73, 81, 82
  susceptibility, 141, 150, 162, 176-193
  variation base station (MVS), 1, 4, 5, 12, 13
Magnetite, 161, 189, 223
Magnetization, 14, 18, 19, 44-65, 74, 76, 146-222
  anomalous, 50
  bulk, 50
  constant, 172
  direction, 58, 62, 63, 65, 66
  effective, 47, 147, 149-172
  heterogeneous, 203
  homogeneous, 54
  ideal, 150
  increment, 46
  intensity, 37, 52, 62, 63
  natural remnant, 150, 161
  negative, 165
  of saturation, 150
  piecewise-constant, 45
  positive, 165
  primary, 178
  regular unique, 45
  remnant, 152-162, 176, 178, 190-205
  remnant saturation, 150
  saturation, 150
  seamount, 48
  susceptibility, 166
  vector, 203, 210
Magnetometer
  component, 2, 3
  fluxgate, 2, 6
  marine, 1, 2
  optically pumped, 1, 2
  proton, 1-3, 6, 8
  scalar, 1, 14
Mantle upwelling, 79
Matrix
  informational density, 40,
  pseudo-inverse, 39, 41
  "resolution defect", 45
  resolution of unknowns, 40, 42
  singular-value decomposition, 38
Matuyama epoch, 194
Mean square error, 10, 14, 18, 40, 72
Measured magnetic field, 24, 25, 28, 31
Measurement
  component, 1, 2
  gradient, 1, 3, 5, 6, 8, 11, 13, 14
  scalar, 1
Membrane tectonics, 168
Metamorphism, 178
  basalt, 162
  greenstone, 161, 175
Method

K/Ar, 175
Q, 184
Mid-oceanic ridge, 7
Moho-boundary, 108, 136
Multi-beam echo-sounder, 158

N

N zone, 230
Noise, 65
   noncorrelated (random), 24, 31, 32
Norm
   of residuals, 39, 41
   of solution, 39-41
Normal
   magnetic field, 70-72
   pseudo solution, 39

O

Observational space, 38, 42
Olduvai event, 201
Olivine, 177
   pyroclastic basalt, 161
Optimal filtering, 11
Oriented specimens, 3
Outcrop, 161
Overlap, 104, 116, 127

P

P-wave velocity, 136, 166
Paleoclimatic data, 172
Paleodynamic reconstruction, 179, 180, 182
Paleolatitude, 166-180, 205-219
Paleomagnetic
   anomaly, 86, 93
   data, 184
   pole, 171-174, 197, 205-219
   reconstruction, 172
Paleontological analysis, 147
Partial melting, 79
Petrochemical techniques, 169
Petrographic
   characteristics, 147, 161
   properties, 161
Petromagnetic, 145
   analysis, 145, 147, 156, 168, 171
   data, 146
   characteristics, 152
Pillow lava, 147, 193
Plagioclase feldspathoid, 175

Poles, virtual, 174
Porphyritic, 161, 177
Possible general solution, 45
Previous amplifier, 7
Problem
   best regression, 42
   direct, 50, 55, 171
   incorrect, 32
   inverse, 38, 50, 52, 203
   potential field transformation, 32
   uniqueness solution, 44
Propagating process, 105, 120, 123
Pseudo-inverse operator, 39-42
Pseudo-solution, 39, 40, 45

Q

Quartz-fayalite magnetite buffer, 224
Quiet magnetic field, 155

R

Radiometric determination, 196, 197
Random irregular noise, 7
Reduction to
   bottom conformable surface, 30, 35-36
   common reference plane, 34-36
   given depth, 32
Refracted waves, 68
Regularization, 178
   adaptive, 24, 32, 41
   level, 38, 40
   parameter, 32, 40, 41,
   procedure, 52
Regularizator, 40
Regularizing family, 32
Reliability, 38
Remnant destructive field, 150
Reparameterization of the model, 38, 41, 45
Residual anomalous magnetic field, 70
Resolution
   matrix, 42
   of unknowns, 40
Rift zones, 205
Ridge regression, 40
Ring system, 2
Reversed polarity, 165

S

Sampling, 145
Sampling time (rate), 2-11, 29

Satellite radio navigation system, 1
Scarps, 152, 155, 158
Schist, 162
Seismic,
  reflection, 68, 77
  refraction, 68
  sounding, 68, 85, 165
  wave velocity, 108
Self-deviation of the sensors, 24, 27
Sensitivity, 1-3, 7
Serpentinite, 83, 156, 162, 169, 176-178
  apoharzburgite, 161
  layer, 168
  outcrops, 161
Serpentinization, 158, 161
Serpentinized
  harzburgite, 175
  hyperbasite, 168
  peridotite, 144, 158, 161, 169
Siltstone, 196
Simulation, 146, 147, 158, 166
Singular-value decomposition of the matrix, 38
Solution space, 38
Space of model, 42
Spectral analysis, 201
Spreading, 176, 187, 202, 204, 215
  axis, 90, 119, 122
  cell, 74, 77, 99
  center, 77, 168
  jumps, 92
  rate, 13, 74-77, 121, 122, 160
  reorganization, 159, 161
  ridge, 158
  trend, 76, 79, 119, 161
Standard deviation, 193, 197
Standard oceanic crust, 13
Stationary magnetic field (EMF), 6-9, 13, 24, 25, 51
Subduction zone, 188
Superimposed volcanism, 18, 165
Superposed
  gradient, 24, 25, 26
  field, 25-26

T

Thermomagnetic analysis, 129
Three-sensor
  differences, 25
  observations, 22
  scheme, 22
Time (temporal) variation, 3-6, 8, 13
Titanoaugite, 215

Titanomagnetite (TiMt), 147, 152, 190, 223
Topography approximation, 55
Total
  field difference, 4
  magnetic field intensity (TMFI), 1, 3, 4
Towing depth, 34
Trachybasalt, 114
Transformation, 194, 199
  Fourier, 203
Transfer function, 64
Transform faults, 47, 74-82, 120-168
Transient zone, 14
Trough, 166, 168
Turbidite layer, 215

U

U-interface, 31
Ulvospinel, 223
Uniqueness of the solution, 38, 44

V

Variance
  observations, 39
  parameters, 39, 40
  solution, 39-41
  white noise, 41
Variation
  difference, 25
  high frequency, 26
  increment, 27, 28
  properties, 26
  temporal, 24-28,
Volatile, 196
Volcanic glass, 168
Volcanoclastite, 215
Volcanite, 200
Volcanism
  basalt, 172, 174, 177, 220, 222
  calc-alkaline, 188
  central-type, 221
  fracture, 201
  intraplate, 202, 213, 215, 221
  superimposed, 18, 165

W

White noise variance, 41

X

Xenolith, 177